MW00846184

Experimental Statistics for Agriculture and Horticulture

Experimental Statistics for Agriculture and Horticulture

Clive R. Ireland

Writtle College
Chelmsford
Essex
UK

www.cabi.org

CABI is a trading name of CAB International

CABI Head Office	CABI North American Office
Nosworthy Way	875 Massachusetts Avenue
Wallingford	7th Floor
Oxfordshire OX10 8DE	Cambridge, MA 02139
UK	USA

Tel: +44 (0)1491 832111
Fax: +44 (0)1491 833508
E-mail: cabi@cabi.org
Website: www.cabi.org

Tel: +1 617 395 4056
Fax: +1 617 354 6875
E-mail: cabi-nao@cabi.org

© Clive R. Ireland 2010. All rights reserved. No part of this publication may be reproduced in any form or by any means, electronically, mechanically, by photocopying, recording or otherwise, without the prior permission of the copyright owners.

A catalogue record for this book is available from the British Library, London, UK.

Library of Congress Cataloging-in-Publication Data

Ireland, Clive R.
   Experimental statistics for agriculture and horticulture / Clive R. Ireland.
     p. cm. -- (Modular texts)
   Includes bibliographical references and index.
   ISBN 978-1-84593-537-5 (alk. paper)
1. Agriculture--Experimentation--Statistical methods. 2. Agriculture--Research--Statistical methods. 3. Horticulture--Research--Statistical methods. I. Title. II. Series.

S540.S7174 2010
630.72'7--dc22

               2010005468

ISBN: 978 1 84593 537 5

Commissioning editors: Stefanie Gehrig, Rachel Cutts
Production editor: Kate Hill

Typeset by SPi, Pondicherry, India.
Printed and bound in the UK by Cambridge University Press, Cambridge.

Experimental Statistics for Agriculture and Horticulture is an independent publication and is not affiliated with, nor has it been authorized, sponsored, or otherwise approved by Microsoft Corporation.

Contents

List of Examples

Preface

The aim of this book is to provide an appropriate training in correct experimental design and appropriate methods of statistical analysis that will allow valid and acceptable conclusions to be drawn from the results of experiments. While the book is primarily aimed at agriculture and horticulture and contains a range of examples drawn from these disciplines, it should still be useful to a broad range of life sciences that involve experimental investigation. The term 'experiment' is, for our purposes, defined in its broadest sense, meaning any type of planned investigation undertaken in order to test a particular hypothesis or answer a particular question. It therefore encompasses everything from the controlled classic scientific experiment through to observational field trials and even questionnaire-based investigations. At the end of the book it is hoped that the reader will be able to achieve the following key tasks:

- Plan an experiment so that it is assured of providing valid data that can be meaningfully analysed.
- Identify an appropriate statistical technique that will permit the analysis of the data that have been collected.
- Fully appreciate the major assumptions and limitations involved in the use of a selected method of data analysis.
- Produce a valid interpretation of the outcome of the data analysis.

In principle it is possible for anyone in possession of a set of instructions to use a technical tool without understanding how that tool actually works. However, when a tool malfunctions an understanding of its mechanism will be vital if it is to be repaired. The analysis of data by statistical tools is no exception to this rule. Therefore, while it is not the intention to present an in-depth account of the underlying theory of statistical analyses, it is the intention to provide sufficient description of the principles to allow the user the best chance of selecting the correct procedure in the first instance and to recognize when an error has been made and to rectify it. The user will hopefully come to recognize that the statistical procedures employed have a very logical and, in many cases, a quite simple theoretical basis.

While the book is designed primarily to assist students and researchers in agriculture and horticulture, it should be helpful to all those biologists who have little or no previous experience in statistical techniques, or who learnt their statistics so long go that all is now rusty and forgotten. While the book confines itself to discussion and demonstration of the main basic techniques in experimental design and data analysis, it is intended to provide a sufficiently advanced understanding of the principles involved that will allow readers to identify when further more specialized techniques are required and then to pursue these through more advanced texts.

In the final chapter of this book some procedures will be outlined for performing a range of the most common data analyses using computer software. The procedures are illustrated using two widely available computer software packages, namely the Data Analysis Toolpak contained within Microsoft Excel® spreadsheet software and the more advanced statistics software package, GenStat. It should be noted that these two particular software packages are employed as examples of modern data handling software and there is no intention either to provide a comprehensive manual of their use or to recommend the use of these rather than any other similar commercial software.

Acknowledgements

I am very grateful to numerous people who have helped me in numerous different ways to complete this book, but I would particularly like to express my thanks to the following: first and foremost my main reviewer Professor Peter Thomson whose detailed but always constructive comments on the chapter manuscripts were absolutely invaluable. To my work colleagues Jon Amory, who made helpful suggestions on a number of chapters, and Peter Hobson, for his skilful artwork. To all the students that have suffered my teaching over the years but who enabled me to develop the ideas and materials for this book. Finally to Fiona for her constant encouragement to keep going. I do not know if this book is of much worth but without these people it would certainly be a lot less worthy.

1 An Introduction to Research by Experimentation

Research: careful search or inquiry; endeavour to discover new or collate old facts, etc., by scientific study of a subject; course of critical investigation.

Experiment: a test, trial; procedure adopted on chance of its succeeding or for testing hypothesis, etc.

(Concise Oxford English Dictionary)

- The philosophical nature of research and different types of research approach.
- An introduction to research by experimentation.
- The principles of experimental design and the main stages in designing an experiment.
- The problem of natural variation.
- The role of statistical analyses in the evaluation of experimental data.

1.1 The Nature of Research and Different Types of Research Approach

Research may be broadly defined as 'the process of finding out and answering questions about the nature of the universe and man's activities within it'. The majority of modern research is, however, undertaken in a precise, carefully planned and systematic way so that it is not liable to changes in direction and the methodology remains consistent and resilient to bias. Thus a more modern and pragmatic definition of research is that it is 'the systematic and objective investigation of a subject or problem with the aim of discovering new relevant information and principles and thereby to increase knowledge'. As such the research process may be conveniently considered to involve four main stages:

1. Identifying a research question. The research question provides the aims and objectives of the research. The broad aim of a large research project may be cited in the form of a generalized question, for example, 'What are the optimum conditions required for producing the maximum yield in tomato?' This, however, will need to be broken down to a series of more specific questions to provide the objectives of individual experiments, for example, 'At what temperature is yield greatest?'; 'In what soil type is yield greatest?'; 'Which variety produces the greatest yield?' etc. It is undoubtedly true that good research arises in the first instance from good questions. However, as will be discussed further, the answering of one research question invariably leads to the posing of a new research question; a process commonly referred to as the 'research wheel'.

2. Collecting data. Data are simply recorded facts, e.g. measurements, counts, descriptions, that on their own have little meaning until processed and placed into context. In the research process data collection is achieved through active investigation that might involve, for example, conducting an experiment, undertaking a survey or making and recording observations.

3. Producing information. Information can be regarded as that which arises from the processing of the collected data and consists of a statement(s) of perceived truth or answer to a particular question. Processing of the data is achieved by techniques of objective data analysis and interpretation. For example, following data processing it may be concluded that a particular tomato cultivar produces a maximum fruit yield when grown in a peat-based compost and held at 25°C.

4. Increasing knowledge. Knowledge arises from the collation and integration of bits of information that eventually allow for an overall understanding of a particular process or subject. For example, through collation of information from a whole range of research, an understanding may be obtained on the overall way that tomato yield responds to temperature and the factors that affect this response.

Research approaches are often considered to fall into one of two mutually exclusive categories; **scientific** and **social scientific**. The scientific approach is traditionally empirical in nature, based on experimentation and observation. Scientific research is conducted primarily through a process of hypothesis testing, evaluation, modifying or producing a new hypothesis as necessary, and then retesting. Social science is the study of the behaviour of human beings, their activities and their organizations and research in social science does not easily lend itself to the classical experimental approach. This is because it is extremely difficult to perform a social science investigation on human activity under constant non-varying conditions and even more difficult to repeat a study under exactly the same conditions. Further, it is extremely difficult to subject human beings to behavioural investigations without them being aware that a study is being conducted. This awareness invariably leads, either consciously or subconsciously, to a change in the behaviour of the subject under investigation. Consequently social scientists use techniques such as questionnaire or interview surveys, case studies and participative action research. Having said that, certain aspects of the experimental approach may often be present in the social science method, in particular, the technique of employing random sampling and the procedure of commencing with a hypothesis that, consequential upon the outcome of the investigative process, is subsequently either accepted or rejected.

A further distinction that is often made is that between **quantitative research**, which involves direct numerical measurement of variables, and **qualitative research**, which involves making descriptions of items and processes. While the analysis of quantitative data is based on numerically precise statistical techniques, analysis of qualitative data is rather more problematical but may involve a 'pseudo-quantitative' approach in which a qualitative variable is converted to a quantitative measure. There are two main ways this can happen. A ranking score might be attributed to items based on a qualitative character, for example, samples of fruit might be given a ranking score based on their flavour. Alternatively, items might be placed into categories based on a qualitative character and the size of categories subsequently compared, for example, the number of people identified from a questionnaire to hold a particular opinion regarding a particular issue. Clearly quantitative research tends to be what scientists do and qualitative research tends to be what social scientists do, but this is a broad generalization and many modern research projects involve elements of both qualitative and quantitative research methodologies.

While an extensive discussion of the different philosophical strands that concern the research process is beyond the scope of this book, it may be strongly argued that the modern researcher should have an awareness of the major alternative, though not necessarily exclusive, research paradigms in order to bring a balanced approach to their research. Further, such awareness may potentially open doors to alternative methodologies for conducting research that otherwise may not have been considered. Therefore, a brief introduction to the philosophies that underlie different research approaches will now be presented. Readers who have no interest in these more esoteric ideas, or who simply need to move rapidly to the practical aspects of experimental design and analysis, are invited to skip the following section.

1.2   An Introduction to the Philosophy of Research

The distinction between the scientific and social scientific research approaches is reflected in two major philosophical paradigms that are concerned with research methodology. The traditional scientific view is that all the processes in the universe operate according to a set of universal laws that is constant and unyielding, this view being encompassed in a philosophy termed **positivism**. Positivists believe that by application of a scientific method that is essentially empirical and based on hypothesis testing, it is possible to elucidate these underlying laws and thereby understand the reality of the universe. Positivists are really only interested in explaining phenomena that can be observed; 'if it is not measurable then it is not important' would be the positivist's point of view. During the last century, however, an alternative philosophy has emerged that is referred to as **post-positivism**. In attempting to explain post-positivism it is perhaps best to start by stating what it definitely is not; post-positivism is not simply the opposite of positivism in that it does not deny absolutely the existence of universal laws. Instead, the post-positivist view is that although such laws may exist they can never actually be determined with certainty. This is because observation and measurement are never infallible and inevitably contain error and bias. Thus, while the traditional scientific positivist view is that any non-objectivity within the observer and non-uniformity in external conditions can effectively be eliminated through the correct application of scientific method, the post-positivist view is that this is impossible and therefore only an estimate of reality can ever be achieved which will always be liable to adaptation and change.

Positivists and post-positivists tend also to employ two different modes of reasoning. The traditional scientific mode of reasoning, termed **deductive**, commences with an idea concerning some particular phenomenon from which a testable hypothesis is formulated. Following testing the hypothesis may be confirmed or rejected and this will lead to formulation of a modified or new hypothesis. It is the contention of the positivist that this continuing process of hypothesis testing will eventually result in the true basis of the phenomenon being resolved. On the other hand, **inductive** reasoning involves a chain of logic that flows in the opposite direction to deductive reasoning. That is, the process begins with a set of neutral independent observations of the phenomenon in question. Through triangulation and integration of observations, patterns and trends become apparent that in turn lead to formulation of hypotheses that then give an explanation of the phenomenon. A light-hearted parallel based on a couple of well-known fictional characters may help to illustrate this distinction (see Fig. 1.1). Detective Lestrade of Scotland Yard commences his investigation into a particular crime by supposing that some particular person is guilty and making an appropriate arrest. Through questioning and appraisal of the available evidence the suspect is either confirmed as the guilty party or is released from custody and an alternative suspect arrested and questioned instead. When a suspect is found for whom no counter evidence can be discerned then a charge of the crime is made. In this mode of operating Detective Lestrade is applying deductive reasoning. Lestrade's famous adversary the private detective Mr Sherlock Holmes is, however, dedicated to inductive reasoning. Sherlock Holmes commences investigations by observing all the clues available while keeping an open mind as to the guilty party. By sifting through the clues and analysing the evidence gradually a picture unfolds of the events that occurred leading to an explanation of the crime and the detection of the guilty person. Thus positivists deduce conclusions about phenomena based on testing of pre-stated hypotheses under carefully controlled conditions, while post-positivists induce explanatory hypotheses based on multiple observations of the phenomena under natural conditions.

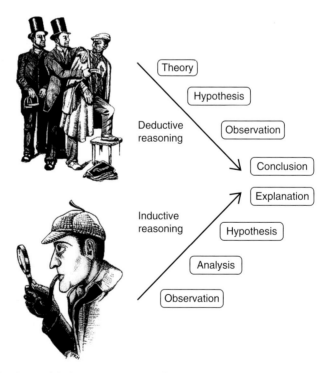

Fig. 1.1. Deductive and inductive reasoning. (Illustrations by Peter R. Hobson.)

While the pros and cons of positivism and post-positivism continue to occupy the time and minds of philosophers, the question for us here is whether these considerations are really of any practical consequence to the modern day science researcher. The application of a positivist approach by scientists over the years has generally proved successful in that it has facilitated the accumulation of a considerable body of knowledge about the universe and the processes that go on within it. It can be convincingly counter-argued, however, that the positivist approach is rather limited, and may cause the researcher to become blinkered to a range of alternative valid explanations of a particular phenomenon once one explanation has been found that readily appears to fit the facts. The post-positivist approach relies more on integrating observations that may be spatially and temporally separated and therefore represents a less confined, more open-ended approach but which may well culminate in more questions being produced rather than a conclusion to a problem being obtained. Of course, modern science research involves aspects of both approaches to solve large problems, for example, while deductive positivist methodology may have led to the precise identification of the active agents in tobacco smoke that induce cells of the human lung lining to become tumorous, it was a post-positivist approach that led to the idea that lung cancer was linked to smoking tobacco in the first place. Therefore, for the individual research scientist working on a highly specific research problem, an awareness of these differing research approaches may be necessary to facilitate the identification of new innovative methodologies that may be applicable and to allow the full range of possible explanations and solutions to the research problem to be recognized.

Having introduced, albeit very superficially, the different strands of research philosophy these must now be put to one side. The aim of this book hitherto is to provide a practical guidance to the design and analysis of experiments in agriculture, horticulture and related scientific disciplines and is thus concerned with the main pillar of the scientific approach, namely hypothesis testing through experimentation.

1.3 Research through Experimentation

In earlier times the scientific research process involved little more than simple observation of a particular phenomenon, followed by the proposition of an explanation of the phenomenon that appeared to fit the facts and then changing the explanation as new observations of the phenomenon were made. Clearly this represented a relatively inefficient and slow approach to the research process so that a phenomenon that is well understood today may in past times have been the subject of a number of alternative explanations that were rigorously defended and argued over by their protagonists for many years, indeed even for many centuries. A good example of this is the identification of the human blood circulation system. The original father figure of human anatomy may be considered to be Galen, a Greek physician of the 2nd century. Galen's writings concerning the structure of the human body were based on dissections of animals, mainly dogs and pigs, and included the assertion that blood was made in the liver. This assertion remained unquestioned for 14 centuries and even Vesalius, who in the early part of the 16th century led a revival of serious interest in the anatomy and physiology of the human body, did not question the role of the liver in this regard. In 1603 Fabricius famously published drawings of the valves in the veins but considered their role was to slow down the flow of blood from the liver to allow it to be more efficiently absorbed by the body's tissues. It was William Harvey who finally described the human circulatory system correctly, showing, through a process of observation and experimentation, that the organ responsible for pumping the blood around the body was the heart and that the role of the valves in the veins was to maintain a one-way flow back towards the heart. This milestone study in human physiology was finally published in 1628.

The evolution of a systematic scientific methodology based on experimentation was slow. Despite the clear lead offered by Galileo in the 16th century and the philosophical pronouncements on experimental method made by Descartes in the 17th century, it was not until the mid-19th century that proper experimental method became a procedure widely adopted by life science researchers. Today, however, investigation by experimentation is the major approach taken by biological researchers in their attempts to explain the mechanisms and causes of biological phenomena. Observational studies are of course very valid and widely undertaken in many areas of science, in particular ecology, but even then observational studies often lead to the formulation of hypotheses that are subsequently tested by experimentation.

What, therefore, do we mean by the term 'experimental approach'? Simply speaking, an experiment is a process through which a phenomenon is observed under a very carefully controlled set of conditions such that, if one condition is varied and all other conditions are held constant, then any observed changes in the phenomenon can be attributed to that one varying condition. Thus the objective of the experiment is to firmly link cause to effect in a way that cannot be achieved through an observational study. Let's take a simple example from biology. It was long ago observed that animals died if placed in a vacuum. With the benefit of scientific hindsight this may seem obvious today but in the 17th century scientists asked why this was. More specifically, it was asked whether animals required all the gaseous components of air in order to live or did they require one particular component? From the initial observation it was not possible to answer this question; however, by subsequent experimentation the answer was quickly resolved. In a series of experiments each one of the gaseous components of air was removed while maintaining all the other components and keeping the experimental conditions constant. It was thereby easily shown that the vital gas necessary for an animal to live is oxygen. The experiment thereby linked a cause to an effect, i.e. it was shown that animals need air because the oxygen contained in the air is essential to the living process. This, however, leads to a new question, namely why is the oxygen necessary? And this of course leads to further experimentation to answer this new question. The experimental approach consists, therefore, of proposing an initial specific hypothesis concerning the cause of a phenomenon, performing an experiment which leads to the hypothesis being accepted or rejected, and then, depending on the conclusion of the experiment, posing a new question/hypothesis that can be similarly tested through a new experiment.

This process may seem second nature to scientific researchers today but it is important to understand how this overall process of finding out by experimentation operates before moving on to discuss experimental design in a more detailed way. In particular, one very important point about experiments should already be clear, that is, that experiments should be designed to answer a very clear and specific question. Planning an experiment to achieve this is frequently much more difficult than it sounds and really successful experiments are only ever achieved through careful planning and design. Even then, once the experiment has been completed and the results collected, it will invariably require detailed and objective analysis of the data before a valid conclusion can be arrived at.

1.3.1 Different types of experimental approach

Experiments in which one or more conditions are deliberately varied while all other conditions are maintained as constant as possible are referred to as **manipulative experiments**. Clearly such experiments are most easily performed under the carefully controlled conditions of a science laboratory. In agriculture and horticulture, however, the growing of crop plants under artificially constant conditions has only a limited practical value. While it might be established through a manipulative laboratory-based experiment that a certain factor has a particular effect on a crop plant, it will usually be necessary to undertake a field experiment to establish if the factor has the same effect on a crop growing under normal field conditions. For example, it may have been discovered that a particular chemical plant growth regulator promotes fruit set in a particular plant when grown under ideal conditions in a glasshouse or growth cabinet; however, it will not be at all certain that this chemical will produce the same effect to the same extent in the plant when grown as a commercial field crop. A **field experiment** will be required to establish this. Environmental biologists and ecologists are of course even more dependent on field experiments; in fact the laboratory experiment may have little relevance to their work at all.

In field experiments it is clearly far more difficult than in the laboratory to manipulate one factor of interest while maintaining all other experimental conditions constant; in fact this may be virtually impossible. In a field experiment it must be expected that external conditions will not be constant and may vary both spatially and temporally. The aim behind the planning of field experiments is, therefore, to ensure that the effects of varying conditions are distributed equally across the experimental subjects, thereby allowing the effect of the manipulated treatment factor to be resolved. This is achieved through the process of **experimental design**, and, while the design of a manipulative laboratory experiment conducted under controlled conditions may be very simple, the design of a field experiment is rarely simple and may become very complex indeed.

As mentioned in the introduction to this chapter, the definition of an experiment may be broadened to encompass some of the types of investigational surveys employed by social scientists. While this might go against the grain with the traditional pure scientist who would consider a social science survey to have more in common with an observational study rather than with a manipulative experiment, a survey using questionnaires and/or interviews can be conducted in a way that employs the principles of experimentally based research. In other words, a survey may set out to establish or disprove a certain pre-stated hypothesis and may be designed to produce quantitative data that can be analysed by statistical techniques, leading to the hypothesis being accepted or rejected. In designing a social science survey many of the same issues need to be addressed as are encountered when designing a scientific field experiment, for example, how can it be ensured that the sample selected represents the population as a whole; how can the effects of secondary influences on the observed response be separated from the particular factor(s) that are being investigated; and how can the collected data be analysed to enable valid objective conclusions to be drawn.

1.4   The Fundamental Rules of Experimental Design

By the term experimental design we mean the total process of planning an experiment including establishing the initial research question, defining the population of subjects to be tested, selecting samples for measurement, identifying and allocating the treatments to be applied, identifying the variables to be measured and identifying the methods of data analysis to be employed.

The design of any experiment can be considered to have three main objectives:

1. To ensure that true treatment effects can be separated from both random variation between subjects and from the effects of non-uniformity in the external experimental conditions.
2. To ensure that observations/measurements can be made without bias.
3. To ensure that the data collected can be subjected to objective and valid analysis.

It may seem obvious but it is none the less important to state that correct and accurate experimental design is essential if an experiment is to result in a valid conclusion being drawn. Many experiments fail because of poor experimental design leading to an inability to attribute an observed result to a particular treatment factor; such experiments are said to be **confounded**. Experiments must be designed to avoid confounding, otherwise they become totally worthless.

The main stages in any experimental design process may be generalized as follows:

- **Statement of the objectives of the experiment.** There must be a clear identification of the particular question that the experiment is to address or the hypothesis that is to be tested.
- **Identification of the resources available for the experiment.** The resources available for the experiment must be clearly identified. Consideration must be given not only to the amount of experimental material available but also to the time and manpower available to carry out the experiment. Many experiments have come to grief due to over-grandiose designs that have led to the experiment proving totally unpractical to perform.
- **Assessment of the location of the experiment and the conditions under which it is to be conducted.** The extent of uniformity in the external conditions must be assessed. The experimental design will need to be extensively modified if the location and/or external conditions are not uniform in nature.
- **Identification of the population of subjects that are to be tested.** It is important to clearly define the population that is to be the subject of the experiment. If samples are to be employed it is essential that the population to which conclusions are to be applied is defined in advance of the experiment and samples are extracted from the correct populations. It is an important rule that statistical analysis does not enable conclusions to be extrapolated from one population to another.

- **Consideration of the amount of variability that is likely to arise within samples.** Where possible, the causes and extent of variation and non-uniformity in the subject material should be identified in advance and consideration should be made concerning how this variability can be minimized.
- **Identification of the type of observation/measurements that are to be made and therefore the type of data that the experiment will generate.** The investigator should be very clear in advance about which variables are to be measured, e.g. plant height, root/shoot ratio, number of stomata per unit leaf area, and thus should be aware of the type of data that will be generated, e.g. continuous measured data, proportional data, discontinuous count data, etc. This is crucial since the type of data determines the method of data analysis to be employed.
- **Identification of treatment groups and assignment of treatments.** There should be a clear specification in advance of the number of treatments and number of replicates required of each treatment. Treatments will then be allocated to subjects or groups of subjects through an identified procedure. A clear unbiased procedure for selecting samples for subsequent measurement must also be formulated.

    Since it is practically impossible to eliminate all variation in experimental material and, therefore, not possible to eliminate all experimental error, it is imperative that errors are maintained at a minimal acceptable level and are spread equally over the whole experiment. This is generally achieved by obeying two golden rules of experimental design:

    Randomize as fully as possible.
    Replicate as much as possible.

- **Identification of the most appropriate technique for analysing the data produced by the experiment.** It is extremely naive to believe that consideration of the statistical analyses to be used need only start once the experiment has been completed and the data collected. The method of statistical analysis to be employed must be identified as part of the experimental design process. The choice of method will depend on the type of measurements being made and the type of data that will be produced. The design of an experiment should ensure that a valid method of statistical analysis can be employed to describe the data and allow conclusions to be drawn from the data. *An experiment that produces data that cannot be objectively analysed is worthless.*

1.5 The Purpose of Statistical Analysis of Experimental Data

The question 'why do we need to apply statistics?' is often asked by students and is certainly a fair question as the answer is not always immediately obvious. Many investigational activities in pure, applied and social science require the collection of numerical data from which a conclusion needs to be drawn. For example, a plant biochemist may need to know the activity of a particular enzyme under a range of different pH conditions, the agricultural scientist may need to know the weight of potato tubers produced when plants were treated with a number of newly formulated fertilizers and the horticultural marketing manager may need to know how many chrysanthemums of different cultivars were sold last year in order to plan for the coming year. In all these cases a large set of numerical data may be available which will require analysis. The branch of mathematics that describes and analyses numerical data is called 'statistics'. Statistical analyses are basically a set of mathematical tools that allow the user to describe numerical data and to objectively evaluate the reliability of conclusions drawn from the data.

The main roles of statistical analysis may therefore be summarized as:

- To describe and summarize data sets.
- To evaluate the validity of using a particular sample(s) to represent a population.
- To identify meaningful trends in data against a background of natural variation.
- To facilitate the drawing of objective conclusions from a data set.

To repeat a very important point, unless objective statistical analyses can be applied to evaluate quantitative experimental data, experimental conclusions drawn from such data are of very limited value. In order to ensure that an experiment is capable of producing data that can be analysed it is very important that *the techniques that are to be used for statistical analysis are identified during the process of experimental planning and before the data are collected*. After all you would not buy a pair of shoes as a present for someone without some knowledge of their foot size, or, if you did, you would be pretty lucky if subsequently they were found to fit!

1.5.1 Describing, summarizing and comparing data sets

One of the primary purposes of the application of statistical analysis is to summarize sets of data so that trends in the data can be discerned. Sets of raw data are difficult to interpret because, apart from anything else, the human brain cannot for the most part instantly remember, sort and recall large banks of numerical values. Furthermore, the more variable the data, the more difficult it is for the brain to analyse and make comparisons. Take, for example, a relatively simple experiment that was designed to inspect the efficacy of a new fertilizer 'Extragrow' by measuring the growth rate of 30 tomato plants grown in pots in a glasshouse that were treated with the new fertilizer and 30 plants grown under the same conditions and treated with a standard fertilizer. The data obtained from such an experiment are shown in Table 1.1. It is clearly extremely difficult, almost impossible, to view the tabulated data and arrive at an immediate conclusion as to the effectiveness of the new fertilizer. The reason for this is twofold. First, there are a relatively large number of data items to appraise and, second, the data values are highly variable. Imagine now that the data had been much simpler, for example, only five plants were measured for each treatment and every 'Extragrow' fertilizer-treated plant had a growth rate of 2.0 g/day and every standard fertilizer-treated plant had a growth rate of 1.0 g/day. Coming to an immediate conclusion about the efficacy of the new fertilizer would now become very straightforward and the application of any statistical analysis would then seem superfluous. Unfortunately experimental data are very rarely as simple as this.

Faced with the data in Table 1.1, it is almost instinctive that we should calculate a mean value for the two sets of plants by summing the values and dividing by the number of observations. In so doing we are applying a simple statistical procedure in an attempt to represent the two data sets each by single values which act as measures of the 'middle value' or **central tendency** of the data sets that can then be easily

Table 1.1. The growth rate (g dry weight/day) of a sample of 30 pot-grown tomato plants treated with a new fertilizer 'Extragrow' and 30 pot-grown plants treated with standard fertilizer (control). Research question: Does the new fertilizer 'Extragrow' improve the growth rate of the plants?

Standard fertilizer treatment			'Extragrow' fertilizer treatment		
0.8	1.3	0.9	1.9	1.7	2.0
1.1	1.4	0.4	2.3	2.2	1.5
0.7	1.8	0.8	1.0	0.8	1.6
0.8	0.9	1.3	1.0	1.1	0.7
1.2	1.1	1.0	1.8	1.2	0.7
1.7	1.6	1.4	1.1	0.9	0.9
1.3	1.0	0.4	1.8	1.5	1.3
1.0	1.2	1.3	2.2	0.7	1.0
1.5	0.7	1.2	1.2	0.9	0.8
0.5	0.5	1.3	1.7	1.1	1.0

compared. When this is done it turns out that the mean growth rate for the 'Extragrow' fertilizer-treated plants is larger than for the standard fertilizer-treated plants as shown below:

	Standard fertilizer	'Extragrow' fertilizer
Mean growth rate of plants:	1.07 g/day	1.32 g/day

So now can we answer the question 'has the new "Extragrow" fertilizer worked?' Despite having the mean values at our disposal, it still remains very difficult to answer this question. This is because the difference between the two means is rather small compared to the amount of variation shown by the samples. The difference between the means could, therefore, be very easily due to the natural variation of the plants rather than to the fertilizer treatment. Since the variability of the data fundamentally affects our ability to interpret the data this variability needs also to be described.

Variability is measured through a number of statistical terms that are collectively known as **measures of dispersion**. When describing a data set statistical measures of dispersion are equally important as measures of central tendency because it is through knowledge of these that the validity of conclusions will ultimately be judged. Measures of dispersion include such statistics as the **range, variance** and **standard deviation** of the data (these will all be discussed in Chapter 2). For example, the standard deviation (which is a measure of the average amount that each value in a data set differs from the mean value) of the two sets of data shown in Table. 1.1 are as follows:

	Standard fertilizer	'Extragrow' fertilizer
Standard deviation:	0.37	0.49

These values are actually rather high compared with the sample means and are indeed considerably higher than the numerical difference between the two means. It becomes apparent, therefore, that trying to draw any firm conclusion from the data in Table 1.1 regarding the efficacy of the new 'Extragrow' fertilizer is actually extremely problematic.

1.5.2 The problem of the variable nature of data in the analysis of experiments

Life for the data analyst would be very much simpler if data did not vary naturally. In fact if data did not vary naturally then there would be virtually no work for the statistician to do. Of course, not every similarly treated plant has the same growth rate; similarly not every leopard (not even twin leopards) has exactly the same number of spots, and not every human of the same sex, age and social background has exactly the same opinion on a particular social question. In other words in our world everything is subject to natural and random variability. Unfortunately, whenever data are obtained, whether it is from a piece of experimental work, from an ecological survey or from a social science survey, the existence of natural variation always clouds the issue when it comes to drawing conclusions. The question is, therefore, how can the observed trends in the data that are due to a particular recognized cause be distinguished from those that are simply due to natural random variation?

In order to further comprehend the problem that faces us let us consider a classical experiment in plant genetics the results of which are represented in Table 1.2. It is suspected that flower colour in a species of pea is controlled by the classical Mendelian inheritance system in which yellow flower colour is genetically dominant and green flower colour is a recessive trait. Consequently, when plants from an F1 generation are crossed it is predicted that yellow and green flowering seedlings will occur in a 3 to 1 ratio. In order to test this idea a cross was carried out and the flower colour in a sample of 100 progeny seedlings was recorded. The question is, assuming that there will be some random variation, what number of yellow and green flowering plants would be expected to occur if the hypothesis that classical Mendelian genetics controls the segregation of flower colour in peas is correct?

In the first trial the results produced a segregation of 75 yellow and 25 green flowering seedlings, i.e. an exact 3:1 ratio, which appears to very clearly support the hypothesis that Mendel's Laws of Inheritance are obeyed in determining flower colour in peas. One would also be fairly happy that the results of the second trial, i.e. a ratio of 72 yellow to 28 green flowering peas, also fits comfortably with the hypothesis.

Table 1.2. The variability of samples from a repeated experiment. Four experimental crosses were made between plants from an F1 generation of pea plants which produce seedlings that flower either yellow or green. The frequencies of the two phenotypes were counted in 100 of the progeny seedlings. The four experiments produced a large variability in the phenotype ratios. Research question: Do the crosses produce a phenotype ratio that adheres to the 3:1 ratio predicted by Mendel's Laws of Inheritance?

	Number of progeny observed		
	Yellow flowering seedlings	Green flowering seedlings	n
Ratio predicted by Mendel's Laws of Inheritance	3 :	1	
Trial cross 1	75	25	100
Trial cross 2	72	28	100
Trial cross 3	65	35	100
Trial cross 4	52	48	100

What, however, would be concluded from the results from the third trial where the observed ratio of 65 to 35 departs appreciably from the expected 3:1 ratio? Now we cannot be so certain of the validity of the conclusion, and in the case of the fourth cross (52:48 ratio) we would probably reject altogether the notion that Mendel's Laws are operating. The question is at what point of departure from the predicted result should the hypothesis be rejected? In fact you might ask yourself again about the 75 to 25 result in trial 1; is it absolutely certain that this result could not have occurred through random variability alone and has nothing to do with Mendel's Laws? The probability of this may seem small but it cannot actually be ruled out.

Natural variability is, therefore, a fundamental problem that besets interpretation of data of all types. One of the main purposes of applying statistical analyses is to provide a means for coping with this problem and to enable real trends and effects to be identified against the background of inherent natural variation. This is achieved in the main part by determining the probability of a particular effect being due to a particular cause while accepting that, due to the existence of natural variability, we can never be 100% certain about the correctness of any conclusion that we may come to. So while cynics may often quote the old adage that 'statistics can be used to prove anything' in fact it is the converse that is true, that is, statistics cannot prove a thing. Instead statistics are used to show the likelihood of an effect having a particular cause. Clearly, if the likelihood is very high, and the same result is found again and again in repeated experiments, then the conclusion will become widely accepted and the outcome will then contribute to the knowledge base.

1.5.3 Using statistical data analysis to evaluate data trends and draw conclusions

To reiterate, researchers have to accept the fact that they can never be 100% certain that an effect observed in an experiment has a particular cause. What they can do, however, is to determine the **probability** that this is the case. Based on statistical measures of the central tendency in relation to the measured variation of sets of data, it is possible to determine the probability that certain trends or differences observed are due to a particular treatment or cause. This is achieved through a statistical procedure called a **statistical test of significance**.

Significance tests operate in what may initially seem to be a rather convoluted manner for, rather than determining the probability that a treatment effect is present, they instead assess the probability that the observed data comply with the negative hypothesis that no treatment effect exists. If it subsequently transpires that the probability of the data agreeing with the negative hypothesis is sufficiently small, for example less than 5%, then it may be concluded that a positive treatment effect does indeed exist and it is this treatment effect that primarily explains the observed data. For an illustration of this procedure let us return to the investigation of the possible effect of the new fertilizer 'Extragrow' described in Table 1.1. By application of an appropriate test of statistical significance the probability of obtaining the observed difference in the two treatment means, or a greater difference, is determined given the assumption that the new fertilizer 'Extragrow' works neither better nor worse than the standard fertilizer. This probability, denoted P, turns out to be only 1.5%, i.e. $P = 0.015$. In this example, therefore, the likelihood that the 'Extragrow' fertilizer failed to cause an effect is very low (<1.5%) and the conclusion that 'Extragrow' must be the primary cause of the effect observed becomes fairly convincing. In this case statisticians would allow a positive conclusion to be stated, although it must be recognized that we are still left with a 1.5% doubt that this positive conclusion is the correct one. (Note that the procedure and calculation of the appropriate test for the fertilizer experiment in Table 1.1 is described in Chapter 6.) Similarly, in Table 1.1 another type of statistical significance test called a 'goodness-of-fit' test can be used to assess the likelihood that the observed frequency of flower colour produced by any one of the trial crosses adheres to a frequency predicted by Mendel's Laws of Inheritance (see Chapter 14).

Statistical significance tests are the basis for the analysis of most types of experimental data and the general concept and principles that govern their use are discussed initially in Chapter 5. The specific principles of operation of a range of different types of significance test are then described in some detail in the subsequent chapters of this book. This is, however, moving ahead too quickly and there is rather a lot of ground to cover first before we can progress to looking at hypothesis testing in detail.

2 Descriptive Statistics

Statistics: numerical facts systematically collected; science of collecting, classifying and using.

(Concise Oxford English Dictionary)

- Measures of the central tendency of data sets: means, median and mode.
- Measures of the dispersion of data sets: range, interquartile range, sum of squares, variance, standard deviation, coefficient of variation.
- Measure of sample reliability: standard error of the mean.
- Graphical representation of descriptive statistics.

2.1 Introduction to Descriptive Statistics

Several basic statistical calculations are used to describe and summarize data sets which are fundamental to all further statistical analyses. These calculations are referred to as **descriptive statistics** because they simply describe the tendency and variability of the data. They cannot, on their own, be used to draw conclusions from collected data. It is assumed that these descriptive statistics will be largely familiar to the reader and are described here in a fairly summary manner. However, their description will serve to introduce the terminology and symbols that will be employed throughout the subsequent chapters. In defining many descriptive statistics, it is necessary to recognize whether the data set in question represents a population, that is the complete set of all possible values, or a sample, that is a limited set of values drawn from a population. (The concept of populations and samples is discussed in detail in Chapter 4.)

2.2 Measurements of Central Tendency

Data sets usually show a central tendency in their frequency distributions, i.e. the majority of the individuals fall around the middle of the range, while extreme values occur relatively rarely. Therefore, a data set can first be described by measuring the middle value around which the data is dispersed. This is most commonly achieved through the mean.

2.2.1 The mean

The mean is the most important and widely used measure of central tendency. Unless otherwise stated it can generally be assumed that whenever the term 'mean' is used it is the arithmetic mean that is being referred to.

The **arithmetic mean** is simply the sum of the individual values divided by the number of values. The mean of a population is denoted by μ (i.e. lower case *mu*) and N denotes the number of individuals in a population; thus in algebraic terms the population mean is given by:

$$\mu = \frac{\sum x}{N}$$

When it is not possible in practice to determine the mean of an entire population, a representative sample may be drawn from the population. The mean of a sample is denoted by $\bar{x}$ (i.e. '*x* bar') and the number of individuals in a sample is denoted by n. Thus:

$$\bar{x} = \frac{\sum x}{n}$$

The sample mean, $\bar{x}$, is then employed to estimate the value of the population mean, μ.

Occasionally and under particular circumstances, two alternative calculations of the mean are employed. The **geometric mean** is considered a better measure of central tendency than the arithmetic mean when the data show an exponential increase, e.g. plant or animal growth, compound interest data, etc. The geometric mean is the *n*th root of the product of the data values where n represents the number of x values in the data set, i.e.:

$$\text{geometric mean} = \sqrt[n]{x_1 \times x_2 \times x_3 \times ... \times x_n}$$

The **harmonic mean** is often considered a better measure than the arithmetic mean when determining the average size of a set of unequal sized samples. The harmonic mean, Hx, is determined by finding the arithmetic mean of the reciprocal of the x values and then taking the reciprocal of this value. The formula simplifies to:

$$\text{harmonic mean } (Hx) = \frac{n}{\sum \dfrac{1}{x}}$$

Neither the geometric nor the harmonic means are employed in any major way in the analysis of experimental data. Therefore, while both statistics have been defined, they need no further consideration here.

2.2.2 The median

The **median** is the value of the middle observation when the data are ranked in an ascending (or descending) order. If there is an even number of observations, then the median becomes the midpoint between the values of the two middle observations. Thus:

when n is odd: median $= x_{[(n + 1)/2]}$; when n is even: median $= (x_{[n/2]} + x_{[(n/2)+1]})/2$

where $x_{[i]}$ is the *i*th value when the x values are ranked.

The median is commonly used to describe the central tendency of data sets in which the values are not evenly dispersed. In data sets containing either a few extremely high or a few extremely low values compared with the majority of values, the median may still give a good indication of the central tendency of the data, whereas the mean tends to be pulled away from the middle towards extreme values. A number of important data analysis techniques, to be described much later, are based on ranking the values in different data sets and then comparing the median values.

2.2.3 The mode

A third measure of central tendency used occasionally is the **mode**. The mode is the most frequently occurring measurement in a data set (i.e. the value at which the frequency distribution reaches a peak). The mode is a useful descriptive statistic only when very large data sets are being described in which there are many repeated values. The mode does not, however, play a role in the more complex quantitative data analysis techniques used to analyse experimental data.

Example 2.1. Calculation of measures of central tendency.

A sample contains 20 observed values ($n = 20$):   3, 7, 10, 0, 2, 8, 3, 11, 2, 4, 4, 9, 7, 1, 18, 3, 13, 17, 3, 10

The arithmetic mean = ($\Sigma x/n = 135/20 =$)   **6.75**

Placed in ascending order the values are: 0, 1, 2, 2, 3, 3, 3, 3, 4, 4, 7, 7, 8, 9, 10, 10, 11, 13, 17, 18

The value of the two middle observations are 4 and 7, thus the **median = 5.5**

(Note that in accordance with the definition of the median, there are an equal number of values, i.e. 10 that are less than 5.5 and greater than 5.5 in the data set.)

The most frequently occurring value is 3 (a total of four observations), thus the **mode = 3**

2.3 Measurements of Dispersion

As with the mean, the dispersion (or variability) of a population and the dispersion of a sample have to be separately defined. Just as a sample mean is used to estimate a population mean, the dispersion of a sample is used to estimate the dispersion of the population from which the sample is drawn.

2.3.1 Range and interquartile range

The **range** is the simplest measure of dispersion, being the difference between the maximum and minimum values in a data set. Although sometimes expressed, it is

---

of relatively little use because it does not account in any way for the frequency distribution of the data. A set of data in which all but one of the values are the same but this one value is much larger than the rest could have exactly the same range as a set of data comprising of different values all widely dispersed between the smallest and the largest values. For example the following two samples have exactly the same range although they clearly have very different patterns of data distribution:

Data set A	Data set B
12, 23, 4, 15, 15, 8, 20, 13, 9, 10	19, 6, 5, 4, 6, 7, 7, 7, 6, 5, 6, 7, 23

In both data sets the range = 23 − 4 = **19**.

In order to describe more adequately the distribution of a data set that contains a limited number of extreme values the **interquartile range** is often expressed. This is the interval between the data value that cuts off the lowest 25% of all values, i.e. the **lower quartile** (Q_L), and the data value that cuts off the highest 25% of all the values, i.e. the **upper quartile** (Q_U).

The most straightforward way to determine the quartiles is to use the median to split the data set into two halves. The lower quartile is then the median of the data in the lower half of the data set and the upper quartile is the median of the data in the upper half of the data set.

Following ranking of the values, the interquartile range of the two data sets shown above can thus be determined:

Data set A	Data set B
⌐4, 8, 9, 10, 12,⌐ ⌐13, 15, 15, 20, 23⌐	⌐4, 5, 5, 6, 6, 6,⌐ 6 ⌐7, 7, 7, 7, 19, 23⌐
median (12.5)	median (6)
Q_L (9) Q_U (15)	Q_L (5.5) Q_U (7)
interquartile range = 15 − 9	interquartile range = 7 − 5.5
= 6.0	= 1.5

The interquartile ranges clearly indicate the much more widely spread data distribution of data set A compared with data set B, a feature that is not apparent when just comparing the ranges.

It should be noted, however, that there are a number of alternative ways of calculating the quartiles precisely and these will often produce slightly varying values. For example, different quartile values from those obtained above will be found if the overall median is included when determining the medians of the lower and upper halves of the data set. Other techniques apply a weighting factor to quartiles when they fall between actual data values. Strange as it seems, there is no universal agreement amongst statisticians, or computer statistical packages, over which approach is best.

2.3.2 Sum of squares (SS)

A more sophisticated approach to describing the dispersion of a data set that takes into account the data distribution frequency is to determine the difference between each individual value and the mean value. The summation of these differences will then provide a relative measure of the variability of the data. Before summing, however, the deviations from the mean are squared in order to remove negative values (otherwise the sum of all deviations from the mean will simply equal zero!). The summation of the squared deviations yields the statistic called the **sum of squares** (denoted **SS**). Again, however, the sum of squares of a sample and of a population must be separately defined:

$$\text{sample SS} = \Sigma\left(x - \bar{x}\right)^2$$

$$\text{population SS} = \Sigma\left(x - \mu\right)^2$$

Once there are more than just a few data values, use of the formulae above for the calculation of SS becomes a very long-winded process. Fortunately, the formula can be mathematically rearranged to yield an alternative formula that, while looking complicated, is much more convenient to use when employing computers and calculators. The alternative formula is:

$$\text{SS} = \Sigma x^2 - \frac{\left(\Sigma x\right)^2}{n}$$

where: Σx^2 = the sum of the squares of all of the values in the sample or population
$(\Sigma x)^2$ = the square of the total of all the values

It may be noted that the squaring of the deviations before summing has another beneficial effect in that, in measuring the variability of a data set, it causes more weight to be placed upon extreme values and thereby more clearly separates highly variable from less variable data sets. The benefit of this is demonstrated in Example 2.2

Example 2.2. The effect of squaring deviations around the mean on the measurement of variation.

Two samples A and B have equal means and similar sum of deviations, $\Sigma[x - \bar{x}]$, but sample A has over twice the sum of squares, $\Sigma[x - \bar{x}]^2$, than sample B, indicating that it has a greater variability.

	Sample A	($\bar{x}$ = 10)			Sample B	($\bar{x}$ = 10)	
	x	$[x - \bar{x}]$	$[x - \bar{x}]^2$		**x**	$[x - \bar{x}]$	$[x - \bar{x}]^2$
	6	4	16		10	0	0
	10	0	0		12	2	4
	14	4	16		8	2	4
	9	1	1		12	2	4
	11	1	1		8	2	4
$\bar{x}$	10				10		
$\Sigma[x - \bar{x}]$		10				8	
$\Sigma[x - \bar{x}]^2$ (= SS)			34				16

2.3.3 Variance (σ^2, s^2)

Although the sum of squares gives a measure of the total variation of the data set, its magnitude will increase as the number of data items increases. This is not very help-ful when comparing the variation of data sets of different size. Therefore, in order to obtain a measure of average variation the sum of squares (SS) is divided by the size of the sample or population. This produces a statistic that might be called the 'mean sum of squares' but is actually referred to as the **variance**.

The variance of a population is denoted by the symbol σ^2. Thus:

$$\sigma^2 \quad = \quad \frac{SS}{N} \quad = \quad \frac{\sum x^2 - \dfrac{(\sum x)^2}{N}}{N}$$

The calculation of the variance of a sample is not quite the same as for a population. To obtain the variance of a sample, SS is divided not by the sample size n but by $n - 1$, where $n - 1$ is the **degrees of freedom** (DF) of the sample. The reason for dividing SS by the degrees of freedom is that it improves the accuracy with which the sample variance acts as an estimate of the population variance. The concept of degrees of freedom is theoretically complex and will be discussed in more detail later.

The variance of a sample is denoted by the symbol s^2. Thus:

$$s^2 \quad = \quad \frac{SS}{n-1} \quad = \quad \frac{\sum x^2 - \dfrac{(\sum x)^2}{n}}{n-1}$$

2.3.4 Standard deviation (σ, s)

There is still a problem with using variance as a measure of variability in that it includes a square function and it is not, therefore, directly related to the magnitude of the actual data and cannot be expressed in the same units as the actual data. This is easily rectified by taking the square root of the variance, a calculation that finally produces the **stand-ard deviation**. To repeat, **standard deviation is the square root of the variance**. In prac-tice, standard deviation is a measure of the average amount that any randomly chosen value within a data set may be expected to differ from the mean value and it constitutes the single most important descriptive measure of data variability.

Population standard deviation is denoted by the symbol σ. Thus:

$$\sigma \quad = \quad \sqrt{\frac{SS}{N}} \quad = \quad \sqrt{\frac{\sum x^2 - \dfrac{(\sum x)^2}{N}}{N}}$$

Sample standard deviation is denoted by the symbol s. Thus:

$$s \quad = \quad \sqrt{\frac{SS}{n-1}} \quad = \quad \sqrt{\frac{\sum x^2 - \dfrac{(\sum x)^2}{n}}{n-1}}$$

2.3.5 Coefficient of variation

As a measure of variability, the magnitude of both standard deviation and variance will depend on the magnitude of the data and cannot, therefore, be used to compare the variability of a measured variable in samples of different types of item. To illustrate this, assume that the average area of a banana leaf is approximately 100 times that of a pea leaf. If the area of leaves is measured in a sample of bananas and peas, although the proportional variability within the samples may be similar, the respective values of s and s^2 will be approximately 100 times and 100^2 times greater in the sample of banana leaves than in the sample of pea leaves. Therefore, neither the standard deviation nor the variance could be used sensibly for comparing the variability of leaf area of bananas with that of peas.

A measure of variability that can be used to compare samples of different items is obtained by dividing the standard deviation by the sample mean. This gives a statistical value termed the **coefficient of variation** and is commonly denoted **CV**. Thus:

$$\mathrm{CV} \quad = \quad \frac{s}{\bar{x}}$$

The coefficient of variation allows the direct comparison of the variation of data sets that comprise of values of different magnitude and thus have very different means.

2.4 Measurement of Sample Reliability: Standard Error of the Mean

Whenever the mean of a single sample is quoted the inference is that the sample mean represents the mean of the entire population from which the sample was drawn. Clearly, therefore, it is important to be able to indicate how well the sample represents the population. The statistical value called the **standard error of the mean**, or more simply the 'standard error' as it is commonly termed, specifically indicates the **reliability** of a sample mean as an estimate of the population mean.

Once the sample standard deviation (s) has been determined, the standard error (SE) is calculated by dividing s by the square root of the number of observations. Thus:

$$\mathrm{SE} \quad = \quad \frac{s}{\sqrt{n}}$$

The theory behind this equation is quite complex and will be discussed later. It will be clear, however, that since the sample size n is the denominator of this equation, albeit as a square root, standard error must automatically increase as sample size is decreased. This tells us something that is fairly obvious but extremely important, i.e. that smaller samples are less reliable. Since it is so important to give an indication of the reliability of a sample, the *standard error should be regarded as a basic descriptive statistic that should always accompany the presentation of a sample mean*. Note that standard error is commonly expressed as a plus/minus value around the sample mean, for example where a sample mean is 1.5 with a standard error of 0.25 then we may express this simply as: $\bar{x} = 1.50 \pm 0.25$.

The calculation of a full set of descriptive statistics for a data set is illustrated in Example 2.3.

Example 2.3. The calculation of statistical measures of data dispersion.

The growth rate (g dry weight/day) of a sample of 30 pot-grown tomato plants treated with a new fertilizer 'Extragrow' and 30 pot-grown plants treated with standard fertilizer (control) (data as in Fig. 1.2).

Growth rate (g dry weight/day)					
Standard fertilizer treatment (control)			'Extragrow' fertilizer treatment		
0.8	1.3	0.9	1.9	1.7	2.0
1.1	1.4	0.4	2.3	2.2	1.5
0.7	1.8	0.8	1.0	0.8	1.6
0.8	0.9	1.3	1.0	1.1	0.7
1.2	1.1	1.0	1.8	1.2	0.7
1.7	1.6	1.4	1.1	0.9	0.9
1.3	1.0	0.4	1.8	1.5	1.3
1.0	1.2	1.3	2.2	0.7	1.0
1.5	0.7	1.2	1.2	0.9	0.8
0.5	0.5	1.3	1.7	1.1	1.0

Simple observation of the raw data provides no indication of the effectiveness, or otherwise, of the fertilizer. Some descriptive statistics are required to summarize the data and to allow some comparison of the data.

		Control	'Extragrow' fertilizer treatment
Number of observations (n)	=	30	30
Total (Σx)	=	32.1	39.6
Mean ($\bar{x}$)	=	1.07	1.32
Range	=	1.4	1.6

Calculation of sum of squares (SS)

Σx^2		= 38.39	59.28
$(\Sigma x)^2$		= 1030.41	1568.16
$SS = \Sigma x^2 - \dfrac{(\Sigma x)^2}{n}$		= 4.04	7.01

Calculation of sample variance (s^2)

$$s^2 = \frac{SS}{n-1} \qquad = \quad 0.14 \qquad\qquad 0.24$$

Note that the data are from samples and thus the denominator for the calculation of variance is $n - 1$, i.e. the sample degrees of freedom, rather than n.

Calculation of standard deviation (s)

$$s = \sqrt{s^2} \qquad = \quad 0.37 \qquad\qquad 0.49$$

Clearly the fertilizer-treated sample is much more variable than the untreated sample.

Calculation of standard error (SE)

$$SE = \frac{s}{\sqrt{n}} \qquad = \quad \pm 0.07 \qquad\qquad \pm 0.09$$

Note that in the calculations above, in accordance with accepted practice, the final solutions have been rounded to just one more decimal place than is used to display the original data.

2.5 Graphical Representation of Descriptive Statistics

When presenting experimental results statistical measures of data dispersion can be shown graphically to important effect.

2.5.1 Error bars

Whenever graphs are used to display sample mean values rather than individual data values, the standard error associated with the sample means should normally be displayed. The recognized way of achieving this is to express the standard error as a plus/minus range around the mean and to plot this range on the graph as a vertical bar through the mean. Such bars are commonly referred to as **error bars**. This can be undertaken on both bar charts and line graphs as shown in Fig. 2.1.

The graphs depicted in Fig. 2.1 show very clearly that in a wheat growth trial, as would be expected, there was a decreasing trend in the amount of light penetrating to the soil surface with increasing canopy area. The standard error bars are, however, noticeably large in the first two samples indicating that the means of these samples are relatively unreliable as an indicator of the true effect and suggest that more replicate measurements should have been made.

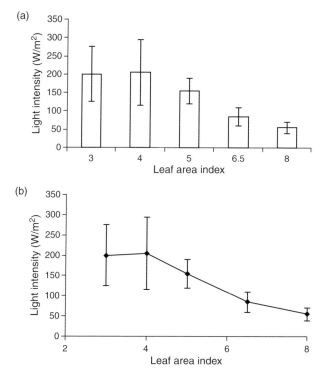

Fig. 2.1. Example of the use of error bars to display the standard error of the mean on (a) a bar chart (b) a line graph. (a) Bar chart showing the mean light intensity at the soil surface within a wheat crop growing at different leaf canopy areas (vertical error bars indicate ± SE, n =10). (b) Line graph showing the mean light intensity at the soil surface within a wheat crop growing at different leaf canopy areas (vertical error bars indicate ± SE, n =10).

2.5.2 Box plots

A box plot (or 'box and whisker' plot as they are sometimes called) is a common graphical technique used to depict the variability of a set of samples and allows a rapid visual comparison of some of the statistical characteristics of the samples.

Using a vertical graphical scale a box (usually a rectangle) is plotted that contains the middle 50% of all the data values of the sample. The base of the box thus represents the lower quartile (Q_L) of the data range and the top of the box represents the upper quartile (Q_U) of the data range. The width of the box can be used to indicate the relative sample size. Within the box the position of the median is shown, usually by a dashed line. From each end of the box single vertical lines are plotted that are called **whiskers**. The 'whiskers' are plotted from each end of the box to the most extreme data value that falls within a further 1.5 × the box length, that is 1.5 times the interquartile range. If the data are perfectly symmetric around the mean then it can be shown mathematically that this range will include 99.3% of all the data values; if the data are non-symmetrically distributed (i.e. skewed) then the proportion of values that lie in this range may be considerably less than this. All data values that occur beyond the end of the whiskers are referred to as **outliers**. These data points are plotted individually, very often using different symbols to indicate those outliers that are within a further 3 × the box length and those far outliers that occur even beyond this point. By comparing the box plots for a number of samples it is then very straightforward to assess their relative variability and the shape of their data distributions.

The construction of a box plot for two contrastingly dispersed data samples is shown in Fig. 2.2. One of the samples (A) has a symmetrical data distribution, i.e. a similar number of values spread each side of the mean, while the other sample (B) is highly non-symmetrical or 'skewed'.

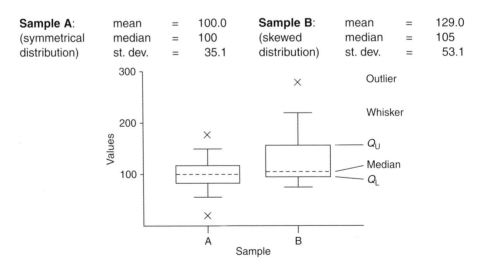

| **Sample A:** (symmetrical distribution) | mean median st. dev. | = = = | 100.0 100 35.1 | **Sample B:** (skewed distribution) | mean median st. dev. | = = = | 129.0 105 53.1 |

Fig. 2.2. Construction of a box plot for two samples. Q_L and Q_U are the lower and upper quartiles. The whiskers include all the data values that lie within a 1.5 × interquartile range beyond the quartiles. More extreme outlier values are individually plotted.

3 Data Distributions and their Use

Datum (pl. data): thing known or granted; fixed starting point of scale etc.; fact of any kind.

(Concise Oxford English Dictionary)

- Introduction to data distributions.
- The main types of data frequency distribution.
  - binomial distribution;
  - Poisson distribution;
  - negative binomial distribution;
  - normal distribution;
  - skewed distributions.
- Mathematical description and characteristics of the normal distribution.
- Use of the normal distribution in testing hypotheses.
- Inspecting for normal distributions.

3.1 Introduction to Data Frequency Distributions

The frequency distribution of a data set is simply a count of the number of times that each value or item occurs within the data set. Data distributions are shown graphically by plotting the frequency of occurrence on the vertical Y-axis against the values or categories on the horizontal X-axis. The frequency distribution of a simple discontinuous data set is plotted in the form of a **bar chart** with discrete values or categories labelled along the X-axis and the frequency of occurrence plotted as columns along the Y-axis. The frequency distribution of a continuous data set is plotted in the form of a **histogram.** In a histogram the data are segregated into class intervals that are plotted on a continuous scale along the X-axis and the number of observed values occurring within each class interval plotted as columns along the Y-axis. The area of each column is, therefore, proportional to the frequency. The frequency distribution of a continuous data set can also be plotted as a **frequency polygon** in which the frequency of observed values is plotted as a point against the midpoint of each class interval and adjacent points may then be joined by lines. These graphical formats are illustrated in Fig. 3.1.

Data sets generally adhere to one of a number of recognized patterns of distribution. It is through a theoretical understanding of these distributions that the probability of occurrence of specified data values can be determined. Since statistical analysis of experimental data is based on determining the probability of obtaining a certain

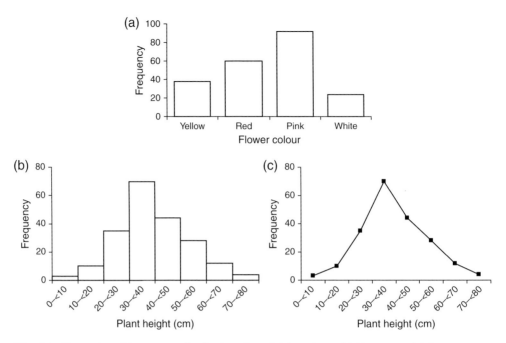

Fig. 3.1. Examples of frequency distribution plots: (a) bar chart; (b) histogram; (c) frequency polygon.

result, it thus becomes clear that the method of data analysis to be used for analysing a particular set of data will be dependent on the type of data distribution to which the data set adheres. It is of great importance, therefore, that the main types of data distribution likely to be encountered are recognized and understood.

3.2 Main Types of Data Frequency Distribution

Four main types of distribution are commonly encountered in life sciences. The **binomial distribution**, the **Poisson distribution** and the **negative binomial distribution** describe possible ways that discrete count data may be distributed. Measured continuous data most commonly adheres to the **normal distribution.** In addition data distributions may be identified as being symmetrical or skewed.

3.2.1 The binomial distribution

Binomial frequency distributions arise when the data refer to counts of items which possess a particular attribute and the attribute in question falls into one of two possible but mutually exclusive categories. For example:

- Counts of individuals that are either male or female (see Fig. 3.2).
- Counts of seeds that are either live or dead.
- Counts of the number of times that a tossed coin comes down heads or tails.

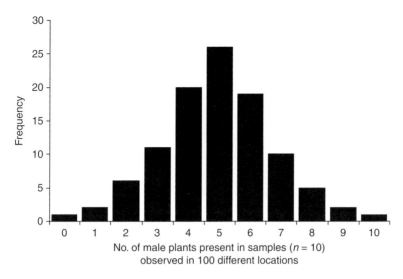

Fig. 3.2. A binomial distribution. Frequency of male plants in wild populations of a dioecious species.

The shape of a binomial distribution is governed by the probability of the occurrence of each observation. Where the probability of each observation is equal, i.e. $P = 0.5$, the peak of the distribution will lie in the middle of the range. An example is shown in Fig. 3.2. If, however, the probability of the two mutually exclusive observations is unequal then the peak will be shifted towards one end of the range or the other and the distribution will become skewed.

If it known that a particular population adheres to a binomial distribution and if every possible sample of given size n was taken then the population mean (μ), population variance (σ^2) and population standard deviation (σ) are given by:

Mean = $\mu = p \times n$
Variance = $\sigma^2 = p \times q \times n$
Standard deviation = $\sigma = \sqrt{p \times q \times n}$

where: p = probability of occurrence of the observation
q = probability of non-occurrence of the observation (note that $q = 1 - p$)
n = sample size

Since by definition the sum of p and q must be 1 then for a binomial distribution *the mean will always be larger than the standard deviation*. In the example shown in Fig. 3.2, if it is assumed that the probability of occurrence of a male plant is 0.5, then the theoretical mean frequency of male plants in all samples of size 20 extracted from the population is $0.5 \times 20 = 10$, and the standard deviation for this frequency is $\sqrt{0.5 \times 0.5 \times 20} = 2.24$.

If the theoretical probability of the occurrence of a particular observation or event is known then the predicted binomial distribution to which the observed data should fit may be calculated. The theoretical basis of the calculation need not concern us here but the formula is a useful one to note:

$$P = \frac{n!}{(x!) \times (n-x)!} \times p^x \times q^{(n-x)}$$

where: P = probability of a given number of observations (x) occurring within a sample (e.g. expected probability of occurrence of a seed batch in which there are x number of live seeds)

p = probability of occurrence of the observation (e.g. probability of a seed being alive)

q = probability of non-occurrence of the observation (e.g. probability of a seed not being alive) (note that $q = 1 - p$)

n = sample size (e.g. total number of seeds per batch)

The binomial equation may therefore be used to predict the outcome of a particular event. For example a seed producer may usefully predict the number of packets of seed that contain a given proportion of viable to non-viable seed as illustrated in Example 3.1.

Example 3.1. Calculation of a binomial probability.

A seed company sells a particular seed in packets of ten. From germination trials it has been determined that the probability of the seed being viable is 80% ($P = 0.8$). For marketing purposes the company needs to know the expected frequency of packets that contain zero and ten viable seeds respectively.

The probability (P) of all ten seeds in any one packet being viable would therefore be:

$$P = \frac{10!}{(10!) \times (10 \times 10)!} \times 0.8^{10} \times (1 - 0.8)^{(10-10)}$$

$$= 0.11 \quad (\text{i.e.} 11\%)$$

The probability (P) of any one packet containing zero viable seeds is similarly calculated:

$$= 1.0 \times 10^{-7} \qquad (\text{i.e.} 0.00001\%)$$

Fortunately it appears to be extremely unlikely that a customer will ever purchase a packet in which no seeds at all will germinate!

The binomial equation also allows us to inspect whether an observed frequency distribution actually conforms to a binomial distribution. This can be achieved by using the binomial formula to determine the expected number of occurrences of each value or category in the data set assuming a binomial distribution and then employing a statistical test called a chi-squared test to examine if the observed frequency and the theoretical binomial frequency are statistically similar. (Chi-squared tests will be discussed in a much later chapter.) This procedure can be important in testing hypotheses concerning the possible underlying causes of a set of data, for example, in genetic studies where it is often required to test whether the observed ratio of phenotypes in a progeny adhere to an expected frequency and therefore whether an expected segregation of genes has taken place.

3.2.2 The Poisson distribution

Poisson frequency distributions arise when the data refer to counts of things that occur within a defined sampling unit, and their occurrence is relatively rare. The things being counted must also be independent of time and of each other, they must not mutually repel or attract each other. Possible examples are:

- The number of weed plants that occur in a specified sample area within an agricultural crop (see Fig. 3.3).
- The number of fungal spores on a leaf.
- The number of goals scored in 90 minutes of a soccer match.

Since the probability p of an event occurring is by definition low and the value zero truncates the distribution, then Poisson distributions are typically asymmetrical with a longer tail extending towards the right-hand side and the peak tending towards the left-hand side of the distribution. A typical Poisson distribution is shown in Fig. 3.3.

A special statistical feature of the Poisson distribution is that *the population mean and the population variance are equal*, i.e. $\mu = \sigma^2$.

As with the binomial distribution, there is a theoretical mathematical formula that describes the Poisson distribution and allows the frequency of occurrence of an observation to be predicted. The formula can be expressed as:

$$P = e^{-\mu} \times \frac{\mu^x}{x!}$$

where: P = probability of x number of observations occurring within a sample (e.g. probability of occurrence of a randomly positioned quadrat in a crop field containing x number of weed plants)

μ = population mean (e.g. mean number of weed plants per quadrat)

e = the exponential mathematical constant (equal to 2.718...)

This equation can be used to predict frequency values and to inspect whether an observed data distribution conforms to a Poisson distribution and thereby whether it adheres to a particular hypothesis about the underlying cause of the data.

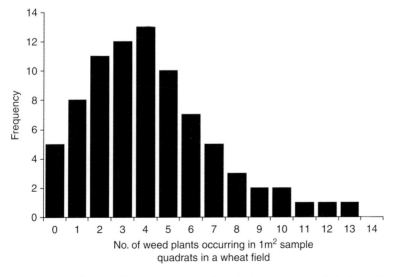

Fig. 3.3. A Poisson distribution. Frequency of weed plants in sample quadrats in a wheat crop.

A further feature of the Poisson distribution worth noting is that when the probability p of an event is very small such that the population mean μ is small, then the Poisson distribution becomes very similar to binomial distributions in which n is large and p is small. Therefore the Poisson equation can be used to estimate the probability of a binomially distributed event occurring where the probability of the event is low, i.e. less than about 0.01.

3.2.3 The negative binomial distribution

The negative binomial distribution (also known as the Pascal distribution) has, as the name suggests, a type of inverse relationship to the binomial distribution. If we have a defined binomial event such as a a randomly selected seed being non-viable, then we can count the number of seeds we need to inspect before we obtain a given number of non-viable seeds. If this process is repeated then the number of sampling trials required to produce the required event a given number of times adheres to a negative binomial distribution.

The main application of the negative binomial distribution in life sciences is to situations where a Poisson distribution might have been applied but is shown to be invalid because the observations have a variance which is larger than the mean, that is the data are **overdispersed**. (Note that for a Poisson distribution the mean = variance but for a negative binomial distribution the mean < variance.) This occurs, in particular, when the data refer to counts of items that occur within a defined sampling unit but the occurrence of individual items is not independent and may exhibit mutual attraction. Consequently the items being counted may occur in clusters. Negative bionomial distributions are relatively rare in experimental investigation but they can successfully describe a range of natural biological situations, for example, the distribution of insect eggs on a leaf and the distribution of antelopes on a grassy plain.

As with other data distributions a theoretical mathematical formula can be derived that describes the negative binomial distribution; however, this is particularly complex. Since its use is only very rarely encountered in experimental investigations the negative binomial distribution will not be considered further here and readers who wish to pursue this topic further are referred to more advanced texts, e.g. Sokal and Rohlf (1994).

3.2.4 The normal distribution

When a continuous variable is measured on all items within an unbiased and reasonably large population or sample and plotted on a histogram as a frequency distribution, the result is commonly a symmetrical bell-shaped curve. This is simply because the majority of the items will tend to have values around the middle of the range so the curve shows a peak in the middle, while there will be progressively fewer values towards the extreme ends of the range which then produce 'tails' on either side of the distribution. Data that follow this pattern adhere to the **normal distribution**. Examples might include measurements of height of plants in a tomato crop, fresh weight yield from potato plants in a field or the concentration of protein in cow's milk in a particular dairy herd. A typical normal distribution is depicted in Fig. 3.4.

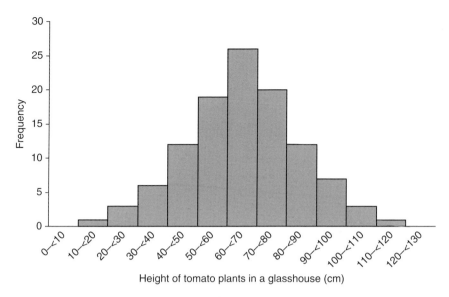

Fig. 3.4. A normal distribution. Distribution of height in a tomato crop.

Clearly if the number of data values plotted is rather small then it would be naive to expect these values to lie exactly on a symmetrical bell-shaped curve but instead will at best only approximate such a curve. However, as the sample or population size is increased the shape of the distribution will become increasingly more defined. With an infinite number of observations the distribution will become exactly symmetrical and will produce a perfect normal distribution curve that can be precisely described mathematically. The normal distribution is very important in statistics and is the basis of an important branch of experimental statistical analysis termed **parametric statistics**. It will be necessary, therefore, to explore the theory of normal distributions in more detail later.

3.2.5 Skewed distributions

When a data distribution displays a disproportionately large number of values towards one end of the range or the other so that the data distribution is not symmetrical the distribution is referred to as being **skewed**. If there is a disproportionately large number of values at the high end of the data range that therefore produces a long right-hand tail, the distribution is termed **positively skewed**. If there are a large number of values at the low end of the range causing a long left-hand tail, the distribution is said to be **negatively skewed**. A very common cause of skewness in data distributions is the existence of bias in the sample selection procedure that favours items of either high or low value. Skewness can, however, have a perfectly natural biological cause. For example, a lack of genetic homogeneity in the material being examined may cause a particular genotypic trait to be favoured that then affects the distribution of the variable being measured.

The extent of skew can be quantitatively described through a descriptive statistic called the **coefficient of skewness**, commonly denoted by the symbol g_1. There are

---

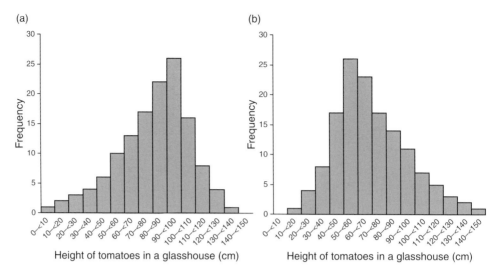

(a)

(b)

Height of tomatoes in a glasshouse (cm)

Height of tomatoes in a glasshouse (cm)

Fig. 3.5. Skewed data frequency distributions. (a) A negatively skewed distribution. The height of tomato plants in a glasshouse that are affected by non-lateral shading causing a disproportionate number of plants to be short. (b) A positively skewed distribution. The height of tomato plants in a glasshouse that are affected by drift of a growth-promoting chemical from one side due to spraying of an adjacent crop causing a disproportionate number of plants to be tall.

several different methods for determining this coefficient and there is no great consistency in the methods used by different statisticians and computer packages. If not perfect, one of the more simple ways to determine a value for this coefficient is to use Pearson's technique:

$$\text{Pearson's coefficient of skewness}, g_1 = \frac{3(\text{mean} - \text{median})}{\text{standard deviation}}$$

A value of zero for g_1 indicates that the distribution is perfectly symmetrical, a negative value indicates the distribution is negatively skewed, i.e. skewed to the left, and a positive value indicates that the distribution is positively skewed, i.e. skewed to the right. In general, values beyond −1 or +1 would suggest a skewed distribution that departs appreciably from the symmetrical normal distribution.

Examples of both a positively and a negatively skewed distribution are shown in Fig. 3.5.

3.3 A More Detailed Description of the Normal Distribution

3.3.1 Mathematical basis of the normal distribution

To recall, if a data set refers to a continuous variable and the sample or population is sufficiently large and unbiased then the data frequency distribution will tend towards the symmetrical bell-shape of the normal distribution curve. The perfect normal distribution curve can be described by a mathematical formula that allows the frequency y for any given value of x to be predicted.

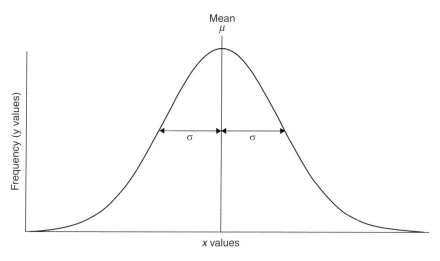

Fig. 3.6. Frequency diagram of a perfect normal distribution.

The equation describing the normal distribution is:

$$y = \frac{1}{\sigma\sqrt{2\pi}} \times e^{-\frac{(x-\mu)^2}{2\sigma^2}}$$

where: μ = mean

σ = standard deviation

e = exponential constant (i.e. 2.718…)

π = pi (i.e. 3.141…)

Fortunately we rarely, if ever, need to use this complex formula in the calculation of statistical tests; however, a whole branch of important statistical tests that we employ is based on this mathematical model. Given that π and e in the above formula are mathematical constants, it should be apparent that the exact shape of a particular normal distribution is dependent upon just two parameters, these are the population mean, μ, and the population standard deviation, σ. For a perfect normal distribution the mean is the data point at which the curve reaches its maximum height and the standard deviation is the horizontal distance from the central axis of the normal distribution curve to the point of inflexion of the curve (i.e. the point where the curve theoretically attains its maximum gradient). This is shown in Fig. 3.6.

An understanding of the theoretical normal distribution is very important because it provides a method for analysing and comparing the variation of different sets of continuous data. If it can be assumed that the measured variable adheres to the normal distribution and that the mean and standard deviation of the population can be accurately estimated by the sample mean and standard deviation respectively, then the probability of an observed data value belonging to a particular normal distribution can be calculated. This in turn allows an assessment of whether two or more different sample means are likely to arise from the same or different populations. In other words, it allows the presence of a **significant difference** to be detected. As previously stated, all statistical analyses that employ the concept of the normal distribution to estimate population parameters, such as the population mean and population standard deviation, from sample statistics are called **parametric analyses.**

3.3.2 Probability values of the normal distribution

Since the shape of the normal distribution is governed by an equation in which the only variables are the population mean μ and the standard deviation σ, it follows that the area under the normal distribution curve lying between the mean and any given number of standard deviations away from the mean is a constant. By using the normal distribution equation, it is possible therefore to determine the theoretical probability that any particular data value from a normal distribution occurs between the mean and a given number of standard deviations from the mean. For example, it can be shown that the portion of the normal curve lying between the mean and plus 1 standard deviation from the mean is 0.341 and the probability of a data item lying within 1 standard deviation either side of the mean, i.e. $\mu \pm 1\sigma$, will therefore be twice this, i.e. 0.682. Conversely the probability of a randomly chosen value from a normal distribution falling outside this range is 0.318. This is illustrated in Fig. 3.7.

The process of converting the scale along the X-axis of the normal distribution plot to equivalent standard deviation values is referred to as the **standardization of the normal distribution,** and the standard deviations plotted are then referred to as z values. The proportion of the normal distribution that lies beyond any given number of standard deviations away from the mean, i.e. beyond μ and $\mu \pm z$, are shown in Table A1.1 in the Appendix.

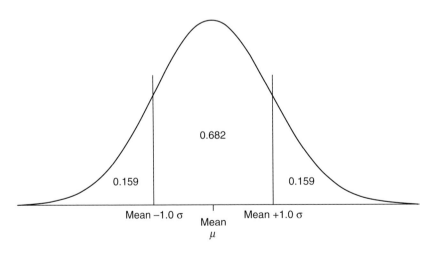

Fig. 3.7. The standardized normal distribution curve showing the proportion of the area below the curve that lies between the mean (μ) and one standard deviation (σ) either side of the mean. These values are equivalent to the probability that any randomly selected value of the X-variable lies within these regions of the distribution.

Suppose that we have obtained a particular data value, then we may ask what is the probability that it comes from a particular population that has a known normal data distribution? In order to determine this, it is first necessary to standardize the value. This is achieved by dividing the difference between the observed value and the population mean by the population standard deviation and thereby converting the observed value to a value of z, i.e.

$$z = \frac{x - \mu}{\sigma}$$

Note that the value z is simply the number of standard deviations that the observed value x is away from the population mean, μ.

The probability *(P)* of the observed data value falling within a given number of standard deviations either side of the population mean can now be determined by referring to the proportions of the normal distribution given in Table A1.1. The format of this table is commonly simplified to show the probabilities associated with some particularly important values of z, termed **critical values**, as in Table 3.1. Table 3.1 shows the probability of obtaining a value for a normally distributed variable that falls between the mean and a given number of standardized deviations (i.e. z values) either side of the mean. For example, 90% of all values of a normally distributed variable are expected to have a value that occurs within the interval between the mean ± 1.645 standard deviations from the mean. Conversely, 10% of all values can be expected to fall outside this range. Similarly, the probability of a normally distributed value occurring within ± 1.96 standard deviations from the mean is 95%. Therefore if it was found that a single observed data value, x, when standardized gave a z value of >1.96, then it would be concluded to have less than a 5% chance of falling within that particular normal distribution. This is a particularly important critical value in parametric statistical analyses because a probability of 95% is generally regarded as the minimum likelihood that must be achieved for an explanatory hypothesis concerning a set of data to be accepted.

Table 3.1. Probabilities *(P)* associated with critical values for the standardized deviations *(z)* of a normal distribution. *P* is the probability of a randomly selected data item from a normal distribution falling either outside or inside the interval defined by the mean ± *z* standard deviations from the mean.

P (outside)	0.1	0.05	0.02	0.01	0.001	0.0001	0.00001
P (inside)	0.9	0.95	0.98	0.99	0.999	0.9999	0.99999
z		1.645	1.960	2.326	2.576	3.291	3.891	4.417

3.3.3 An example of the use of the probability values of the normal distribution

To help place the theory of the normal distribution into a practical context, let us now consider an example. In a survey of milk quality produced by an apparently homogeneous population of dairy cattle, one cow was found to yield milk with a particular high level of the protein casein. The casein content in milk from this cow was 37 mg/g whereas the mean casein content determined from a very large population of cows of the same breed was 32 mg/g. The question is, therefore, whether there is something especially different about this one cow or is this high value for milk casein concentration simply the consequence of random variation? It is of course impossible to answer this question definitively based on just one observation. However, if we are able to assume that casein content in milk is a normally distributed variable then what we can do is determine the **probability** of obtaining the high casein value in this one cow given the hypothesis that this cow is no different from the other cows in the study. If this probability proves to be sufficiently low then we would be justified in rejecting this hypothesis and claiming that the cow was truly different from the other cows. The hypothesis of no difference that we inspect in this process is referred to by statisticians as the **null hypothesis**.

In order to determine the probability that the observed data conform to the hypothesis of no difference, in addition to knowing the population mean (32 mg/g) we also need to know the population standard deviation; let us assume that this has been previously recorded and is 2.5 mg/g. The observed difference between the milk casein content of the single cow and the population mean of the other cows is standardized to produce a z value; this might be referred to as $z_{observed}$ since it is based on the observed data, thus:

$$z_{observed} = \frac{x - \mu}{\sigma} = \frac{37 - 32}{2.5} = 2.0$$

The value of $z_{observed}$ can now be referred to the critical z values shown in Table 3.1. In order for there to be, at the very least, a 5% probability that the observed cow is part of the same normal distribution to which the main population of cows conforms, the value of $z_{observed}$ would need to be no greater than 1.96. The observed value is, however, 2.0. This indicates that the probability of obtaining the high milk casein value in this particular cow, were the cow to be homogeneous with the main population, is less than 5%. Of course this does not mean that the cow with the high milk casein content is definitely different in some way from the other cows but it does provide substantial statistical evidence that this is likely to be the case and justifies further investigation.

In fact what we have undertaken here is a type of **statistical significance test** that in this case is called a **z-test**. Significance tests are used to test hypotheses about the data collected, in this case the hypothesis that the cow with the particularly high milk casein content was truly different from the other cows examined. Before we can explore significance testing in further detail we need, however, to understand rather more about the nature of samples and populations, which is the subject of the next chapter.

3.3.4 Testing for a normal distribution

It should now be apparent how important it is to be able to inspect whether a sample of continuous data adheres to a normal distribution, or, more pertinently, whether it is likely to have been extracted from a normally distributed population. An assumption of normality is required by all parametric statistical analyses and if the assumption proves invalid then the use of such tests becomes invalid. For example, in the z-test illustrated above it was necessary to assume that milk protein is a normally distributed variable in the population of cows under examination. In many cases it may be obvious that the sample data are taken from a normal distribution and most biometrical data such as plant height, crop yield, milk protein concentration, if collected from an unbiased sample held under uniform conditions, should be normally distributed. There will be occasions, however, when it is not so clear. A set of data may appear to be slightly asymmetrical or 'skewed' for example, or it may be symmetrical but have a shape that is relatively short and fat, or tall and thin, or it may appear to contain more than one peak, i.e. multi-modal. Under such circumstances it may be necessary to test whether the observed distribution departs significantly from the normal distribution or whether it remains close enough for analyses requiring the assumption of normality still to work well.

There are several procedures that can be employed to test for normality, none of which is definitive in its own right but which together should provide sufficient evidence one way or the other:

1. The first obvious step is to graphically plot the distribution and simply see if it looks like a normal distribution. Frequently this will be all that is required but it does of course depend on having sufficient data values to facilitate a valid graphical plot.
2. Secondly, the standardized proportions of the distribution may be examined. From the theory explained 'Probability values of the normal distribution', in section 3.3, it would be expected that for any normally distributed sample approximately 68% of data values should fall between the mean and 1 standard deviation either side of the mean. It is not possible to put exact limits on this but, if the proportion of data values that fell in this range was between about 60 and 75% then this would be reasonable evidence of a normal distribution.
3. Thirdly, the statistical coefficients of skewness and kurtosis may be examined. As already explained (section 3.2, 'Skewed distribution') the extent of skew can be measured by the **coefficient of skewness**, g_1. A value of zero indicates that the distribution is perfectly symmetrical, while values beyond +1 or –1 would suggest appreciable departures from symmetry. While remaining symmetrical a distribution may also depart from the theoretical normal distribution by becoming excessively tall at the peak and narrow at the base, or conversely by becoming flattened at the peak but broad at the base. The term **kurtosis** is used to describe the shape of the peak of the curve and a **coefficient of kurtosis**, g_2, can be calculated as a descriptive statistic. A value of g_2 of zero indicates that the distribution is perfectly normally shaped while values much beyond +1 or –1 suggest appreciable departures from normality. The calculation of the coefficients of skewness and kurtosis are generally complex (although note that a relatively simple approach to determining g_1 was given in section 3.2, 'Skewed distribution'); however, the descriptive statistics function in most computer software packages, including the Excel® spreadsheet software Analysis Toolpak and Genstat, do calculate these statistics. (See Zar, 2009, for a detailed explanation of these statistics.)
4. A rather more complex graphical technique is provided through a **normal probability plot**, although most statistics software packages will produce these in one form or another. The simplest approach involves ordering the data from lowest to highest and determining the cumulative probability p for each value in order. (The

formula for the cumulative probability is $p_i = (i - 0.05)/n$ where i is the rank of the data value and n the number of data values in the data set). If the data are normally distributed a plot of the ordered data values against their corresponding probability values will produce an 'S'-shaped curve. However, it becomes much easier to appraise if the probability values are converted to standardized z scores based on the normal distribution, the values being obtained from the table of proportions of the normal distribution (see Table A1.1 in the Appendix). Where the data are normally distributed a plot of the z scores against the ordered data values will then produce a straight line. Example 3.2 demonstrates the determination and interpretation of normal probability plots. Because of the random variation in data we cannot expect such plots to produce perfect straight lines but in practice we only need to assess whether major discrepancies from linearity are present. Normal probability plots are used in particular to test the adherence of data to a normal distribution prior to undertaking linear regression analysis to determine the best-fit line through data points on a scatter graph and their use in this context is described in Chapter 12.

Example 3.2. Testing for normality.

The growth rate (g dry weight/day) of a sample of 30 pot-grown tomato plants treated with a new fertilizer 'Extragrow' and 30 pot-grown plants treated with standard fertilizer (control). The descriptive statistics for these data were determined in Example 2.3; however, before further analysis can be performed, it is required to inspect the extent to which the data sets adhere to a normal distribution.

Growth rate (g dry weight/day)					
Standard fertilizer treatment (control)			'Extragrow' fertilizer treatment		
0.8	1.3	0.9	1.9	1.7	2.0
1.1	1.4	0.4	2.3	2.2	1.5
0.7	1.8	0.8	1.0	0.8	1.6
0.8	0.9	1.3	1.0	1.1	0.7
1.2	1.1	1.0	1.8	1.2	0.7
1.7	1.6	1.4	1.1	0.9	0.9
1.3	1.0	0.4	1.8	1.5	1.3
1.0	1.2	1.3	2.2	0.7	1.0
1.5	0.7	1.2	1.2	0.9	0.8
0.5	0.5	1.3	1.7	1.1	1.0

continued

Example 3.2. Continued.

Frequency histograms

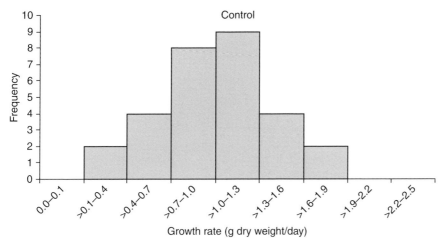

Control

Growth rate (g dry weight/day)

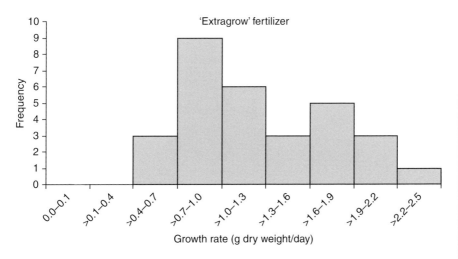

'Extragrow' fertilizer

Growth rate (g dry weight/day)

Conclusion: frequency plots suggest that while the control data approach a normal distribution the 'Extragrow' fertilizer data appear non-symmetrical.

Proportion of values lying in the range $\mu \pm 1.0\sigma$
Using the sample mean ($\bar{x}$) and sample standard deviation (s) as estimates of population parameters μ and σ then:

	Control	'Extragrow' fertilizer
$\bar{x} \pm s$	0.697 – 1.443	0.829 – 1.812
No. of values in range $\bar{x} \pm s$	22	20
Proportion	22/30 = 0.73	20/30 = 0.67

continued

Example 3.2. Continued.

Conclusion: estimated proportion of values falling in range $\mu \pm 1.0\sigma$ suggests that while the 'Extragrow' fertilizer data closely approach a normal distribution (proportion very close to 0.68) the control data appear less well normally distributed.

Coefficients of skewness and kurtosis
Values determined using Genstat statistics program

	Control	'Extragrow' fertilizer
Coefficient of skewness	−0.09	0.53
Coefficient of kurtosis	−0.68	−0.96

Conclusion: coefficient of skewness and kurtosis values do fall in the range −1 to +1 for both data sets but values are much closer to zero in the control data suggesting a closer adherence to a normal distribution.

Normal probability plots
The data values are ordered according to magnitude and the corresponding cumulative probability (p) for each value determined by $p_i = (i - 0.05)/n$ where i is the rank position of the value in the ordered data set. The z value corresponding to each value of p is obtained from the table of standardized proportions of the normal distribution (Appendix Table A1.1).

Growth rate (g dry weight/day)							
Standard fertilizer treatment (control)				'Extragrow' fertilizer treatment			
Sorted values	Rank[a]	p	z value	Sorted values	Rank	p	z value
0.4	1	0.02	−2.128	0.7	1	0.02	−2.128
0.4	2	0.05	−1.645	0.7	2	0.05	−1.645
0.5	3	0.08	−1.383	0.7	3	0.08	−1.383
0.5	4	0.12	−1.192	0.8	4	0.12	−1.192
0.7	5	0.15	−1.036	0.8	5	0.15	−1.036
0.7	6	0.18	−0.903	0.9	6	0.18	−0.903
0.8	7	0.22	−0.784	0.9	7	0.22	−0.784
0.8	8	0.25	−0.674	0.9	8	0.25	−0.674
0.8	9	0.28	−0.573	1	9	0.28	−0.573
0.9	10	0.32	−0.477	1	10	0.32	−0.477
0.9	11	0.35	−0.385	1	11	0.35	−0.385
1	12	0.38	−0.297	1	12	0.38	−0.297
1	13	0.42	−0.210	1.1	13	0.42	−0.210
1	14	0.45	−0.126	1.1	14	0.45	−0.126
1.1	15	0.48	−0.042	1.1	15	0.48	−0.042
1.1	16	0.52	0.042	1.2	16	0.52	0.042
1.2	17	0.55	0.126	1.2	17	0.55	0.126
1.2	18	0.58	0.210	1.3	18	0.58	0.210
1.2	19	0.62	0.297	1.5	19	0.62	0.297
1.3	20	0.65	0.385	1.5	20	0.65	0.385
1.3	21	0.68	0.477	1.6	21	0.68	0.477

continued

Chapter 3

Example 3.2. Continued.

Growth rate (g dry weight/day)							
Standard fertilizer treatment (control)				'Extragrow' fertilizer treatment			
Sorted values	Rank[a]	p	z value	Sorted values	Rank	p	z value
1.3	22	0.72	0.573	1.7	22	0.72	0.573
1.3	23	0.75	0.674	1.7	23	0.75	0.674
1.3	24	0.78	0.784	1.8	24	0.78	0.784
1.4	25	0.82	0.903	1.8	25	0.82	0.903
1.5	27	0.88	1.192	2	27	0.88	1.192
1.6	28	0.92	1.383	2.2	28	0.92	1.383
1.7	29	0.95	1.645	2.2	29	0.95	1.645
1.8	30	0.98	2.128	2.3	30	0.98	2.128

[a]Note that equal data values are awarded successive rather than equal rank positions.

The ordered values are now plotted on the Y-axis against the corresponding z values plotted on the X-axis to produce the normal probability plot.

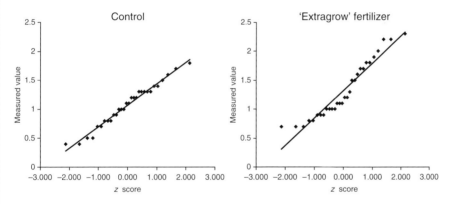

Conclusion: data points fall close to a straight line for the control data suggesting reasonable adherence to a normal distribution but clearly deviate from linearity for the 'Extragrow' fertilizer data.

Overall a combination of evidence would indicate that the control data show good adherence to a normal distribution but the 'Extragrow' fertilizer data fail to do so.

5. It is possible to apply a statistical significance test to inspect if the data fit a normal distribution. The most common of these tests, which determine the probability that a set of data values falls within the expected standardized proportions of the normal distribution, are the 'goodness-of-fit' chi-squared test and the Kolmogorov–Smirnov test. We are not in a position to examine these tests any further at this point, except to note that, while they are employed widely within statistical software packages, there is much disagreement over their complete validity among statisticians. (See sections 14.3.3 and 14.7 for further discussion of these tests.)

4 Populations, Samples and Sample Reliability

Population: total number or quantity of things in a given place or region; total group of items under consideration.

Sample: small separated part of something illustrating the qualities of the mass.

(Concise Oxford English Dictionary)

- The concept of employing samples to represent populations.
- Selection of samples.
  - the need for randomization;
  - the need for replication.
- Determination of required sample size.
- Analysis of sample reliability.
  - standard error of the mean;
  - confidence limits around the mean.
- Interpretation and graphical representation of standard errors and confidence limits.
- The problem of analysing small samples.
- Using Student's *t* distribution to analyse small samples.

4.1 Populations and Samples

Complete reliability in an experiment can only be obtained when every possible individual item in the investigated population can be measured. In practice, however, entire populations are invariably too large to allow this and instead a sample of the population is selected to represent the population and measurements are then confined to the selected sample. Samples may also be necessary when the measurements to be made are destructive, for example, the change in biomass of a growing crop can only be determined from a succession of samples that are harvested from the crop. The ultimate aim of the investigator will be to draw conclusions that, although based on the sample data, can be extrapolated to the whole population. For example, if it is concluded that a particular fertilizer increases the dry weight of a sample of wheat plants it would be nice to be able to extend this conclusion to a much larger population of wheat plants, not just the few plants that make up the selected sample. In order to do this it will be necessary to establish how accurately the sample represents the population. Sample data are, therefore, employed to represent the population data and the calculated sample statistics are used to estimate the population parameters. Further statistical techniques are then employed to evaluate how representative the sample is of the population from which it is drawn and thereby how much reliance can be placed on any conclusions drawn from the sample data but subsequently applied to the population.

4.1.1 The problem of using samples to represent populations

The validity of conclusions drawn from an experiment which uses samples will be completely dependent on how well the selected sample represents the population. The question that therefore arises is how can we possibly judge how well a sample represents the population when we do not know what the population parameters are? (If we knew the values of the population parameters we would not of course be bothering to select samples in the first place!)

The problem is illustrated in Fig. 4.1. A volume of milk is collected from one particular cow and it is required to determine the concentration of casein, the major milk protein, in the sample. The volume of milk collected is far too large to permit the protein analysis without adopting a sampling procedure. A series of five small volumes are extracted from the bulk volume that then represent the **sampling units**. The casein concentration of each sample is measured by spectrophotometric analysis and the values turn out to vary considerably from each other. This may be due to a number of causes, for example, the protein was not evenly distributed throughout the bulk solution, the spectrophotometer performance fluctuated, etc. The question is how can we know which of the five measurements most accurately estimates the actual

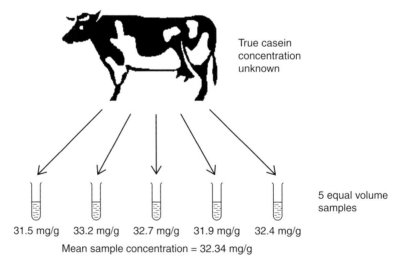

Fig. 4.1. An example of the variability of samples: determination of the casein protein content of milk. In order to determine the concentration of casein protein in a bulk sample of milk obtained from one particular dairy cow, five subsamples of equal volume are obtained. The samples subsequently show variability in the measured concentration of the protein, i.e. they display 'sample error'. Two questions therefore arise: How can the most accurate sample be identified? How near to the actual protein concentration in the cow's milk is the sample mean?

concentration of casein in the milk? Further, if we take a statistical approach and determine the mean of the five measurements, how close can we then expect this sample mean to be to the actual but unknown casein concentration of the bulk volume?

4.1.2 The selection of samples

The procedure for selecting samples to ensure that they represent the population from which they are extracted as accurately as possible is a critical part of the experimental design process. There are two main obstacles to valid sample selection that must be overcome: biased selection by the investigator and the presence of a systematic variation in the population from which the sample is to be selected. Biased sample selection occurs readily if the investigator fails to carefully consider the sampling approach. For example, a plant sample might be chosen from the edge of a field simply because these are the most accessible plants and physically the easiest to harvest. Plants growing at the edge of a field are, however, likely to be subjected to different growing conditions from the majority of the crop and selection of such plants would not produce a sample which accurately reflects the entire crop. This type of problem is easily avoided by taking a common-sense approach. Systematic variability in the population is, however, much more difficult to deal with, not the least of the problems being the need to recognize when a systematic variation is present in the first place.

The term **systematic variation** refers to a trend of variation that runs through the population. In most areas of field investigation, it is rare that an experiment is undertaken under absolutely uniform conditions. Staying with the example of a crop trial, most field-grown crops will be subject to a whole range of climatic and environmental conditions that will cause systematic variation in growth and production of the crop across the field. For example, there may be shade and wind protection from trees along one edge, there may be a slope across the field causing uneven drainage, the soil type might vary across the field, and so on and so on. Both the allocation of experimental treatments to items and selection of samples of items for measurement must be undertaken in such a way that the underlying variation is accounted for and can be separated from the specific treatment effects.

The problem and importance of sample selection are illustrated in Fig. 4.2, which shows the plan of a crop field trial. The trial consists of 20 rows of potatoes, each row with 40 plants. Employing a very simple experimental design, one-half of the trial (i.e. 20 plants per row) is treated with a new fertilizer formulation and the other half left untreated. (This is actually a very poor approach to experimental design for reasons that will become clear very soon.) The potential yield of each plant is shown on the plan. However, let us assume that, for practical reasons, it is not actually possible to measure the entire crop but that a sampling procedure needs to be employed. How then should the samples be selected? How large should the samples be? How many samples would be required? The example in Fig. 4.2 shows three different approaches to sample selection that could have been used and, not surprisingly, each provides a different estimate of the population mean; the sample means are considerably variable and none are particularly close to the actual population mean.

Control plot

Plant No.

Row	1	2	3	4	5	6	7	8	9	10	11	12	13	14	15	16	17	18	19	20
1	0.50	0.76	0.41	0.67	0.77	0.43	0.40	0.59	0.85	0.48	0.33	0.48	0.52	0.57	0.42	0.21	0.47	0.55	0.52	0.45
2	0.82	0.56	0.38	0.52	0.76	0.59	0.84	0.28	0.42	0.54	0.57	0.62	0.56	0.46	0.37	0.54	0.46	0.52	0.43	0.23
3	0.72	0.61	0.34	0.53	0.48	0.62	0.41	0.69	0.56	0.56	0.53	0.61	0.68	0.49	0.49	0.37	0.47	0.57	0.41	0.48
4	0.44	0.73	0.49	0.37	0.44	0.69	0.49	0.19	0.63	0.55	0.61	0.50	0.42	0.40	0.45	0.55	0.50	0.53	0.70	0.52
5	0.77	0.39	0.60	0.75	0.50	0.54	0.72	0.41	0.34	0.30	0.50	0.55	0.64	0.55	0.60	0.57	0.52	0.56	0.65	0.53
6	0.45	0.68	0.48	0.51	0.44	0.59	0.26	0.59	0.66	0.67	0.50	0.57	0.56	0.58	0.54	0.52	0.58	0.55	0.70	0.51
7	0.53	0.57	0.25	0.77	0.62	0.83	0.37	0.57	0.51	0.62	0.43	0.48	0.52	0.51	0.58	0.57	0.53	0.57	0.54	0.48
8	0.67	0.27	0.49	0.45	0.52	0.31	0.57	0.53	0.41	0.47	0.61	0.49	0.52	0.32	0.50	0.56	0.55	0.53	0.55	0.59
9	0.53	0.57	0.55	0.66	0.53	0.35	0.52	0.62	0.52	0.31	0.57	0.53	0.52	0.50	0.52	0.55	0.58	0.54	0.58	0.53
10	0.45	0.61	0.62	0.49	0.50	0.71	0.59	0.63	0.59	0.62	0.53	0.57	0.50	0.56	0.54	0.58	0.56	0.50	0.62	0.65
11	0.55	0.71	0.59	0.68	0.55	0.53	0.42	0.60	0.52	0.55	0.58	0.51	0.61	0.57	0.56	0.52	0.52	0.50	0.63	0.68
12	0.52	0.58	0.50	0.57	0.51	0.57	0.56	0.52	0.50	0.54	0.54	0.49	0.55	0.52	0.55	0.60	0.57	0.59	0.52	0.55
13	0.62	0.53	0.61	0.48	0.60	0.59	0.38	0.58	0.49	0.53	0.56	0.58	0.58	0.54	0.57	0.57	0.59	0.67	0.54	0.61
14	0.58	0.38	0.53	0.54	0.42	0.51	0.61	0.63	0.53	0.39	0.55	0.60	0.62	0.55	0.61	0.55	0.60	0.53	0.58	0.59
15	0.49	0.56	0.57	0.58	0.56	0.50	0.54	0.57	0.52	0.60	0.71	0.57	0.61	0.55	0.55	0.50	0.64	0.63	0.53	0.58
16	0.61	0.56	0.58	0.50	0.58	0.54	0.58	0.55	0.55	0.78	0.53	0.59	0.49	0.60	0.54	0.54	0.57	0.60	0.50	0.60
17	0.53	0.61	0.63	0.53	0.52	0.69	0.68	0.79	0.58	0.52	0.65	0.70	0.68	0.63	0.56	0.71	0.69	0.59	0.50	0.53
18	0.71	0.52	0.58	0.61	0.54	0.58	0.59	0.58	0.52	0.55	0.59	0.51	0.67	0.58	0.60	0.57	0.52	0.79	0.55	0.58
19	0.63	0.54	0.60	0.63	0.55	0.61	0.62	0.53	0.54	0.58	0.69	0.58	0.68	0.68	0.64	0.69	0.81	0.73	0.87	0.76
20	0.60	0.51	0.52	0.55	0.67	0.68	0.53	0.62	0.67	0.50	0.58	0.64	0.72	0.83	0.62	0.68	0.67	0.80	0.63	0.62

Treated plots

Row	1	2	3	4	5	6	7	8	9	10	11	12	13	14	15	16	17	18	19	20
21	0.47	0.52	0.42	0.55	0.59	0.49	0.45	0.67	0.70	0.54	0.49	0.54	0.59	0.63	0.68	0.46	0.32	0.48	0.52	0.41
22	0.60	0.65	0.49	0.81	0.74	0.68	0.63	0.54	0.59	0.60	0.58	0.67	0.60	0.63	0.56	0.58	0.53	0.32	0.26	0.55
23	0.55	0.61	0.73	0.49	0.60	0.58	0.55	0.64	0.52	0.60	0.59	0.71	0.57	0.61	0.57	0.62	0.55	0.43	0.45	0.55
24	0.65	0.54	0.52	0.48	0.62	0.70	0.52	0.68	0.48	0.60	0.65	0.64	0.62	0.68	0.54	0.65	0.61	0.54	0.63	0.60
25	0.72	0.64	0.61	0.49	0.58	0.74	0.81	0.65	0.62	0.71	0.67	0.65	0.68	0.64	0.53	0.68	0.65	0.53	0.66	0.53
26	0.52	0.74	0.56	0.63	0.48	0.71	0.79	0.62	0.56	0.67	0.60	0.69	0.64	0.65	0.60	0.66	0.66	0.64	0.68	0.64
27	0.48	0.52	0.61	0.59	0.59	0.47	0.64	0.60	0.90	0.58	0.59	0.64	0.66	0.70	0.54	0.62	0.63	0.58	0.63	0.65
28	0.67	0.45	0.39	0.56	0.62	0.68	0.74	0.72	0.61	0.48	0.69	0.59	0.69	0.59	0.62	0.68	0.59	0.68	0.64	0.67
29	0.64	0.64	0.79	0.81	0.72	0.52	0.71	0.61	0.68	0.62	0.71	0.63	0.73	0.62	0.64	0.64	0.57	0.64	0.57	0.60
30	0.59	0.61	0.60	0.49	0.38	0.71	0.39	0.45	0.62	0.43	0.64	0.58	0.66	0.68	0.65	0.65	0.53	0.67	0.58	0.63
31	7.00	0.60	0.59	0.54	0.60	0.65	0.65	0.63	0.61	0.55	0.69	0.68	0.65	0.63	0.60	0.61	0.68	0.68	0.53	0.68
32	0.68	0.65	0.62	0.48	0.47	0.62	0.60	0.58	0.61	0.46	0.64	0.64	0.80	0.65	0.62	0.60	0.59	0.66	0.66	0.66
33	0.62	0.63	0.63	0.60	0.63	0.54	0.74	0.64	0.63	0.68	0.67	0.65	0.63	0.62	0.68	0.54	0.69	0.63	0.70	0.80
34	0.50	0.64	0.61	0.63	0.66	0.60	0.62	0.61	0.60	0.39	0.74	0.65	0.77	0.73	0.65	0.62	0.70	0.66	0.74	0.76
35	0.55	0.60	0.62	0.60	0.66	0.55	0.82	0.57	0.58	0.42	0.54	0.69	0.68	0.44	0.69	0.52	0.64	0.78	0.64	0.64
36	0.61	0.58	0.68	0.49	0.59	0.59	0.68	0.53	0.63	0.65	0.59	0.66	0.64	0.64	0.64	0.68	0.67	0.72	0.78	0.64
37	0.63	0.59	0.63	0.65	0.60	0.63	0.59	0.60	0.62	0.60	0.67	0.71	0.67	0.68	0.78	0.73	0.65	0.57	0.72	0.59
38	0.58	0.61	0.61	0.63	0.61	0.66	0.60	0.62	0.60	0.71	0.64	0.68	0.65	0.78	0.70	0.68	0.78	0.66	0.60	0.67
39	0.74	0.68	0.64	0.72	0.60	0.62	0.60	0.60	0.61	0.59	0.78	0.73	0.66	0.66	0.91	0.76	0.82	0.66	0.70	0.64
40	0.63	0.76	0.78	0.70	0.70	0.87	0.70	0.68	0.66	0.67	0.76	0.81	0.68	0.78	0.87	0.82	0.71	0.64	0.70	0.60

A field trial consists of 40 rows of potatoes with 20 plants per row. Half the rows are treated with standard fertilizer (control) and the other half are treated with a new fertilizer. The yield (kg) of each plant is shown in the plan. Assuming that it is not possible to measure all the treated plants, samples must be selected. Three possible sample selection procedures are shown together with the mean values that arise from these samples.

Population means:

Control mean: 0.558 kg   'Extragrow' fertilizer treatment mean: 0.641 kg

Assuming a sample size of 15 plants, three approaches to sample selection are shown

(a) ☐ selection of a batch of plants from the middle of each plot.
This produces the following results:

Control sample mean 0.524 kg   'Extragrow' treatment sample mean: 0.603 kg

(b) ▓ selection of evenly spaced plants right across each plot.
This produces the following results:

Control sample mean 0.607 kg   'Extragrow' treatment sample mean: 0.643 kg

(c) ▉ totally random selection of plants but discarding plants in edge rows.
This produces the following results:

Control sample mean 0.553 kg   'Extragrow' treatment sample mean: 0.605 kg

Fig. 4.2. An example of the effect of different approaches to sample selection.

4.1.3   The need for randomization

The first golden rule of sample selection is that, in order to prevent biased sample selection, *samples must be as random as possible*. The application of statistical techniques to estimate population parameters from sample data is only valid when randomized samples have been employed. If no systematic variation in the population can be identified then samples should be chosen fully randomly; an experiment conducted on this basis is then referred to as a **completely randomized design** (CRD). There are a number of ways that random selection can be made. The most straightforward today is to use the random number function available on electronic calculators and computers to generate the components of a code which identifies individuals in a population, e.g. the row and plant number of plants growing in a field trial. Alternatively, tables of random numbers published in books of statistical tables can be consulted. It should be noted, however, that, even in completely random designs, items on the edge of an experimental plot would not normally be included in order to remove the possibility of 'edge effects' that bias the data.

Where an identified systematic variation is present in a population a fully randomized procedure for selection of samples is, however, not justified. This is because full randomization does not guarantee that individuals selected for a sample are widespread across the population and there is always the possibility that the majority of a sample, by bad luck, originates from a similar position within the experiment. If this happens and at the same time that there is a systematic variation running through the experiment, then the sample will become biased. Under such circumstances what is required is a random sample selection procedure that at the same time assures that samples represent the whole experimental plot. This appears to be a contradiction in terms; however, the solution is to employ a **randomized block design** (RBD). In a randomized block designed experiment the population is divided into equal sized blocks. It is subsequently ensured that a full set of treatments is randomly allocated within each block but with a different randomization employed for each block. This technique ensures that treatment samples include items spread across the whole population but at the same time involves a sufficient degree of randomization to remove the possibility of biased selection. Randomized block designs, together with the techniques for their analysis, are discussed in more detail in Chapter 9; in the meantime a simple randomized block design is illustrated in Fig. 4.3.

4.1.4   The need for replication

The need for replication in sample selection is fairly obvious. If we could be certain that every single individual in a population had exactly the same measurement and, following imposition of an experimental treatment, every single individual then gave the same new measurement, then no replication would be required and single before and after measurements would suffice. However, in all populations there is inherent natural variation between individuals and this inherent variation will be exacerbated by the presence of any non-uniformity in the experimental conditions. In order to obtain a good estimate of the actual measurement we need to measure, therefore, a number of individuals from the population.

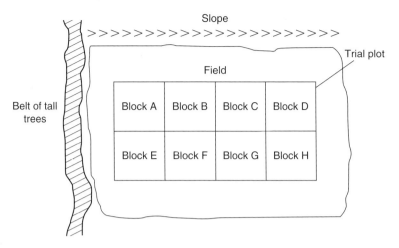

Fig. 4.3. An example of a randomized block designed field trial. Due to the systematic variation across the field caused by shading and a slope the trial plot is divided into equal sized blocks. A full set of treatments is randomly allocated to each block and each treatment sample will be represented by an equal number of plants selected randomly from each block. Each sample is then assured of including plants from right across the field.

There is another good reason why we need to make replicate measurements. Following the performance of an experiment we are usually left in a position of wishing to know whether there is a truly significant difference between two or more differently treated populations, that is, a difference that can be attributed to the experimental treatments rather than to random variation. In order to obtain an answer to this question we perform a statistical analysis on sample measurements made from each population. Such analyses, called 'tests of statistical significance', depend upon a quantitative knowledge of the amount of variation that is present within each sample. The amount of variation within a sample can of course only be determined if the sample consists of a number of replicate observations. Furthermore, the statistical technique for determination of sample reliability, i.e. how close the sample mean may be expected to be to the population mean, also depends on the sample consisting of a number of replicates. This may seem obvious but it is surprising how often this is ignored. For example, in the scenario depicted in Fig. 4.1 in order to estimate the casein concentration in the milk it would be tempting to make a single measurement on just one sample extracted from the bulk solution. Clearly, however, this runs the considerable risk that dispersion of the protein in the bulk solution is not homogeneous or that there is some, albeit small, random fluctuation in the spectrophotometer readings. Under these circumstances unless replicate measurements are made it is impossible to determine the reliability of the sample concentration as an estimate of the true concentration in the bulk solution.

4.1.5 Determination of the required sample size

Having understood the need for replication, the question that arises is how many replicate observations or measurements per sample should be made? This might seem a simple question but it is actually one of the most problematic questions to answer when designing an experiment. Simple logic tells us that the more measurements made the more reliable will the sample become and the simple answer is, therefore, that *a sample should be as large as possible*. However, since time and available resources will generally be the limiting factor to sample size perhaps a more useful way of posing the question is 'What is the smallest number of replicates necessary to provide a reliable estimate of the population?' Unfortunately there is no easy way to answer this question. A common conception is that the size of a sample should be some fixed proportion of the population; however, it can be easily demonstrated that this is not the case. For example, so-called 'exit polls' following a governmental general election in the UK are based on asking a random sample of around 5000 people how they voted. This represents less than 0.025% of the voting population; however, because the sample is very large in absolute terms ($n = 5000$) exit polls do prove to be very accurate in predicting the overall voting pattern of the population. On the other hand, if the population consisted of only 100 people then 0.025% of the population would represent only 3 people (rounded up) and clearly such a small sample is most unlikely to indicate how a population of 100 people voted as a whole. In the case of such a small population a meaningful exit poll sample might need to be at least 10% of the population in order to reliably predict the election result. Therefore, it is the absolute size of the sample that is important irrespective of the proportion of the population that it represents.

I have previously stated that the greater the variability between items, the larger the sample will need to be, but of course we do not know how variable the sample is until we have selected and measured it in the first place. Therefore common sense needs to be applied together with some understanding of the nature of the measurements being made. Previous experience of the material under investigation and possibly a previous 'look-see' type of experiment will often have provided the investigator with a guide to the amount of variation that might be expected. For example, an animal pathologist will know that mammalian blood is extremely homogeneous and percentage variation between cell counts of samples is very small, maybe less than 0.1%. On the other hand, a crop researcher will know that yield per plant within a crop is highly variable, maybe greater than 10%. The human pathologist may therefore consider that relatively few replicate measurements of blood cell count are required while the crop researcher will require rather more replicate yield values to be collected before any confidence can be placed in the conclusions.

Statistical analysis may provide some assistance in choosing an appropriate sample size. Sample reliability is determined through the statistic termed the standard error (originally introduced in section 2.4 and discussed in detail in the next section). In any experiment the standard error, which naturally decreases with increasing sample size, needs to be maintained as low as possible. One approach is, therefore, to consider in advance the maximum level of the standard error that is acceptable. For example, it might be decided that the maximum expected sample standard error should be no greater than a certain proportion of the sample mean, say 10%. In a preliminary trial measurements are made on samples of increasing size until this criterion is satisfied and the required sample size thereby identified. Another rather more complicated approach is to set the sample size to give a minimum required level of probability that the conclusion produced by a particular statistical analysis (i.e. a particular significance test) is the genuinely correct conclusion. This is achieved by determining the so-called **power** of the statistical test, a concept that will be discussed in more detail later. While random variation can never be eliminated and sample reliability can never be 100%, the application of such approaches does at least confer a degree of rigour to the sampling process.

4.2 Analysis of Sample Reliability

To reiterate, samples are used to estimate the values of population parameters, e.g. the sample mean is used as an estimate of the unknown mean of the population from which the sample was drawn. Therefore whenever sample statistics are quoted it is only fair to try and address the question 'how reliable is the sample as an estimate of the population?' However good the sample selection procedure has been, it is always important that some evidence of the sample reliability is provided. This is achieved by determination of two statistical values called the **standard error of the mean** and the **sample confidence limits**.

4.2.1 The standard error of the mean (SE)

The standard error of the mean (or more simply the 'standard error') was originally introduced in Chapter 2 as a descriptive statistic that indicates the reliability of a sample, that is, a measure of how close the sample mean is to the true population mean. However, to understand fully what this statistic tells us about a sample it is necessary to examine this important sample statistic in more detail.

Consider an exercise in which a large number of samples of equal size are extracted from a population and the means of these samples plotted as a data distribution. What will this distribution of sample means look like? It is fairly obvious that if the population is normally distributed then the distribution of sample means will also be normally distributed. What is less obvious is the fact that even if the population is not normally distributed, e.g. skewed, the distribution of sample means still tends towards a normal distribution and this is a tendency that becomes stronger the larger the sample size. This rule is known as the **central limit theorem**. The other important feature of the distribution of sample means is that it will be much more tightly packed around the population mean than are the individual data values. This is because by taking sample means the most extreme individual values in the population are eliminated from consideration. Consequently, the variance and standard deviation of a series of sample means must be smaller than the variance and standard deviation of the population from which the samples were drawn. Clearly it follows that the smaller the standard deviation of the sample means, the closer the samples means are to the population mean and, therefore, the more reliable any single sample is likely to be. It is the standard deviation of the means of all the possible samples of a given sample size, n, that are drawn from the population that is termed the standard error of the mean. This concept is illustrated in Fig. 4.4.

It is logical to argue that the greater the spread of the population the less likely it is that any given sample mean will give an accurate estimate of the true mean. However, as the sample size increases the chances of the sample mean being a close estimate of the true mean improve and the variance of the sample means will therefore

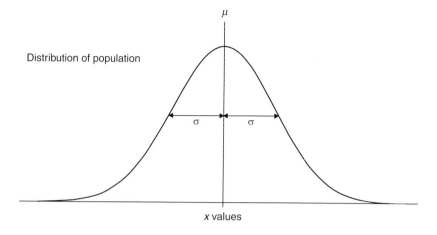

Distribution of population

σ   σ

x values

All possible samples of size = 10

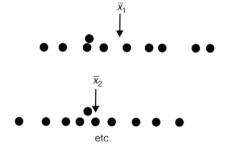

$\bar{x}_1$

$\bar{x}_2$

etc.

Distribution of means of all possible samples of size = 10

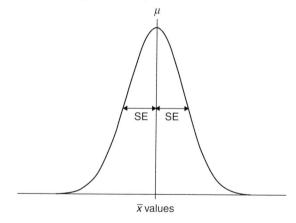

SE   SE

$\bar{x}$ values

Fig. 4.4. The distribution of sample means around a population mean. The standard deviation of sample means is smaller than the population standard deviation and is termed the standard error of the mean (SE).

decrease. In fact it can be shown that the variance of all possible sample means ($\sigma_{\bar{x}}^2$) of a given size is mathematically proportional to the population variance (σ^2) and inversely proportional to the sample size (n), i.e.

$$\sigma_{\bar{x}}^2 = \frac{\sigma^2}{n}$$

Standard deviation is the square root of variance, thus the standard deviation of the sample means $\sigma_{\bar{x}}$ is given by:

$$\sigma_{\bar{x}} = \sqrt{\frac{\sigma^2}{n}} = \frac{\sigma}{\sqrt{n}}$$

It is the standard deviation of the samples means that is termed the **standard error of the mean** and given the symbol **SE**.

However, we still have a problem. While we know the value for n, we do not know the value for the population standard deviation σ. However, the spread of the observations in the sample should reflect the spread of the variable in the population. Therefore we can quite legitimately use the sample standard deviation s as an estimate of σ. Thus for a single sample the formula for calculation of standard error simply becomes:

$$SE = \frac{s}{\sqrt{n}}$$

where: s = sample standard deviation
$\quad\quad\quad n$ = sample size

Comparison of the use of standard error and standard deviation

Inspection of the respective formulae for the determination of the standard error and the standard deviation reveals a very important difference between the two statistics. In the formula for standard deviation the sample size n is a component of both the term above the division line, i.e. the numerator, and the term below the division line, i.e. the denominator. In the formula for standard error, however, n is a component of the denominator only. Therefore an increase in sample size decreases standard error but it has no systematic effect on standard deviation. Thus, while the standard deviation gives an indication of the reliability of a single random measurement in estimating the population mean, it is the standard error which indicates the reliability of a sample mean in indicating the population mean.

As previously noted in Chapter 2, the standard error may be considered as a basic descriptive statistic that should accompany the presentation of any sample mean. While its absolute value may not in itself validate or invalidate a given sample mean, clearly the smaller the standard error the more reliable the sample and its calculation does allow the relative reliability of different samples to be compared.

One final very important point regarding standard error; the above arguments hold for all samples extracted from all types of data distribution and the formulae do not rely on the population being normally distributed. Consequently, the *standard error of the mean can be calculated for any sample of numerical data independent of the data distribution.*

4.2.2 Confidence limits for the sample mean (CL)

While the standard error provides a relative measure of sample reliability, what would be even more useful would be to be able to determine quantitatively how close we might expect the sample mean to be to the true population mean. While the population mean is a parameter which cannot be determined, it is possible to determine the range around the sample mean which has a defined probability of including the population mean. Clearly the smaller this range the closer the sample mean must be to the population mean. The limits around the sample mean which have a defined level of probability of including the population mean are termed the **confidence limits** (**CL**) and the numeric distance between the upper and lower confidence limits around the mean is termed the **confidence interval** (**CI**). The confidence limits and confidence interval thus provide a more exact quantitative measure of the reliability of a sample mean than the standard error.

Before continuing it must be pointed out that, unlike the standard error, *the calculation of the confidence limits depends on the frequency distribution of the population data*. The most usual situation under which confidence limits are determined is when the variable being measured is normally distributed and its calculation then employs *z* values derived from the standardized normal distribution.

Calculation of confidence limits for a sample taken from a normally distributed population

If it is known that a particular population shows a normal distribution then it is possible to calculate the probability of any one data item falling within any given number of standard deviations around the mean. For example, there is a 95% probability that any one item will fall in the range defined by the mean $\pm 1.96\sigma$, where 1.96 is the *z* value derived from the table of standardized normal deviations (Table 3.1). Similarly, if we have a number of samples extracted from this normally distributed population then the means of these samples will also be normally distributed and their values will also fall with a predictable distribution around the true population mean. Recalling that the standard error is the standard deviation of a set of sample means, then in fact 95% of sample means will occur in the range $\mu \pm 1.96$ standard errors around the mean.

In practice we will normally have only a single sample mean at our disposal, so what can we say about a single sample mean? From the above it should be clear that we are now in a position to define the probability of a sample mean falling within a certain range around the true mean. For example, there is a 95% probability that a measured sample mean falls within the limits $\mu \pm 1.96 \times SE$. This is of little use in itself because we cannot measure the true population mean, μ. However, we can turn the argument round and define the range around a sample mean within which the true population is likely to fall. By applying the standard error interval to the sample mean instead, i.e. $\bar{x} \pm 1.96 \times SE$, it follows that there is a 95% probability that this range around the known sample mean will include the true population mean, μ. This is illustrated in Fig. 4.5.

By referring to the appropriate value for *z* the confidence limits may be calculated for any given level of probability desired although the 95% limits are those most usually

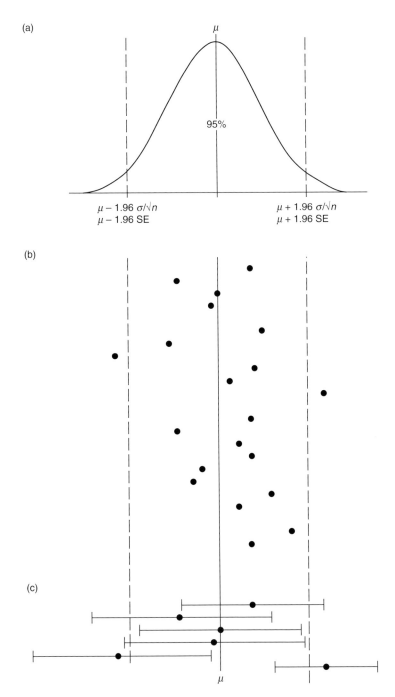

Fig. 4.5. Determination of 95% confidence limits: the range around a sample mean which has a 95% probability of including the population mean μ. (a) The distribution of sample means around the true mean of a normally distributed population. (b) Means of 20 randomly chosen samples of size n. (c) Sample means $\pm 1.96 \, s/\sqrt{n}$. The horizontal bars represent 95% CL obtained from repeat samples so that, on average, 95% of these intervals would include the true but unknown population mean.

Chapter 4

quoted. The confidence limits are then simply calculated by multiplying the standard error by the standardized normal deviation z for the required level of probability, thus:

CL (at P probability $= \pm$ SE $\times$ Z (at P probability)

The upper confidence limit is obtained by adding the CL value to the mean and the lower confidence limit obtained by subtracting it from the mean. The range between the upper and lower confidence limits is the confidence interval.

4.2.3 Determination of confidence limits for a sample from non-normally distributed populations

The standardized normal deviation z can only be used for calculation of confidence limits for samples that are drawn from normally distributed populations and cannot be employed for samples drawn from other types of data distribution, such as binomial, Poisson and skewed distributions. In particular it should be remembered that data in the form of counts, scores, proportions or percentages cannot form a normal distribution because of the discontinuous nature of the data and the truncation of the tails of the distribution. There are two approaches that can be taken to determination of confidence limits for non-normally distributed data sets, either undertaking a data transformation or applying a different data distribution model.

Data transformation

Data transformation is simply the mathematical conversion of the values of a variable by applying some constant function to each value. By doing this it is often possible to convert the data so that they then approximate to a normal distribution. The type of mathematical conversion required depends on the type of data that are to be transformed. Some examples of functions often used to transform data include logarithmic (primarily base 10 and e), square root and arcsin. Following transformation of the data subsequent statistical analysis is then performed on the transformed values. In presenting the results the type of transformation employed should of course be made clear and both the untransformed and transformed treatment means and standard errors should be given. The most common transformations and the circumstances of their use are shown in Table 4.1.

Conversion of numerical values to their equivalent root or log values is straightforward but the arcsin transformation, also termed the angular transformation, needs

Table 4.1. Common transformations used to convert data to fit a normal distribution.

Type of data	Possible transformations
Skewed	$\sqrt{x}$; log x
Counts	log x
Counts that include zero values	log $(x + 1)$
Counts from a Poisson distribution	$\sqrt{x}$
Counts from a Poisson distribution that include zero values	$\sqrt{(x + 0.5)}$
Percentages based on counts	arcsin
Binomial proportions	arcsin

a little further explanation. As stated in Table 4.1 the arcsin transformation is commonly used for count data that have been expressed in the form of percentages. It can also be used for data that adhere to a binomial distribution. To perform the transformation percentage values are initially divided by 100 to convert to a proportion. The required transformed value is then the angle that has a sine value equal to the square root of this proportion and can be obtained from tables. The original data value is thereby converted to an angle between 0 and 90°. An example of the transformation calculation is shown in Example 4.1.

Example 4.1. Calculation of the arcsin transformation of % data based on counts.

X (%) (data value to be transformed)	P (X/100)	$\sqrt{P}$	arcsin (α) (where sine $\alpha = \sqrt{P}$)
1	0.01	0.10	5.74
10	0.10	0.32	18.43
25	0.25	0.50	30.00
50	0.50	0.71	45.00
75	0.75	0.87	60.00
90	0.90	0.95	71.57
99	0.99	0.99	84.26

It is relatively straightforward to perform the transformation using an appropriate computer software package that includes an arcsin function. Note, however, that many packages including Microsoft Excel® spreadsheet software return the arcsin value in radians rather than degrees. Where this occurs the radians need to be converted to degrees by multiplying by 57.296, although Excel does have a function for achieving this (called 'DEGREES').

Application of other distribution models

If the data are known to adhere to a different type of distribution, such as a binomial or Poisson distribution, then alternative techniques for the determination of the confidence limits can be employed. The calculation is based on the theoretical mathematical description of the distribution model concerned and can be rather complex. The details of these methods will not be dealt with here and readers requiring further details are referred to alternative texts, e.g. Heath (1995).

4.2.4   The use and graphical presentation of confidence limits

The confidence limits define the range around a measured sample mean that has a given probability of containing the true but unknown population mean and are literally, therefore, a measure of the confidence that one can place in a sample mean. The smaller the confidence limit interval at a given level of probability the more accurately the sample mean estimates the true mean.

The other use that is often made of confidence limits is to provide a simple assessment of whether the observed difference between sample means is likely to represent a true difference due to some underlying cause or treatment or whether it is simply due to natural random variation. This is achieved by making a simple comparison of the confidence intervals around the sample means of interest. If the confidence interval around two sample means numerically overlap then it suggests that the means are different estimates of the same population mean. If, however, the confidence intervals do not overlap it becomes more likely that the two sample means estimate two different population means and the samples are, therefore, likely to be significantly different. This is shown in Example 4.2. This is, however, a rather controversial area and non-overlapping confidence intervals must not be used as a definitive test of significant differences. Thus *the presence of non-overlapping confidence intervals does not permit any statement to be made regarding the probability of a significant difference being present.*

Whenever sample means are graphically plotted, the standard error of the mean or the confidence limits around the mean should also be shown. Either can be

---

Example 4.2. Example of the use of confidence limits for testing for significant differences between sample means.

The calculated mean and 95% confidence limits of three samples, A, B and C, were as follows:

A: Mean = 3.0   95% CL = ±2.0 : 95% confidence interval = 1.0 to 5.0
B: Mean = 4.0   95% CL = ±1.5 : 95% confidence interval = 2.5 to 5.5
C: Mean = 7.5   95% CL = ±1.0 : 95% confidence interval = 6.5 to 8.5

Comparison of 95% confidence intervals:

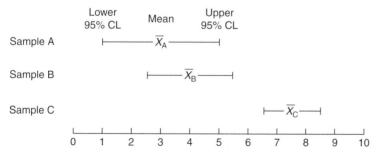

The 95% confidence intervals for Sample A and Sample B means overlap, suggesting that both sample means are estimates of the same population mean. Sample A and Sample B are, therefore unlikely to be significantly different from each other. The 95% confidence interval for Sample C mean does not overlap with that of either Sample A or Sample B mean, suggesting that it is an estimate of a different population mean. Sample A and Sample B are, therefore, likely to be significantly different from Sample C. (Note that this is not a definitive test of presence or absence of significant differences and does not allow any statements regarding the probability of such differences being present.)

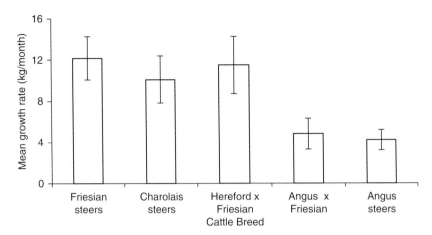

Fig. 4.6. Example of the use of vertical bars to indicate the 95% confidence limits around graphically plotted sample means. Growth rate of different breeds of cattle in a beef production breeding trial (vertical bars indicate 95% confidence intervals).

represented by an error bar plotted vertically through each sample mean and scaled against the Y-axis as described in Chapter 2 (see Fig. 2.1). Presentation of the confidence limits in this way becomes a particularly useful method for showing graphically those means that can be inferred to be significantly different. An example of this is shown in Fig. 4.6. Note that it is a common fault to depict vertical error bars but to fail to clearly indicate whether they represent standard errors or confidence limit intervals. Clearly it is important to state in the figure legend which of these the error bars represent.

4.3 Analysis of Small Samples

4.3.1 The problem of analysing small samples

When using the standardized normal deviation z to calculate the confidence limits for a sample mean, a rather large assumption is made. The normal deviation (z) is derived, not from the observed sample data distribution, but from the theoretical consideration of the normal distribution of a population. In employing z to calculate the confidence limits for a sample, we have to assume, therefore, that the sample is drawn from a normally distributed population and the sample data distribution does not appreciably differ from this distribution. If the sample is sufficiently large, this is a reasonable assumption; however, as the sample size decreases, the amount of sample error increases and at some point the discrepancy between the sample data distribution and population data distribution will become too large to ignore. To put this in another way, if the sample is large then the sample variance s can be expected to be very close to the population variance σ and we can use the former to estimate the latter. The smaller the sample becomes, however, the less valid becomes this estimate. Consequently it then becomes invalid to use the normal deviation z to calculate the confidence limits around the mean of a small sample.

4.3.2 Student's *t* distribution – a solution to the problem of analysing small samples

The problem of how to analyse small samples was solved in 1908 by William S. Gosset, publishing under the name of 'Student'. Gosset showed that the analysis of small samples should not be based on the normal distribution but instead on a modified data distribution that he called the *t* distribution. The major difference between the normal distribution and the *t* distribution is that, unlike the normal distribution, the shape of the *t* distribution varies with sample size. Consequently, there is not a single *t* distribution but a series of *t* distributions covering all possible sample sizes. The shape of the *t* distribution is shown in Fig. 4.7.

Exactly how the *t* distribution is mathematically derived from the normal distribution really need not worry us here; however, two important features of the *t* distribution should be noted: first, that as sample size decreases the spread of the *t* distribution becomes both wider and flatter; and, secondly, that at infinite sample size ($n = \infty$) the *t* distribution is exactly equivalent to the normal distribution. The *t* distribution therefore describes a trend in which the sample data distribution increasingly departs from the normal distribution as the sample size decreases.

Since the *t* distribution is to be used to analyse small samples, the obvious question is what size a sample must for it to be classified as 'small'. Student himself suggested that at a sample size of 30 ($n \leq 30$) the departure of the sample distribution from the population distribution became too large to ignore. Subsequently it has been agreed by consensus that *all samples of 30 items or fewer must be considered to be small* and should be analysed using Student's *t* distribution. Since, however, modern computers make running complex statistical analyses very straightforward, many statisticians today use the more complex but accurate *t* distribution for all analyses irrespective of sample size so that the *z* test is actually becoming rather redundant.

The theoretical values of *t* are given in the Student's *t* table (see Appendix: Table A2). Because the *t* distribution depends on sample size, instead of a single set of

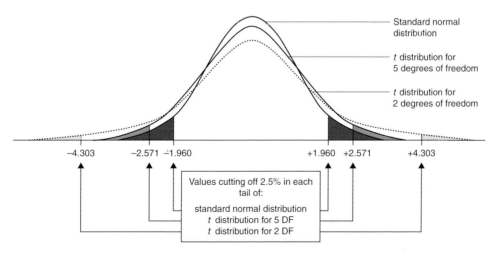

Fig. 4.7. The normal distribution and the Student's *t* distribution for two sample sizes. Note that the normal distribution is equal to the *t* distribution at DF = ∞

---

values for each level of probability the table shows a whole series of t values for each probability. Therefore the Student's t table must be consulted at both the required probability and the required sample size. When samples are being statistically analysed, sample size is measured not by n but by the **degrees of freedom** (**DF**) of the sample. The *degrees of freedom of a single sample are simply given by n – 1.* Therefore to obtain the required value for t the value is read from the table, reading along the rows to locate the required probability, and reading down the columns to locate the required degrees of freedom. It may be noted that at infinite degrees of freedom ($n - 1 = \infty$) for any given level of probability the value of the t statistic is the same as the z statistic. However, as the degrees of freedom decrease, t becomes increasingly larger than the equivalent value of z.

A note on degrees of freedom (DF)

As indicated above, in the statistical analyses of samples, sample size is described not by n but by the sample degrees of freedom, $n - 1$. The reason for this is highly theoretical but it can be shown that it is more accurate to employ the degrees of freedom as a measure of sample size whenever samples are being used to represent populations. Their use has the effect of increasing the value of the calculated sample variance, standard error and confidence limits and thereby produces safer estimates of the population parameters.

One way of looking at degrees of freedom is to consider them as a measure of the number of items within a sample that are theoretically free to take any value. Remembering that a sample mean is used to estimate the population mean, then, if all but one of the values in a sample are known, the value of the final item will be automatically fixed by the population mean. In other words, the value of the final measurement is not free to have any value and, therefore, the number of items that are free is $n - 1$. It should be noted that, while the degrees of freedom of a single sample are always given by $n - 1$, there are a number of situations in statistical analyses when the degrees of freedom for a parametric statistical test are given by a calculation other than $n - 1$.

4.3.3 Determination of confidence limits for a small sample taken from a normally distributed population

In the case of small samples the Student's t statistic is used to calculate the confidence limits. This is simply done by replacing z in the confidence limit formula by the appropriate value of t read from the Student's t table at the appropriate degrees of freedom and the required level of probability. Thus for small samples the confidence limits around the sample mean are given by:

CL (at P probability = $\pm$ SE $\times$ t (read at P probability and DF = $n - 1$)

It is important to remember that statistical tables generally present values for the probability of a data item falling outside the required distribution. Therefore, when consulting the Student's t table to obtain the value of t for the calculation of the 95% confidence limits, it is the 5% P value which is consulted, i.e. at $P = 0.05$. Example 4.3 demonstrates the calculation of 95% confidence limits for a small sample.

Example 4.3. Example of the calculation of the 95% confidence limits and confidence interval for a mean of a small sample.

If a sample of size $n = 20$ has a mean of 15 and a standard error of 0.5, then the 95% confidence limits are given by:

95% confidence limits	$= \pm 0.5 \times 2.093$	$= \pm 1.046$
Upper 95% CL	$= 15 + 1.046$	$= 16.046$
Lower 95% CL	$= 15 - 1.046$	$= 13.954$

where the value of t read from the Student's t table (at $P = 0.05$ and DF $= 19$) $= 2.093$.

Conclusion: it is 95% probable that the data interval 13.954 to 16.046 contains the population mean μ.

4.4 Summary of the Use of Sample Standard Error and Confidence Limits around the Mean

Statistical analyses therefore provide us with two main measures that enable us to judge how close we might expect a set of sample data to represent the actual population values. The **standard error of the mean** is a relative measure of sample reliability that potentially allows the most reliable of a number of available samples to be chosen. The **confidence limits**, stated for a particular probability, give a precise numerical range which has a defined likelihood of including the true but unknown population mean. To illustrate their usefulness let us return to Fig. 4.1 where five small samples are used to estimate the concentration of casein protein in a bulk volume of milk. From the five sample measurements the mean concentration, the standard error of the mean and the 95% confidence limits are as follows:

Sample mean (mg/g)	Standard error of the mean	95% confidence limits
32.34	±0.30	±0.83

What do these values now tell us? Since the standard error of the mean is relatively very small compared with the actual mean (approximately 1%) this implies strong sample reliability, i.e. the sample mean is an accurate measure of the actual casein concentration in the milk. In other words, if we took a further similar sample we would not expect it to vary from this sample mean by more than 0.30 units. More precisely, the confidence limits tell us that it is 95% certainty that the actual casein concentration of the bulk milk solution occurs within plus or minus 0.83 mg/g of the sample mean. Or, to put it another way, we can be 95% confident that the actual casein concentration lies between 31.51 and 33.17 mg/g. In this case, therefore, it is reasonably assured that the sample mean is providing an accurate indication of the true casein concentration of the milk.

While there are no firm rules about how large or small a standard error or confidence limit should be, the presentation of these sample statistics do provide an objective measure of the validity of a set of sample data. In order that the reader can come to their own view concerning the reliability of a set of presented sample data, it is universally recognized good practice to always return the standard error and/or the confidence limits whenever a sample mean is expressed.

5 Inferential Statistics and Hypothesis Testing

Hypothesis: supposition made as basis for reasoning or as starting point
for investigation.

(Concise Oxford English Dictionary)

- Introduction to the principles of inferential statistics and significance testing.
- The null hypothesis.
- The main procedural steps in significance testing.
- Setting the critical level of probability for accepting or rejecting the null hypothesis.
- The concept of one- and two-tailed significance tests.
- Errors and fallacies in significance testing.

5.1 Introduction to Inferential Statistics

Whenever an experimental investigation is carried out, whether it is based on a classical manipulative experimental approach or upon observational studies, the researcher commences the investigation with a basic hypothesis. The hypothesis may take the form of a suggestion of how some recognized factor affects the items being investigated, or it may be a possible explanation of a particular mechanism. Once the experiment has been conducted and the data collected, the branch of statistics which is concerned with testing the validity of the stated hypotheses and drawing conclusions based on the observed data is called **inferential statistics**.

As discussed extensively in the previous chapters, the main problem in analysing trends and differences in research data is that these arise not just from the treatments applied but also from the natural random variation that exists between all items. Consequently in trying to draw conclusions from data we always need to ask one fundamental question: are the trends or differences observed in the data due to a recognized cause or are they simply due to natural random variation between the items being observed? In other words, are differences in the data 'significant'? In order to answer this question, a **test of statistical significance** must be performed and such tests are the basis of inferential statistics. Since, however, we can never be 100% certain that natural random variation is not the cause of the differences observed in the data, statistical significance tests cannot totally prove or disprove a hypothesis. Instead, statistical significance tests determine the **probability** that differences in the data arise due to a recognized cause or treatment.

© C. Ireland 2010. *Experimental Statistics for Agriculture and Horticulture* (C. Ireland)

5.2 Testing for Statistical Significance

Deciding whether a significance test is to be used and, if so, deciding which type of test is an integral part of the experimental planning process. A great many different types of statistical significance test are available enabling data arising from a large range of experimental designs to be analysed. The first major problem that therefore has to be resolved is how to identify which type of test is the most appropriate to analyse a particular set of data. The choice of test depends on a number of considerations, the most important being: the precise question being asked (e.g. is there a difference or a relationship between data sets?); the type of data involved (e.g. measured or count data?); and the type of distribution to which the data adheres (e.g. a normal or non-normal data distribution?). Figure 5.2 at the end of this chapter indicates the decision-making process that leads to the selection of the major types of significance test that researchers in agriculture and horticulture are likely to encounter.

5.2.1 The principles of statistical significance tests and the null hypothesis

All statistical significance tests are based on the same set of major procedural principles; once these are fully understood the performance of any significance test becomes a much more logical and straightforward process.

The first step in any test of significance is to state the **null hypothesis** for the particular experiment being conducted. The null hypothesis, which is commonly abbreviated H_0, states that there are no treatment effects present that influence the data observed. To put this another way, the null hypothesis proposes that any apparent differences or trends in the data are due solely to natural random variation. The alternative **positive hypothesis**, which is commonly abbreviated H_1, proposes that there is a definite treatment effect present that influences the observed data set. It is important to understand that these hypotheses apply to the populations of items under investigation and not just to samples that may have been drawn from the populations. Thus, if the differences between the means of a number of differently treated sets of measured items were being investigated in relation to the application of some experimental treatments, then we would state the null and positive hypotheses as follows:

H_0 = there is no difference between the population means: $\quad \mu_1 = \mu_2 = \mu_n$
H_1 = there is a difference between the population means: $\quad \mu_1 \doteqdot \mu_2 \doteqdot \mu_n$

The objective of the subsequent significance test is to determine the probability that the observed data support the null hypothesis as the correct explanation of the results obtained. The final conclusion drawn will then depend on this calculated level of probability. If the probability of the observed data supporting the null hypothesis is sufficiently low, then the null hypothesis may be rejected and the alternative positive hypothesis can be accepted as the more likely explanation of the results obtained. If, however, the probability of the data supporting the null hypothesis is not sufficiently low, then the null hypothesis must be accepted and no significant treatment effect can be concluded to be present.

At this stage you may be wondering why it is that the probability of the data supporting the null hypothesis is examined rather than testing the probability that the data support the positive hypothesis. There are two reasons for this; the first is philosophical and the second is technical. First, by testing the null hypothesis it is ensured that no prior assumptions are made about the effectiveness of any applied treatment. In this way it is less likely that bias in favour of obtaining a positive type of conclusion will be introduced. Second, while it is relatively straightforward to mathematically calculate the probability that data support the null hypothesis, it is extremely problematical to calculate the exact probability of the alternative positive hypothesis being correct. This is because the null hypothesis is more numerically exact. If, for example, we were examining the difference between two differently treated populations, then the null hypothesis clearly states that the numerical difference between the population means is zero. On the other hand, the positive hypothesis cannot be exact since it is not possible to state how large the difference between the sample means should be before a real treatment effect can be assumed to be present. The positive hypothesis is therefore tested indirectly by examining in the first place whether the difference is sufficiently greater than zero for the null hypothesis to be rejected. If the null hypothesis is rejected, the alternative positive hypothesis may then be accepted in its place.

5.2.2 The basic procedure for conducting statistical significance tests

The basic procedure for conducting statistical significance tests arises mainly from the work of Sir Ronald Fisher, Jerzy Neyman and Ego Pearson in the 1930s; see Quinn and Keough (2002) for a historical and philosophical account of the development of statistical testing. To reiterate, the principle behind all significance tests is that if the probability of the observed data supporting the null hypothesis is sufficiently small (remember it can never be zero) then the null hypothesis may be rejected as the correct explanation of the results. The probability value at which the null hypothesis is rejected is termed the **critical probability** (α) and of course the lower the value at which this is set the safer will be the final conclusion reached.

The main procedural steps for conducting all significance tests are as follows:

1. **Choice of an appropriate test of significance.** There is wide range of significance tests available. The test of significance chosen for analysing a particular set of results will depend on the type of experimental design, the type of data involved and the type of comparison being made (see Fig. 5.2).

2. **Statement of the appropriate null hypothesis for the experiment being analysed.** The null hypothesis, H_0, is that there are no treatment effects present that influence the experimental data.

3. **Calculation of the value of the test statistic.** The probability that the observed data comply with the null hypothesis being correct is determined by calculating from the observed data, using appropriate statistical formulae, the value of a single statistic value called the 'test statistic'. The numerical value of the test statistic generally depends on the variability of the data within and between the samples. Further, the test statistic will have a known probability distribution so that the probability of

obtaining any given value for the test statistic on a completely random basis is known.

4. Determination of the probability (*P*) of obtaining the observed value for the test statistic in the event that H$_0$ is actually true. A statistical probability table is consulted which gives the probability of obtaining the observed value for the test statistic in a situation where the null hypothesis is actually correct and there is only natural random variation present in the data. (Probability values for performing different significance tests are commonly published in table form such as those presented in the Appendix of this book and in more complete form in Rohlf and Sokal, 1994. Note, however, that in modern computer statistics packages the probability values are determined by the software and the user does not usually need to consult statistical tables directly.)

5. Decision to reject or accept the null hypothesis. The probability *P* of obtaining the observed value of the test statistic is compared with a **pre-chosen** critical level of probability α (commonly 5%) for the rejection of H$_0$. If the value of *P* is smaller than (or equal to) the critical value, then H$_0$ is rejected and it can be concluded that the data support a significant effect being present at the given level of probability. If the value of *P* is larger than the critical value, then H$_0$ must be accepted as the more likely explanation of the data. It is important to realize, however, that accepting the null hypothesis on this basis does not mean that it is certainly true. It simply means that the data inspected do not provide sufficient statistical evidence to allow a significant effect to be claimed.

5.2.3   Setting the critical level of probability

One of the major problems in significance testing is deciding at what level the critical probability for rejecting H$_0$ should be set. There is no purely objective mathematical solution to this question and an answer can only be obtained through a reasonable consensus of experienced statisticians and investigators. It is the general rule that *the probability of the data concurring with the null hypothesis being correct must be no greater than 5% for the null hypothesis to be rejected.* Conversely, the probability of the data supporting a significant effect being present must be at least 95% for the positive hypothesis to be accepted. A 5% critical value for rejection of H$_0$ has generally proved, in the history of experimental investigation, to be neither too small so that many erroneous claims have been made, nor too large so that many valuable discoveries have been left undiscovered for too long. It is, in effect, a compromise value that maintains the incidence of invalid conclusions at an acceptably low level. There will be occasions, however, when the use of a 5% probability level for rejection of H$_0$ is not considered sufficient. For example, medical researchers will need to be more than 95% certain that a new drug does not have harmful side effects and may wish to set a much lower critical value for rejection of the null hypothesis. The consequences of getting it wrong in land-based research are rather less critical and 5% is commonly regarded as an acceptable maximum probability for rejecting the null hypothesis in agriculture and horticulture experiments.

Obviously, the lower the probability that the observed data support the null hypothesis, the greater the confidence one can have in any positive conclusions that are drawn from the data. Therefore it is extremely important that the critical level of probability is set before any experimental investigation is carried out and is not subsequently altered in the light of the results obtained. Only by doing this can objectivity be maintained. If the critical level of probability is set after the experiment is completed, then there may arise a strong temptation to set a critical level of probability that will give a certain desired result. It goes without saying that this is a course of action that must be firmly resisted.

Finally, whenever the conclusion to a statistically analysed experiment states that a significant effect is present it is very important that the critical level of probability employed to facilitate this conclusion is clearly indicated. This is commonly achieved by simply confirming within parentheses that the calculated probability (P) is equal to or less than the critical probability set for the experiment. For example, the expression '($P \leq 0.05$)' indicates that the probability of the data supporting the null hypothesis is equal to or less than the critical probability that was set at 0.05. More usefully, if the exact value of P has been determined – and many modern computer statistics packages will do this – then this exact value can be presented, e.g. '($P = 0.012$)'. In effect P represents the likelihood that you got it wrong.

5.3   One- and Two-tailed Tests of Significance

Significance tests that inspect the differences between two data samples may be performed as either a **one-tailed test** or a **two-tailed test**. If it is required to test whether a difference is significant irrespective of which sample mean is the larger, then a two-tailed test is employed. If, however, we wish to test whether one of the sample means is specifically larger than the other, then a one-tailed test may be employed instead. To understand this we need to consider carefully how a significance test works and I shall use a simple example to help explain this.

Suppose that a new breed of dairy cow has been developed and it is necessary to know whether the protein content of the milk of the new breed is in any way different from that of an old existing breed. An appropriate significance test is performed to compare the means of samples of the two breeds and this produces an observed value for a particular test statistic (there is no need to be concerned about how this value is calculated at this stage). The test statistic will have a known probability distribution on the basis that there is no difference between the sample means, i.e. that the null hypothesis is correct. What we then need to determine is the probability that the observed value for the test statistic derived from the milk protein data belongs to this same distribution. If the value is sufficiently extreme that its probability of belonging to the distribution is lower than the preset critical probability α, then we will conclude that the data fail to support the null hypothesis and the positive hypothesis may be accepted in its place. Since, however, it has not been specified in which direction we are looking for a difference between the means, we need to consider that the observed value of the test statistic might occur on either side of the probability distribution. Therefore, if the critical probability α for rejecting H_0 has been set at 5%, what we are actually interested in is whether the observed value falls within either of the tails of the probability distribution, each of which contain only 2.5% of all possible values. This is illustrated in Fig. 5.1(a). Such a test is therefore referred to as a **two-tailed test**. Thus, if we have no previous notion of

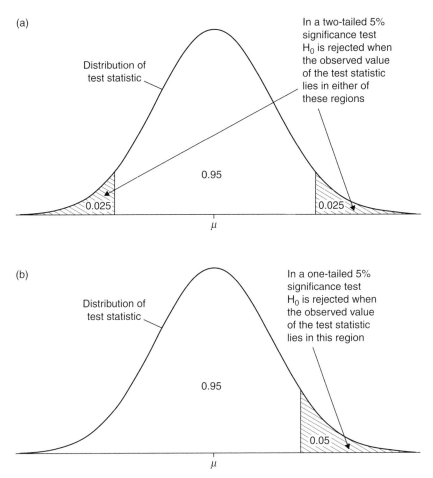

(a)

Distribution of
test statistic

In a two-tailed 5%
significance test
H_0 is rejected when
the observed value
of the test statistic
lies in either of
these regions

0.95

0.025

0.025

μ

(b)

Distribution of
test statistic

In a one-tailed 5%
significance test
H_0 is rejected when
the observed value
of the test statistic
lies in this region

0.95

0.05

μ

Fig. 5.1. Probability distribution of a test statistic (e.g. Student's t) showing the regions where the null hypothesis will be rejected in (a) a two-tailed 5% significance test and (b) a one-tailed 5% significance test. Value shown in the hatched tail is the probability of a random value occurring within that tail.

whether the milk protein level may be higher or lower in the new breed compared with the old breed, the analysis is conducted as a two-tailed test in order to inspect for a difference in either direction.

If, however, the new breed of dairy cow had been deliberately developed to try to increase milk protein, then a **one-tailed test** may be performed in order to detect if the milk protein had been significantly increased. In this case the possibility that the milk protein may have decreased is of no interest to the researcher and is not specifically tested for. The calculation of the test statistic is performed in exactly the same way for a two-tailed test of significance, the distinction lies in the critical value of the test statistic that is then consulted in order to determine the probability that the data support the null hypothesis. In a one-tailed 5% significance test we are concerned with only one tail of the distribution at either the lower or higher end, and we will be interested in the value of the test statistic that cuts off 5% of all values in that single tail. If the calculated value of the test statistic is larger than this critical value, then the null hypothesis can be rejected and one of the samples can be concluded as being significantly larger than the other. *The critical value of the test statistic at a particular probability for a one-tailed test is, therefore, equal to the critical two-tailed value at twice that probability.* The difference in the critical probability values for a one- and two-tailed test of significance used for a normally distributed variable such as milk protein level are illustrated in Fig. 5.1.

Since the absolute critical value of the test statistic will be lower for a one-tailed test than for a two-tailed test, it might seem that the chance of obtaining a significant result from a one-tailed test is considerably enhanced compared with a two-tailed test. This, however, is not the case as long as the direction of the difference we are interested in is defined beforehand. Remember that a one-tailed test is only concerned with one tail of the distribution, as shown in Fig. 5.1. If a difference arises but is in the opposite direction to that which we are interested in, then this must be considered as a non-significant effect, however large the difference might be. Returning to the dairy cow example, if it is decided to test the one-tailed hypothesis that the protein level of milk from the new breed is greater than from the old breed but we find subsequently that the new breed has a lower milk protein level than the old breed, this observation must be, for all intents and purposes, ignored and no significant difference can be claimed. Therefore before conducting a one-tailed test it is vital that the investigator is very sure that they really will have no interest if the result goes in the opposite direction to the one that they might be anticipating.

Finally it is very important to understand that the decision to undertake either a one-tailed or a two-tailed test must be made before the experiment is conducted and it is totally invalid to swap from one to the other later on in order to produce a significant result. An investigator must, therefore, have very good *a priori* reasons for performing a one-tailed test and if there is any doubt as to which should be performed then performance of a two-tailed test is the safe default option. It might also be pointed out that the statistics editors of a number of scientific journals today do not accept the use of one-tailed tests, presumably because they consider them prone to misuse and misinterpretation.

5.4   Errors in Significance Testing and the Power of Significance Tests

Since conclusions arrived at by statistical significance testing are based on probabilities, it is inevitable that sometimes the conclusion reached will actually be the wrong one. Occasionally the null hypothesis will be rejected even when it is actually true. In fact, if the critical level of probability is set at 5% an erroneous conclusion maybe expected to be obtained one time in every 20 times that the same experiment is performed. The rejection of the null hypothesis when it is actually true is termed a **Type I error**. On the other hand, the null hypothesis may sometimes be accepted when there really are significant differences present in the population data. This is termed a **Type II error**. Clearly as the critical level of probability α for rejecting H_0 is reduced the chance of making a Type I error is reduced but conversely the chance of making a Type II error must simultaneously increase. In practice this means that as α is reduced the more difficult it becomes to obtain a significant result but the greater the confidence one can have in any significant result that is obtained. It can be shown that in practice the only way to reduce the probability of making both Type I and Type II errors simultaneously in an experiment is to increase the number of replicate measurements made.

The **power of a significance test** is a means of indicating how likely it is that a particular significance test will lead to the correct conclusion being drawn. More formally defined, the power of a significance test (termed α) is the probability of the test resulting in the correct rejection of the null hypothesis when it really is not true. In mathematical terms, if the probability of a significance test producing a Type I error is β, then the power of the test is $1-\beta$. By identifying in advance of an experiment the power required from a particular significance test this can assist the investigator in the experimental design and in the decision on the appropriate sample size to use. This concept will be discussed further in the following chapters covering particular significance tests.

Finally, it should always be remembered that *statistical tests of significance are not infallible*. When drawing conclusions the following three points need always to be remembered:

- Just because a null hypothesis has been rejected it does not mean for absolute certainty that the data are significant, it only means that they are significant with a certain calculated level of probability.
- Just because a null hypothesis is accepted it does not mean that it is certain there are no significant effects, only that it is unlikely given the evidence currently available.
- Just because a null hypothesis has been rejected it does not mean that the null hypothesis is definitely false, it just means that a certain set of observed data do not support it.

In fact, given that the null hypothesis states that two or more populations are unequal, then because of the existence of random variation it is actually most unlikely that the null hypothesis could ever be truly wrong. While we can leave most of the more theoretical aspects of hypothesis testing for the theoretical statisticians to argue over, it is important that researchers are aware that there still remains much discussion, indeed controversy, over many aspects of statistical significance testing. Arguably, the most important function of the application of statistical analyses to research data is that they force all researchers to work by the same set of rules and to apply the same standards when deriving conclusions from their experimental data.

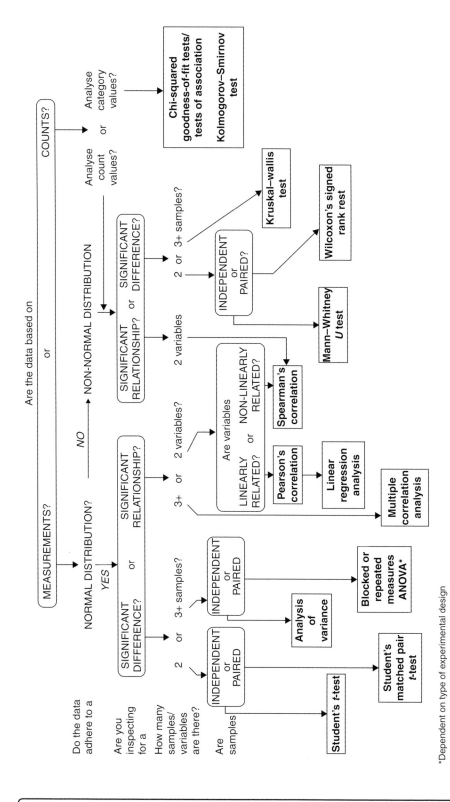

Fig. 5.2. Flow chart indicating the decision-making process required to select the major statistical significance tests used to make inferences from experimental data.

*Dependent on type of experimental design

6 Single-sample and Two-sample Parametric tests

Test: critical examination or trial of person's or thing's qualities.

(Concise Oxford English Dictionary)

- Introduction to parametric statistical tests for comparing samples.
- The *z*-test of statistical significance for examining whether the mean of a large sample is significantly different from a hypothesized population mean.
- The *z*-test for analysing the difference between two large samples taken from two normally distributed populations.
- Student's *t*-test of statistical significance for examining whether the mean of a small sample is significantly different from a hypothesized population mean.
- Student's *t*-test for analysing the difference between two small samples taken from normally distributed populations.
- Paired-sample *t*-test for analysing the difference between two non-independent samples taken from normally distributed populations.

6.1 Introduction to Parametric Single-sample and Two-sample Comparison Testing

One of the most common and indeed most simple types of experimental design is where two contrasting treatments are applied to two independent populations and single samples are then selected from each of the two populations for measurement. It is then required to test for a significant difference between the two populations using the sample data to estimate the population parameters. This situation arises particularly frequently where one population represents a treatment and the other population represents an untreated control, e.g. the measured yield of a sample taken from a population of tomatoes treated with a new fertilizer compared with the measured yield of a sample from an untreated population of tomatoes. Less common but still fairly familiar is the situation where a single sample has been extracted from a population and it is required to know whether the sample mean differs significantly from a theoretical or hypothesized mean of a different population. For example, whether the mean milk yield of a sample of cows of a new breed differs significantly from the already known mean yield of an existing breed. In both cases, if the measured samples are extracted from populations that can be safely assumed to adhere to a normal distribution, then a parametric significance test may be employed.

Two parametric tests exist for comparing two samples, namely the *z*-test, where the analysis is based on large samples ($n > 30$), and **Student's *t*-test**, which was originally devised to analyse small samples ($n \leq 30$) but can in fact be employed for any sized samples. These two tests will be described in some detail in this chapter. If, however, the two samples to be compared are extracted from populations that cannot be assumed to adhere to a normal distribution, and it is not desired to apply a data transformation to convert the data to fit to a normal distribution, then alternative non-parametric techniques are required, as described in Chapter 10.

6.2 Analysis of Large Samples using the *z*-test of Statistical Significance

If two samples are extracted from normally distributed populations, and if the samples are sufficiently large for the sample variances to be relied on to provide accurate estimates of the variances of the respective populations, then the z statistic may be employed to test for a significant difference between the samples. This so-called 'z-test' may also be employed to examine whether the mean of a large sample ($n >$ 30), is significantly different from a given hypothesized population mean. This latter use of the z-test is perhaps the most straightforward of all parametric tests of significance and its detailed description will serve to introduce many of the important principles of parametric statistical testing.

6.2.1 Analysis of the difference between the mean of a large sample and a hypothesized population mean using the *z*-test

The null hypothesis for this form of the z-test is that the mean of a population, which is estimated by the mean of a large sample of measured values, is not significantly differently from a given hypothesized value. We have in fact already come across an example of the use of the z-test statistic in Chapter 3 (section 3.3, 'An example of the use of the probability values of the normal distribution') to inspect whether the level of the protein casein in the milk of a particular cow was significantly different from the known mean of a particular population of cows. To describe the z significance test further we will modify this example. Suppose that, in an attempt to increase the casein content of cow's milk, a new breed of dairy cow has been developed. A large sample of cows from this breed was subsequently observed to have a mean milk casein content of 37 mg/g, while the mean milk casein content of an existing commercial breed maintained under exactly the same conditions was known to be 32 mg/g. In order to evaluate whether the new breed displays a genuine change in milk casein content, it is necessary to test whether the difference between the sample mean, which estimates the population mean for the new breed, and the hypothesized population mean of the existing breed is statistically significant. The null hypothesis for this analysis is that the population mean for the new breed and the hypothesized population mean for the commercial breed are equal.

Once the concept of the normal distribution is understood, the theory behind this z-test is quite straightforward. The aim is to determine the probability that the observed sample mean and the hypothesized population mean are actually representing the same population or whether they represent two different populations. Recalling the theory of the normal distribution (Chapter 3, section 3.3) we know that the probability of occurrence of a particular value within a normally distributed data set is mathematically related to the number of standard deviations (z) between the

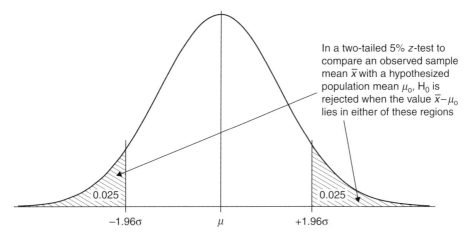

In a two-tailed 5% z-test to compare an observed sample mean $\bar{x}$ with a hypothesized population mean μ_0, H_0 is rejected when the value $\bar{x} - \mu_o$ lies in either of these regions

0.025　0.025

-1.96σ　μ　$+1.96\sigma$

Fig. 6.1. The region of the normal distribution used for rejecting the null hypothesis in a two-tailed 5% z-test comparing an observed large sample mean $\bar{x}$ with a hypothesized population mean μ_o.

population mean and the value. For example, the probability of a value within a normally distributed population falling more than 1.96 standard deviations above the population mean is less than 2.5% and similarly the probability of it falling more than 1.96 standard deviations below the population mean is less than 2.5%. This principle is illustrated in Fig. 6.1.

A test statistic, which we will call $z_{observed}$, is therefore calculated by taking the numerical difference between the observed sample mean $\bar{x}$ and the hypothesized population mean μ_o and dividing by the population standard deviation σ, thus:

$$Z_{observed} = \frac{\bar{x} - \mu_o}{\sigma}$$

The problem is, however, that we cannot determine $z_{observed}$ from the above equation because we do not have a value of the population standard deviation σ. The only way we can obtain a value for σ is to estimate it from the standard deviation of samples extracted from the population. As we are dealing here with large samples, then it may be assumed that any such estimate is accurate and valid. Recalling that the standard deviation of the distribution of sample means is called the standard error of the mean (SE), then the formula for the test statistic, $z_{observed}$, becomes:

$$z_{observed} = \frac{\text{difference between the sample mean and population mean}}{\text{SE}}$$

Since standard error is given by sample standard deviation (s) divided by $\sqrt{n}$, the equation becomes:

$$z_{observed} = \frac{\bar{x} - \mu_o}{s/\sqrt{n}}$$

where: $\bar{x}$ = the sample mean;　　　　s = the sample standard deviation;
μ_o = the hypothesized population mean;　n = sample size.

If the value of the test statistic $z_{observed}$ is greater or equal to the z value equivalent to the critical probability α that has been selected for testing the null hypothesis, then the probability of the sample mean and the hypothesized population mean belonging to same population must be less than α and the null hypothesis may be rejected. The critical probability values that need to be consulted are given in the table of standardized normal deviations (Table 3.1). Thus for a two-tailed z-test conducted at a critical level of probability of 5%, if the calculated value of $z_{observed}$ is greater or equal to 1.96 then the null hypothesis may be rejected and a significant difference between the observed sample mean and the hypothesized population mean can be claimed. The regions of the normal distribution used for rejecting the null hypothesis in a two-tailed z-test are illustrated in Fig. 6.1.

Often a z-test is performed as a one-tailed test where the null hypothesis states that the sample mean is not specifically larger (or smaller) than the known population mean. In this case we need only to concern ourselves with one tail of the distribution, either the lower or upper tail, whichever is specified by the one-tailed hypothesis. To inspect for 5% significance the critical z value now becomes 1.645, i.e. the value of z that cuts off 5% of values occurring in one tail of the distribution (see Fig. 6.2). The one-tailed null hypothesis can then be rejected when $z_{observed}$ exceeds this value.

In Example 6.1 the calculation of a z-test used to compare a sample mean with a known population mean is described employing the example of the casein content of cows discussed above.

Remember that when stating the conclusion of a significance test the critical probability used for rejection of the null hypothesis must always be quoted. In Example 6.1 the conclusion is qualified by stating '($P < 0.05$)' indicating that the probability of obtaining the observed value of the test statistic $z_{observed}$ assuming that H_o is correct is less than 5%. In effect this is stating that the chance of the conclusion being wrong is less than 5%.

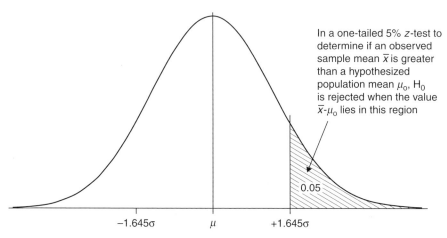

In a one-tailed 5% z-test to determine if an observed sample mean $\bar{x}$ is greater than a hypothesized population mean μ_0, H_0 is rejected when the value $\bar{x}$-μ_0 lies in this region

0.05

−1.645σ   μ   +1.645σ

Fig. 6.2. The region of the normal distribution used for rejecting the null hypothesis in a one-tailed 5% z-test to determine if an observed large sample mean $\bar{x}$ is greater than a hypothesized population mean μ_0.

Example 6.1. A z-test of significance for analysing the difference between the mean of a large sample and a hypothesized population mean.

A two-tailed test to determine whether dairy cows of a new breed have a significantly altered level of the protein casein in their milk compared with an existing commercial breed.

Suppose that the mean milk casein content (μ_o) of a main commercial breed of dairy cow is known to be 3.2 mg/g. A new breed of cow has been developed and breeders wish to know whether a significant change in milk casein content is present in this new breed compared with the commercial breed. In a sample of 100 cows of the new breed the mean casein level ($\bar{x}$) was 3.7 mg/g with a standard deviation (s) of 0.55 mg/g.

Null hypothesis: there is no difference between the mean of the population of the new breed and the hypothesized population mean ($H_0 : \mu = \mu_o$).

For the sample of the new breed of dairy cow:

$$z_{observed} = \frac{\bar{x} - \mu_o}{s/\sqrt{n}} = \frac{3.7 - 3.2}{0.55/\sqrt{100}} = 9.09$$

The critical z value at $\alpha = 0.05$ for a two-tailed test is 1.96. The observed value of z is greater than 1.96; therefore the null hypothesis may be rejected.

Conclusion: cows of the new breed have a milk casein content which is significantly different from the mean milk casein content of the main commercial breed of dairy cow ($P < 0.05$).

One-tailed test: this analysis could have been undertaken as a one-tailed z-test. In this case H_0 will have been stated in the form: the observed sample mean is not greater than the known population mean. The critical one-tailed z value at $\alpha = 0.05$ is 1.645. Since $z_{observed} > 1.645$ then H_0 is rejected and it is concluded that cows of the new breed have a milk casein content which is significantly greater than that of the existing commercial breed ($P < 0.05$).

6.2.2 Analysis of the difference between the means of two large samples by the z-test

Very often two populations are encountered which may have been subjected to different conditions and we wish to know whether there is a significant difference between their means. The simple example already referred to is where the yield of crop plants treated with a new type of fertilizer needs to be compared with that of crop plants treated with a standard fertilizer, the latter representing a control treatment. Since it is not feasible to measure all the crop plants present, a large sample is randomly selected from each treatment group and the difference between their sample means, i.e. $\bar{x}_1 - \bar{x}_2$, is then used to estimate the difference between the means of the differently treated populations, i.e. $\mu_1 - \mu_2$.

If two populations are both normally distributed, then it is valid to presume that if numerous pairs of large samples were to be extracted from these populations then the differences between the means of these pairs of samples, i.e. $\bar{x}_1 - \bar{x}_2$, will also be normally distributed. It then follows that the further away the observed value of $\bar{x}_1 - \bar{x}_2$ is from the mean of the normal distribution of $\bar{x}_1 - \bar{x}_2$, the more likely it becomes that the observed difference between $\bar{x}_1$ and $\bar{x}_2$ represents a true treatment effect. So the question we can now ask is whether the actual observed value of $\bar{x}_1 - \bar{x}_2$ is sufficiently far away from the mean of the theoretical normal distribution of sample mean differences for it to become probable that $\bar{x}_1$ and $\bar{x}_2$ actually represent the means of two significantly different populations. In order to answer this question we need to determine a value for the test statistic z_{observed} which can be used to test against critical values of z and thereby determine the likelihood that the observed data comply with the null hypothesis being correct.

In this case z_{observed} is obtained by dividing the difference between the observed sample means by the standard error of the difference (SED) and the known formula for this is:

$$z_{\text{observed}} = \frac{\bar{x}_1 - \bar{x}_2}{\sqrt{\dfrac{s_1^2}{n_1} + \dfrac{s_2^2}{n_2}}}$$

where: $\bar{x}_1, \bar{x}_2$ = mean of samples 1 and 2 respectively;
$\quad\quad\; s_1^2, s_2^2$ = variance of samples 1 and 2 respectively;
$\quad\quad\; n_1, n_2$ = size of samples 1 and 2 respectively.

The null hypothesis (H_0) for the test is that there is no difference between the two population means. The value of the test statistic z_{observed} is then compared with the theoretical values for z given in the the table of standardized normal deviations (Table 3.1, Chapter 3). As we have already seen, if the critical level of probability is set at 5% for rejection of H_0 ($\alpha = 0.05$), then z_{observed} needs to be ≥ 1.96 in order to claim a significant difference. This is illustrated in Fig. 6.3.

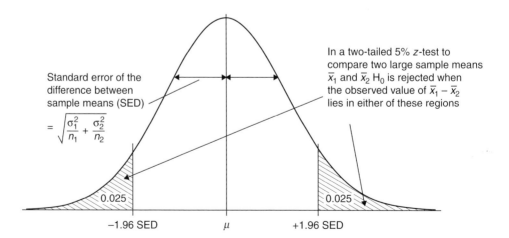

Fig. 6.3. The region of the normal distribution used for rejecting the null hypothesis in a two-tailed 5% z-test examining the difference between two sample means $\bar{x}_1$ and $\bar{x}_2$.

The full calculation of the z-test for comparing two large independent samples is shown in Example 6.2 and the reader is invited to move directly to this example if they wish. For those who require a better understanding of this test, the following section explains the derivation of the test statistic $z_{observed}$.

6.2.3 Derivation of $z_{observed}$ for the comparison of two large samples

It should be recalled from section 6.2, 'Analysis of the difference between the mean of a large sample and a hypothesized population mean', that the test statistic $z_{observed}$ is given by the difference between the observed sample mean and the theoretical population mean but expressed in standard deviations. The mean of the distribution of $\mu_1 - \mu_2$ values is of course unknown but can assumed to be zero based on the premise that if all possible pairs of samples of a given size are randomly selected from two populations then as many negative differences will arise between them as positive differences and these will cancel each other out. Therefore, for a two-sample comparison test $z_{observed}$ is given by:

$$z_{observed} = \frac{[\text{difference between sample means}] - [\text{zero}]}{\text{standard deviation of the distribution of } \bar{x}_1 - \bar{x}_2}$$

In statistical terminology the standard deviation of the differences between sample means is referred to as the **standard error of the difference (SED)**. The formula therefore simplifies to:

$$z_{observed} = \frac{\text{difference between sample means}}{\text{SED}}$$

So now we need to know how to determine the standard error of the difference (SED) between two sample means.

Statistical theory shows us that when we have two normally distributed populations that have known variances σ_a^2 and σ_b^2 then the differences between pairs of randomly selected single values from each population will also follow a normal distribution. Furthermore, it can be shown that the variance of the differences between these pairs of values will be equal to the sum of the variances of each individual population, i.e. $\sigma_{a-b}^2 = \sigma_a^2 + \sigma_b^2$. Similarly, if a large number of random samples of constant size are selected from the two populations the variance of the difference between pairs of sample means, i.e. $\sigma_{\bar{x}_1 - \bar{x}_2}^2$, is the sum of the sample variances divided by their respective sample sizes, i.e.:

$$\sigma_{\bar{x}_1 - \bar{x}_2}^2 = \frac{\sigma_1^2}{n_1} + \frac{\sigma_2^2}{n_2}$$

Remembering that standard deviation is given by the square root of variance, then the standard deviation of the differences between sample means, i.e. $\sigma_{\bar{x}_1 - \bar{x}_2}$, is the square root of the above term. It is this square-root value which is called the standard error of the difference (SED), thus:

$$\text{SED} = \sigma_{\bar{x}_1 - \bar{x}_2} = \sqrt{\frac{\sigma_1^2}{n_1} + \frac{\sigma_2^2}{n_2}}$$

where: n_1, n_2 = size of samples 1 and 2 respectively.

Of course, we cannot know the true population variances and cannot therefore determine $\sigma^2_{\bar{x}_1 - \bar{x}_1}$ definitively, but because we are dealing with large samples we can assume that the sample variances, s^2_1 and s^2_2, will provide accurate estimates of these. Thus in practice SED is obtained by:

$$\text{SED} = s_{\bar{x}_1 - \bar{x}_2} = \sqrt{\frac{s^2_1}{n_1} + \frac{s^2_2}{n_2}}$$

Therefore the formula for calculating the *z-test* statistic for comparing the means of two large samples finally becomes:

$$z_{\text{observed}} = \frac{\bar{x}_1 - \bar{x}_2}{\sqrt{\dfrac{s^2_1}{n_1} + \dfrac{s^2_2}{n_2}}}$$

Example 6.2. A two-tailed z-test of significance for analysing the difference between the means of two large samples.

A significance test to determine whether pot-grown tomato plants in a glasshouse treated with a new type of fertilizer 'Extragrow' have a significantly different growth rate compared with plants treated with a standard fertilizer.

The growth rate (g dry weight gain/day) of a randomly selected sample of 50 'Extragrow' treated plants was compared with that of a sample of 100 standard fertilizer treated plants. (Note that the sample sizes do not have to be the same for this test.) The following information was obtained:

Sample A:	$\bar{x} = 1.32$	Sample B:	$\bar{x} = 1.07$
'Extragrow'	$s = 0.49$	standard	$s = 0.37$
fertilizer	$n = 50$	fertilizer	$n = 100$

Null hypothesis: the mean growth rate of plants treated with 'Extragrow' fertilizer is the same as the mean growth rate of standard fertilizer treated plants ($H_0 : \mu_1 = \mu_2$).

The test statistic z_{observed} is now calculated:

$$z_{\text{observed}} = \frac{\bar{x}_A - \bar{x}_B}{\sqrt{\dfrac{s^2_A}{n_A} + \dfrac{s^2_B}{n_B}}} = \frac{1.32 - 1.07}{\sqrt{\dfrac{0.49^2}{50} + \dfrac{0.37^2}{100}}} = \textbf{3.18}$$

The two-tailed critical z value at $\alpha = 0.05$ is 1.96. The observed value of z is greater than 1.96; therefore, the probability of obtaining the observed value of z given the null hypothesis is true is less than 5%. H_0 is therefore rejected.

Conclusion: plants from the crop treated with the new fertilizer 'Extra-grow' show a significantly different growth rate from plants from the crop treated with standard fertilizer ($P < 0.05$).

The test shown in Example 6.2 is performed as a two-tailed test since it was required to determine only whether a significant difference was present irrespective of which sample was the greater. The fertilizer was of course developed in an attempt to specifically improve plant growth rate and it might have been decided, therefore, to undertake a one-tailed test to ascertain if the new fertilizer significantly increased growth rate compared with a standard fertilizer. In the case of a one-tailed z-test the critical value of z at probability α is equivalent to the two-tailed value at twice α. Thus for a 5% significance one-tailed z-test the critical value of z becomes 1.645. Example 6.3 shows the performance of the appropriate one-tailed z-test.

Example 6.3. A one-tailed z-test of significance for analysing the difference between the means of two large samples.

A significance test to determine whether pot-grown tomato plants in a glass-house treated with a new type of fertilizer 'Extragrow' have a significantly greater growth rate compared with plants treated with a standard fertilizer (same data as those in Example 6.2).

The growth rate (g dry weight gain/day) of a randomly selected sample of 50 'Extragrow' treated plants was compared with that of a sample of 100 standard fertilizer treated plants. (Note that the sample sizes do not have to be the same for this test.) The following information was obtained:

Sample A:	$\bar{X} = 1.32$	Sample B:	$\bar{X} = 1.07$
'Extragrow'	$s = 0.49$	standard	$s = 0.37$
fertilizer	$n = 50$	fertilizer	$n = 100$

Null hypothesis: the mean growth rate of plants treated with 'Extragrow' fertilizer is not greater than the mean growth rate of standard fertilizer treated plants (H_0: $\mu_1 \leq \mu_2$).

The test statistic $z_{observed}$ is calculated exactly as in Example 6.2 except that the direction of the difference now needs to be acknowledged, thus:

$z_{obsered} = + \textbf{3.18}$

For a one-tailed test the critical value of z at the 5% level of probability for rejecting H_0 is 1.645. Taking into account the sign of the observed value of z, this is greater than 1.645; therefore, the probability of obtaining the observed value of z given the null hypothesis is true is less than 5%. H_0 is therefore rejected.

Conclusion: plants treated with the new fertilizer 'Extragrow' show a significantly greater growth rate than plants treated with standard fertilizer ($P < 0.05$).

6.3 Analysis of Small Samples using Student's *t*-test of Statistical Significance

In Chapter 4 it was explained that it is unsafe to assume that small samples (i.e. $n \leq 30$) adhere to a normal distribution even when the samples are extracted from known normally distributed populations. Consequently it becomes invalid to use the z statistic, which is based on the normal distribution, in the analysis of small samples. Instead Student's t distribution is a more valid descriptor of the way that small samples behave and Student's *t* values are therefore used in the calculation of the confidence limits for small samples. For exactly the same reason it is invalid to employ the *z* statistic when examining the difference between a pair of small samples in a significance test. Consequently Student's *t* distribution is utilized when inspecting for significant differences between two small samples and this gives rise to one of the most familiar of all significance tests, namely **Student's *t*-test.**

Student's *t*-test is based on exactly the same working principles as the *z*-test, the essential difference being that the probability of significance is obtained by consulting Student's *t* distribution rather than the normal distribution. Since, for a given level of probability and sample size the Student's *t* value is greater than *z*, in practice the two means being compared need to show a larger numerical difference for a Student's *t*-test to reveal a significant effect than when a *z*-test is employed. Furthermore, the numerical difference that is required for significance to be detectable by a Student's *t*-test becomes larger as the samples become smaller and this, therefore, ensures that a more reliable conclusion is reached. In fact Student's *t*-test can be used with any sized samples, both small and large, and many statisticians never use the *z*-test, preferring always to employ the more cautious but safer Student's *t*-test.

6.3.1 Analysis of the difference between the mean of a small sample and a hypothesized population mean using Student's *t*-test

We will start our examination of Student's *t*-test by seeing how it is used to compare the mean of a population that is estimated using a small sample with a single hypothesized population mean. The aim of the test is to determine whether the probability of obtaining the observed value for the difference between the hypothesized population mean μ_o and the sample mean $\bar{x}$ is smaller than the selected critical probability for rejecting the null hypothesis (H_0). Thus for a 5% significance Student's *t*-test H_0 will be rejected if $\bar{x} - \mu_o$ falls in either of the 2.5% tails of the *t* distribution, as illustrated in Fig. 6.4. In order to determine whether this is the case, the distance that

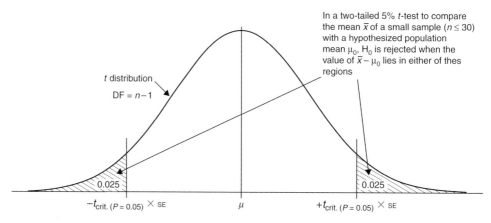

In a two-tailed 5% *t*-test to compare the mean $\bar{x}$ of a small sample ($n \le 30$) with a hypothesized population mean μ_0, H_0 is rejected when the value of $\bar{x} - \mu_0$ lies in either of thes regions

t distribution
DF = *n*−1

0.025

0.025

$-t_{\text{crit.} (P = 0.05)} \times {}_{\text{SE}}$

μ

$+t_{\text{crit.} (P = 0.05)} \times {}_{\text{SE}}$

Fig. 6.4. The region of Student's *t* distribution used for rejecting the null hypothesis in a two-tailed 5% Student's *t*-test comparing an observed small sample mean $\bar{x}$ with a hypothesized population mean μ_o.

the value of $\bar{x}$ is away from the population mean needs to be expressed in terms of the number of standard deviations, σ, that this distance represents. In other words the value of $\bar{x} - \mu_o$ needs to be standardized. Thus, if $\bar{x} - \mu_o$ is divided by σ, in theory we obtain a test statistic and, since this test statistic will be referred to the Student's t distribution rather than the normal distribution, it will be referred to as $t_{observed}$.

The problem once again is that, since we are dealing with samples, we do not know the value for the population parameters and the population standard deviation σ can only be estimated through the standard deviation, s, of samples extracted from the population. Since the standard deviation of a set of sample means is given by the standard error of the mean, $t_{observed}$ is obtained in practice by dividing the difference between the population and sample means by the sample standard error (SE). Recalling that SE is given by $s/\sqrt{n}$ then:

$$t_{observed} = \frac{\bar{x} - \mu}{s/\sqrt{n}}$$

where: $\bar{x}$ = the sample mean
μ = the population mean
s = the sample standard deviation
n = the number of observations in the sample.

It may not have escaped the reader's notice that, when comparing a sample mean with a single hypothesized population mean, the formulae for the test statistics $t_{observed}$ and $z_{observed}$ are actually the same. The distinction is that when analysing small samples the value of $t_{observed}$ is then compared with critical values from Student's t distribution (as given in the Student's t table; see Table A2 in the Appendix) rather than with critical z values based on the normal distribution. As Student's t distribution is dependent upon sample size, the Student's t table has to be consulted at the sample degrees of freedom which are given by $n - 1$. If $t_{observed}$ is larger than the critical value of t, the probability of obtaining the observed difference in means assuming that the null hypothesis is true must be less than the critical probability selected for conducting the test and the null hypothesis may then be rejected. Conversely, if $t_{observed}$ is smaller than the critical value of t, then the null hypothesis must be accepted on the basis that there is insufficient evidence to permit a claim of a significant difference between the means.

As with the z-test, Student's t-test may be performed as either a one-tailed or a two-tailed test, as illustrated in Example 6.4.

Example 6.4. A Student's t-test of significance for analysing the difference between the mean of a small sample and a hypothesized population mean.

A two-tailed test to determine whether dairy cows of a new breed have a significantly altered level of the protein casein in their milk compared with an existing commercial breed.

continued

Example 6.4. Continued.

Suppose that the mean milk casein content of a main commercial breed of dairy cow is known to be 3.2 mg/g. A new breed of cow has been developed and breeders wish to know whether a significant change in milk casein content is present in this new breed compared with the commercial breed. In milk samples from 15 cows of the new breed the mean casein level was 3.7 mg/g with a standard deviation, s, of 0.55 mg/g.

Null hypothesis: there is no difference between the mean of the population of the new breed and the hypothesized population mean (H_0: $\mu = \mu_0$).

A value for the test statistic t is calculated:

$$t_{observed} = \frac{\overline{x} - \mu_0}{s/\sqrt{n}} = \frac{3.7 - 3.2}{0.55/\sqrt{15}} = \textbf{3.52}$$

Since the sample size, n, is 15, the degrees of freedom for the test are $15 - 1 = 14$.

For a 5% significance test the critical value of t for a two-tailed test read from tables ($\alpha = 0.05$, DF = 14) is 2.145. The observed value of t is greater than 2.145; therefore the null hypothesis is rejected.

Conclusion: cows from the new breed of dairy cow have a milk casein content that is significantly different from the mean milk casein content of cows of the commercial breed ($P < 0.05$).

One-tailed test: this analysis could also be performed as a one-tailed Student's t-test to inspect whether cows of the new breed of dairy cow have a greater milk casein content than cows of the commercial breed. In this case H_0 is stated in the form: the population mean of the new breed is not greater than the hypothesized population mean (H_0: $\mu \leq \mu_0$). The critical one-tailed value at $\alpha = 0.05$ is 1.761. Since $t_{observed} > 1.761$, then H_0 is rejected and it is concluded that cows from the new breed have a milk casein content which is significantly greater than the mean milk casein content of cows of the existing commercial breed ($P < 0.05$).

6.3.2 Analysis of the difference between two small independent samples by Student's t-test

One of the most commonly used statistical significance tests is the Student's t-test for analysing the difference between two small independent samples where both samples can be safely assumed to have been drawn from normally distributed populations. This test can be used for comparing samples of any size but must be used if either of the samples has size ≤ 30. The principles of the test are exactly the same as those for the equivalent z-test for comparing two large samples (section 6.2) and the test statistic for the Student's t-test, $t_{observed}$, is thus given by the difference in the two sample means divided by the standard error of the difference (SED):

$$t_{observed} = \frac{\text{difference between sample means}}{\text{SED}}$$

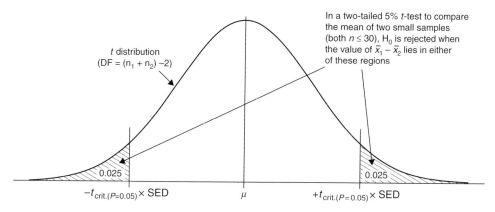

In a two-tailed 5% *t*-test to compare the mean of two small samples (both $n \leq 30$), H_0 is rejected when the value of $\bar{x}_1 - \bar{x}_2$ lies in either of these regions

t distribution
(DF = $(n_1 + n_2) - 2$)

0.025

0.025

$-t_{crit.(P=0.05)} \times$ SED μ $+t_{crit.(P=0.05)} \times$ SED

Fig. 6.5. The region of Student's *t* distribution used for rejecting the null hypothesis in a two-tailed 5% *t*-test examining the difference between two sample means $\bar{x}_1$ and $\bar{x}_2$.

Having determined $t_{observed}$, the value is referred to the critical value of *t* that cuts off the required tails of the *t* distribution. For example, if a 5% significance two-tailed test is to be conducted, then a value of $t_{observed}$ that is greater than the critical value of *t* read from the Student's *t* table at a probability of 0.05 would indicate that the observed value falls in one of the 2.5% tails and therefore the null hypothesis of no significant difference may be rejected. This is illustrated in Fig. 6.5.

The method of determination of SED depends, however, on whether or not the variances of the two populations from which the samples are drawn can be assumed to be equal. This is in fact a reasonable assumption under the majority of experimental situations since applying different treatments to a population of individuals may alter the mean of one set of treated individuals compared with another without affecting the distribution of the values around the mean. If the two populations can be assumed to have equal variances, then during the calculation of the SED the two sample variances can be pooled to produce a more accurate estimate of the population variance. If, however, the two population variances cannot be assumed to be equal, then a different mathematical approach is required.

6.3.3 Student's *t*-test assuming statistically equal population variances

Where the two small samples can be assumed to have been drawn from two populations that have equal variances, the test statistic $t_{observed}$ is given by:

$$t_{observed} = \frac{\bar{x}_1 - \bar{x}_2}{SED} = \frac{\bar{x}_1 - \bar{x}_2}{\sqrt{s_p^2 \times \left(\dfrac{1}{n_1} + \dfrac{1}{n_2} \right)}}$$

where: $\bar{x}_1, \bar{x}_2$ = means of samples 1 and 2 respectively;

n_1, n_2 = sizes of samples 1 and 2 respectively;

s_p^2 = pooled variance of both samples

$$= \frac{s_1^2 (n_1 - 1) + s_2^2 (n_2 - 1)}{(n_1 - 1) + (n_2 - 1)};$$

s_1^2, s_2^2 = variances of samples 1 and 2 respectively.

Since this calculation involves combining the two sample variances to produce a pooled statistic, i.e. s_p^2, the degrees of freedom (DF) for the test are the sum of the degrees of freedom of the two samples, i.e.:

$$DF = (n_1 - 1) + (n_2 - 1)$$

The observed value for t is then compared with the critical value read from the Student's t table (Table A2 in the Appendix) at the appropriate probability and degrees of freedom. *If the observed value is greater or equal to the critical value, then the null hypothesis can be rejected at the chosen level of probability.* Note that, if the test is being conducted as a two-tailed test, then the sign of $t_{observed}$ (i.e. whether it is positive or negative) can be ignored. If however, the test is being conducted as a one-tailed test, then it is important that the hypothesized smaller mean is subtracted from the larger when calculating $t_{observed}$ and that the sign of $t_{observed}$ is subsequently taken into account when comparing with the critical one-tailed value of t read from the Student's t table.

For those interested in the derivation of the formulae for SED and $t_{observed}$, these will now be explained more fully, although the reader may wish to move directly to Example 6.5, which provides a fully worked example of the Student's t-test for comparing two small samples. The reader may note that the data in this example have been previously encountered in Chapter 1 (section 1.5, 'Describing, summarizing and comparing data sets') and so finally it can be revealed whether or not the new fertilizer 'Extragrow' really worked.

6.3.4 Derivation of the formula for calculating $t_{observed}$ assuming homogeneous variances

It has previously been shown that the standard error of the difference between sample means (SED) is given by:

$$SED = \sqrt{\frac{\sigma_1^2}{n_1} + \frac{\sigma_2^2}{n_2}}$$

where: σ_1 and σ_2 are the population variances.

Yet again we are faced with the dilemma that we do not know the values of the population parameters and in this case we only have the variances of two small and, by definition, unreliable samples to estimate the population variances. However, the problem is reduced if it can be assumed that the variances of the two populations from which the samples are drawn are statistically equal, i.e.:

$$\sigma_1^2 = \sigma_2^2 = \sigma^2$$

Under these circumstances it can be argued that, as long as the sample sizes are equal, both sample variances act as equally good estimates of the population variance, i.e.:

$$s_1^2 = s_2^2 \approx \sigma^2$$

Therefore the best estimate that we can obtain for σ^2 is an average of the two sample variances but weighted to take into account the possibility that they may be of different sizes and therefore provide different amounts of information. This weighted average is termed the **pooled sample variance**, s_p^2. The weighted variances are obtained by multiplying each sample variance by its respective sample degrees of freedom (a calculation which in effect returns the sum of squares, SS, for each sample). The pooled sample variance is then calculated by summing the two weighted sample variances and dividing by the sum of the two sample degrees of freedom:

$$s_p^2 = \frac{s_1^2 (n_1 - 1) + s_2^2 (n_2 - 1)}{(n_1 - 1) + (n_2 - 1)}$$

where: s_1^2, s_2^2 = variances of samples 1 and 2 respectively;
n_1, n_2 = size of samples 1 and 2 respectively.

The value for the pooled sample variance, s_p^2, can now be substituted into the equation for SED:

$$SED = \sqrt{\frac{s_p^2}{n_1} + \frac{s_p^2}{n_2}}$$

This can be further simplified to:

$$SED = \sqrt{s_p^2 \times \left(\frac{1}{n_1} + \frac{1}{n_2}\right)}$$

The test statistic, $t_{observed}$, is, therefore, finally calculated by:

$$t_{observed} = \frac{\bar{x}_1 - \bar{x}_2}{SED} = \frac{\bar{x}_1 - \bar{x}_2}{\sqrt{s_p^2 \times \left(\frac{1}{n_1} + \frac{1}{n_2}\right)}}$$

where: $\bar{x}_1$, $\bar{x}_2$ = means of samples 1 and 2 respectively.

Example 6.5. A Student's t-test of significance for analysing the difference between the means of two small samples (assuming equal variances).

(a) A two-tailed test to determine whether a small sample of plants treated with a new fertilizer and a small sample of plants treated with a standard fertilizer have significantly different growth rates.
　　Pot-grown tomato plants randomly positioned in a glasshouse were treated either with a new fertilizer 'Extragrow' or with a standard fertilizer (control). The growth rates of a randomly selected small sample ($n = 30$) of each of the 'Extragrow' treated plants and standard fertilizer treated plants were determined. (Note, however, that the sample sizes do not necessarily have to be equal for this test.) The following data and sample statistics were obtained:

continued

Example 6.5. Continued.

The growth rate (g dry weight/day) of a sample of 30 plants					
'Extragrow' fertilizer			Standard fertilizer		
1.9	1.7	2.0	0.8	1.3	0.9
2.3	2.2	1.5	1.1	1.4	0.4
1.0	0.8	1.6	0.7	1.8	0.8
1.0	1.1	0.7	0.8	0.9	1.3
1.8	1.2	0.7	1.2	1.1	1.0
1.1	0.9	0.9	1.7	1.6	1.4
1.8	1.5	1.3	1.3	1.0	0.4
2.2	0.7	1.0	1.0	1.2	1.3
1.2	0.9	0.8	1.5	0.7	1.2
1.7	1.1	1.0	0.5	0.5	1.3

Sample A: $\bar{X} = 1.32$
'Extragrow' $s = 0.49$
fertilizer $n = 30$

Sample B: $\bar{X} = 1.07$
standard $s = 0.37$
fertilizer $n = 30$

Null hypothesis: the mean growth rate of 'Extragrow' fertilizer treated plants and the mean growth rate of standard fertilizer treated plants are the same (H_0: $\mu_1 = \mu_2$). The pooled variance, s_p, for both samples is first calculated:

$$s_p^2 = \frac{s_a^2(n_1-1)+s_b^2(n_2-1)}{(n_a-1)+(n_b-1)} = \frac{0.49^2(30-1)+0.37^2(30-1)}{(30-1)+(30-1)} = 0.19$$

The test statistic, $t_{observed}$, is now calculated:

$$t_{observed} = \frac{[\bar{X}_a - \bar{X}_b]}{\sqrt{s_p^2 \times \left(\frac{1}{n_a}+\frac{1}{n_b}\right)}} = \frac{[1.32-1.07]}{\sqrt{0.19 \times \left(\frac{1}{30}+\frac{1}{30}\right)}} = 2.22$$

From Student's t table the two-tailed critical t value at $\alpha = 0.05$ and DF = 58 is 2.002. The value of $t_{observed}$ is greater than 2.002; therefore, the probability of obtaining the observed value of t given the null hypothesis is true is less than 5%. H_0 is therefore rejected.

Conclusion: there is a significant difference in the growth rate of 'Extragrow' fertilizer treated plants and standard fertilizer treated plants ($P < 0.05$).

(b) A one-tailed test to determine whether a small sample of plants treated with a new fertilizer 'Extragrow' have a significantly greater growth rate than a small sample of plants treated with standard fertilizer (same data as in (a) above).
    The growth rate of a randomly selected small sample ($n = 30$) of 'Extragrow' fertilizer treated tomato plants and a similar sample of standard fertilizer treated tomato plants was determined. The following data and sample statistics were obtained:

Sample A: $\bar{X} = 1.32$ Sample B: $\bar{X} = 1.07$
'Extragrow' $s = 0.49$ standard $s = 0.37$
fertilizer $n = 30$ fertilizer $n = 30$

continued

Example 6.5. Continued.

Null hypothesis: the mean growth rate of 'Extragrow' fertilizer treated plants is not greater than the mean growth rate of standard fertilizer treated plants (H_0: $\mu_1 \leq \mu_2$).

The test statistic $t_{observed}$ is calculated exactly as in the two-tailed test (a) above, although the sign of the statistic must be noted:

$$t_{observed} = +2.22$$

From Student's t table the one-tailed critical t value at $\alpha = 0.05$ and DF = 58 is 1.672 (i.e. equivalent to the two-tailed t value at $\alpha = 0.10$). Taking into account the sign, the value of $t_{observed}$ is greater than 1.672; therefore, the probability of obtaining the observed value of t given the null hypothesis is true is less than 5%. H_0 is therefore rejected.

Conclusion: the 'Extragrow' fertilizer treated tomato plants do have a significantly greater growth rate than the standard fertilizer treated plants ($P < 0.05$).

6.3.5 Student's t-test assuming unequal population variances

The assumption of homogeneous population variances will not always be valid. We often presume that when a treatment is applied this will affect the central tendency of the data without having an effect on the uniformity of the data. This is, of course, not always the case and many types of treatment may actually affect the dispersion and therefore the variance of the treated populations. Furthermore two population variances may also be different due simply to different amounts of random variation being present (otherwise known as bad luck!). In such cases we cannot use a pooled variance from the two samples as an estimate of a single homogeneous population variance σ^2. Instead the two sample variances each make an independent estimate of two different population variances and both are therefore directly involved in the determination of the standard error of the difference between means (SED). Therefore where variances are unequal $t_{observed}$ is calculated by the following calculation:

$$t_{observed} = \frac{(\bar{x}_1 - \bar{x}_2)}{SED} = \frac{(\bar{x}_1 - \bar{x}_2)}{\sqrt{\left(\dfrac{s_1^2}{n_1}\right) + \left(\dfrac{s_2^2}{n_2}\right)}}$$

where: $\bar{x}_1, \bar{x}_2$ = means of samples 1 and 2 respectively;
s_1^2, s_2^2 = variances of samples 1 and 2 respectively;
n_1, n_2 = size of samples 1 and 2 respectively.

While this may seem a relatively simple calculation of $t_{observed}$, there is a problem in that values of $t_{observed}$ calculated in this way do not adhere perfectly to a Student's t distribution. As a result, this tends to increase the probability of making a Type I statistical error, i.e. rejecting the null hypothesis when it is actually true. For large samples (n ≥ 30) the error is sufficiently small and can be ignored, but as samples become smaller the error becomes appreciable. In order to correct for this, a very complex adjustment can be made to the degrees of freedom which shifts the value of $t_{observed}$ back towards a Student's t distribution. To achieve this the degrees of freedom are estimated by the following equation:

$$DF \approx \frac{\left(\dfrac{s_1^2}{n_1} + \dfrac{s_2^2}{n_2} \right)^2}{\dfrac{\left(\dfrac{s_1^2}{n_1} \right)^2}{n_1 - 1} + \dfrac{\left(\dfrac{s_2^2}{n_2} \right)^2}{n_2 - 1}}$$

The above equation (which today is readily determined by most statistical computer packages) will usually produce a non-integer for the degrees of freedom, in which case the next smallest integer should be used. Once $t_{observed}$ and DF have been determined, the $t_{observed}$ value can be referred to the table of critical Student's t values to assess the probability of the null hypothesis being correct and the test then proceeds in the same way as the equal variance Student's t-test described above.

6.4 Analysis of the Difference between Two Sample Variances

It was explained in section 6.3 that the method of calculation of Student's t-test depends on whether or not the variances of the two populations from which the samples are drawn are statistically equal. Very often inspection of the experimental design and experience from previous investigations will allow a sensible judgement to be made regarding the likely homogeneity of the population variances. If, however, it is not certain that an assumption of statistically equal population variances is valid, then the assumption will need to be formally tested. Furthermore, in some experiments the effect of a treatment on the variation of the material will occasionally be of prime interest in itself. For example, a treatment may have been applied to batches of seed with the intention of improving the uniformity of germination rather than the quantity of germination. In this case examination of the difference in variance of germination rate between the seed batches will be more relevant than an analysis of the difference between the means.

The simplest test available to examine the difference between two variances is the **F-ratio test**. Again, however, we are unlikely to know the population variances so that the test is conducted on the sample variances making the assumption that the sample statistics provide reliable estimates of the parameters of the populations from which they are drawn. If two samples have similar variation, then the ratio of their variances will be close to a value of 1; the larger the difference between the variances the further removed from unity this ratio will become. Without going into further theory for the moment, the ratio of one sample variance to another provides a test statistic called the F-ratio. Once determined, the F-ratio value can be tested against critical values derived from the known probability distribution of F in order to assess whether two variances are statistically different. (The theory of the F probability distribution and the F-ratio test statistic will be described in detail in the coming chapters dealing with multiple sample analysis.)

In this case the F-ratio is determined by dividing the larger variance by the smaller, i.e.:

$$F_{ratio} = s_a^2 / s_b^2$$

where: s_a^2 = variance of the sample with the larger variance;
  s_b^2 = variance of the sample with the smaller variance.

The test is conducted as a two-tailed test and therefore a table of two-tailed critical values of F needs to be consulted (see Table A5.3 in the Appendix) at the required probability (usually $\alpha = 0.05$) and at the degrees of freedom of both samples, where V_1 = DF of the sample with the larger variance, V_2 = DF of the sample with the smaller variance. The null hypothesis for this test is that there is no difference between the population variances (H_0: $\sigma_1^2 = \sigma_2^2$). If the observed F-ratio $\geq F_{critical}$, then the null hypothesis must be rejected and a difference between the variances must be assumed. If, on the other hand, the F-ratio $< F_{critical}$, then the two variances can be considered as statistically equal. Where the difference between two small samples is to be subsequently inspected, the appropriate form of the Student's t-test can then be selected, i.e. a test assuming either equal or unequal variances.

Example 6.6. An F-ratio test to determine if the variances of two small samples are statistically homogeneous.

A two-tailed test to determine whether the variance in growth rate of tomato plants treated with a new fertilizer 'Extragrow' and standard fertilizer treated plants are significantly different (same data as in Example 6.5).

Sample A:	$\bar{X}$ = 1.32	Sample B:	$\bar{X}$ = 1.07
'Extragrow'	s = 0.49	standard	s = 0.37
fertilizer	s^2 = 0.242	fertilizer	s^2 = 0.137
	n = 30		n = 30

Null hypothesis: the variances of the 'Extragrow' and standard fertilizer treated plants are equal (H_0: $\sigma_1^2 = \sigma_2^2$).

$$F_{ratio} = s_{larger}^2 / s_{smaller}^2 = 0.242 / 0.137 = 1.77$$

The two-tailed critical value of F (read from the F table at $\alpha = 0.05$ and V_1 degrees of freedom of 29 and V_2 degrees of freedom of 29) is 2.10. As the observed F-ratio is $< F_{critical}$, the null hypothesis must be accepted.

Conclusion: there is no significant difference between the variances in growth rate of the 'Extragrow' treated plants and the standard fertilizer treated plants. The difference between the sample means may be examined by a Student's t-test assuming equal population variances.

6.5 Confidence Limits for the Difference between Two Sample Means

As a result of the application of a Student's t-test, a conclusion may be reached that there is a significant difference between two sample means. Since the sample statistics are being employed to estimate population parameters, this is providing good evidence that a true difference exists between the two populations from which the samples are drawn. What we do not know, however, is what the true size of the difference between the population means might be. This can be addressed by determining the confidence limits for the difference between the sample means.

In the case of a two-sample comparison, the confidence limits provide the numerical range around the difference between the sample means that has a defined probability of including the true difference between the population means. Now recall that the confidence limits for a single sample mean are given by multiplying the sample standard error by the appropriate Student's t value. Similarly the confidence limits for a difference between two sample means $CL_{(\bar{x}_1, \bar{x}_2)}$ are given by multiplying the standard error of the difference (SED) by the appropriate Student's t value. Thus:

$$CL_{(\bar{x}_1 - \bar{x}_2)} = \pm SED \times t$$

Student's t is the critical two-tailed value read from tables at the level of probability required and the combined degrees of freedom of the two samples.

By way of illustration, the 95% CL around the difference in growth rate between the sample means of the tomato plants treated with a new fertilizer or with standard fertilizer, shown in Example 6.5, can now be determined. The difference between the treated and untreated sample means was 0.25 g/day, the pooled sample variance, s_p^2, was 0.19 and both samples consisted of 30 plants ($n_1 = n_2 = 30$). The SED is then given by:

$$SED = \sqrt{s_p^2 \times \left(\frac{1}{n_1} + \frac{1}{n_2} \right)} = \sqrt{0.19 \times \left(\frac{1}{30} + \frac{1}{30} \right)} = 0.113$$

The two-tailed 5% Student's t value is 2.002 (based on a combined DF = 58); therefore the 95% CL for the difference between means are:

$$95\% CL_{(\bar{x}_1 - \bar{x}_2)} = 0.113 \times 2.002 = \pm 0.23$$

And the 95% confidence interval is therefore:

$$0.25 - 0.23 \text{ to } 0.25 + 0.23 = 0.02 \text{ to } 0.48 \text{ g/day}$$

It is therefore concluded that there is a 95% probability that the interval 0.02 to 0.48 g/day includes the true difference beween the population means, $\mu_1 - \mu_2$.

6.6 The Power of Student's t-test and Determination of Appropriate Sample Size

It was mentioned in Chapter 5 (section 5.4) that the ability of a significance test to detect a significant effect is given by a measure called the 'power of the test'. The power of a significance test is the probability that a test will result in a correct rejection of the null hypothesis when it really is not true. While the determination of this probability for a two-sample Student's t-test is rather theoretical (see, for example, Zar, 2009), it is easily shown that *the power of a test improves with increased sample size*. Furthermore, *the power of a Student's t-test is greatest when the sample sizes are equal*, which is another good reason for employing equal sample sizes wherever possible.

Rather than actually quoting the power of a test, a more pertinent question for the investigator to ask is what size the samples should be to allow the efficient detection of a significant difference. This question can be answered if the investigator has in mind the smallest difference between means which is required to be assessed for significance and has some notion of the existing variance in the populations under examination.

To help discuss this further, let us return to Example 6.5 concerning the effect of our newly developed fertilizer 'Extragrow' on the growth rate of a crop but imagine that, due to the cost of the fertilizer, the grower is only interested if the effect produced is an increase in the mean of at least 0.2 g dry weight/plant/day. We can now ask what sample size we would need in order to be sure that a Student's t-test would have sufficient power to enable a difference between means of 0.2 g dry weight/day to be detectable as being significant (i.e. at $P \leq 0.05$). If from previous experience of working with the same material, or through a preliminary experiment, the variance of the material can be reliably estimated, then the equation for determining the confidence limits for the difference between means can be exploited to allow this question to be answered. If the samples each contain the same number of measurements (n) and if the difference between means that is required to be detectable is denoted as δ, then the required sample size will be the smallest integer value of n that satisfies the following condition:

$$\delta > t_{crit} \times \sqrt{\frac{2\sigma^2}{n}}$$

where: σ^2 = the estimated population variance;

t_{crit} = critical value of Student's t read from tables at DF $n-1$ and at $P = 0.05$. This therefore becomes an iterative process in which decreasing values for n are successively tested until the equation is satisfied for the smallest integer value of n.

Let us now return to the example of the fertilizer trial where we wish to be able to detect a difference between means of 0.2 g dry weight/day as being significant at a 5% level of significance. Suppose now that previous trials had indicated that a good estimate of the variance in growth rate in standard fertilizer treated crop plants was 0.05. An initial value for n has now to be guessed at and tested; a reasonable first guess might be $n = 15$. Then:

$$t_{crit} \times \sqrt{\frac{2\sigma^2}{n}} = 2.131 \times \sqrt{\frac{2 \times 0.05}{15}} = 0.17$$

A sample size of 15 would thus allow us to detect a difference of 0.17 g dry weight/day as being significant ($P \leq 0.05$). Smaller values of n may then be tested and it is found that:

where $n = 12$: $\quad t_{crit} \times \sqrt{\dfrac{2\sigma^2}{n}} = 2.179 \times \sqrt{\dfrac{2 \times 0.05}{12}} = 0.199$

where $n = 11$: $\quad t_{crit} \times \sqrt{\dfrac{2\sigma^2}{n}} = 2.201 \times \sqrt{\dfrac{2 \times 0.05}{11}} = 0.210$

Therefore, in order to determine that a difference between means of 0.20 dry weight/day is significant, a sample size of at least 12 would be required.

It remains true, of course, that in terms of accurately representing populations the larger the sample the better. However, taking this statistical approach to the determination of the smallest sample size that still permits a valid statistical analysis can be particularly useful where the amount of experimental resources available is limited and it is necessary to use as small a quantity of material as possible. It is also true that if samples are employed that are very much larger in size than is strictly necessary then this may lead to an increase in the magnitude of the variance since there will then be an increased likelihood of both making measuring errors and sampling from non-homogeneous units.

6.7 Analysis of the Difference Between the Means of Two Non-independent Samples using the Paired Samples *t*-test

The discussion so far has concerned the comparison of two independent randomly selected samples. It is quite common, however, for a comparison to be made between two sets of measurements that are derived from the same sample of individuals. In this case the two sets of data cannot be independent and are described as being 'paired'. (Note that the term 'matched pairs' is also often used but this more specifically refers to a situation where the pairing of the measurements has arisen through a deliberate experimental design procedure on behalf of the investigator.) If the samples can be assumed to be extracted from normally distributed populations, then significant differences between paired samples can be examined by a **paired samples *t*-test**.

Paired samples arise in two main ways. In the first case, two separate sets of measurements may be made on the same variable at different times. For example, the respiratory rate of ten apples in storage might be measured at the end of week 1 and the respiratory rate of the same ten apples might be remeasured at the end of week 2. In this case the individual apple that displayed the highest respiration rate in week 1 is also very likely to show the highest respiration rate in week 2, while the individual with the lowest respiration rate in week 1 is also likely show the lowest respiration rate in week 2. Clearly, therefore, the two sets of respiration data cannot be regarded as being independent. The second way in which paired samples commonly arise is where a single sample of items provides two sets of simultaneous measurements. Such data are often termed **bivariate**. For example, the number of hairs on the upper side of a sample of ten leaves and the number of hairs on the underside of the same ten leaves might be recorded at the same time. Again the two sets of measurements cannot be treated as being independent but are related through having been derived from the same single sample of leaves.

The two-tailed null hypothesis for the paired sample t-test is that the means of the populations from which the samples are extracted are equal. The first stage of the test is to determine the numerical difference between each pair of measurements, for example, between the respiration rate in week 1 and in week 2 for each apple in a store room. If the overall means of the two sets of measurements are exactly equal, it follows that the mean of the differences between each data pair must be zero. The greater that the mean of the differences deviates from zero, the greater is the probability that a significant difference exists between the two samples. Therefore, as long as the data are from normally distributed populations, a Student's t-test can be employed to test whether the mean of the differences between the pairs of measurements $(\bar{d})$ differs significantly from a value of zero. The test therefore becomes the equivalent of a Student's t-test for comparing a sample mean with a hypothesized mean μ_o (see section 6.3.1). Since in this case μ_o is zero, then the equation for calculating the test statistic becomes:

$$t_{observed} = \frac{\bar{d}}{s/\sqrt{n}}$$

where: $\bar{d}$ = the mean of the differences between paired measurements;
μ = the theoretical mean (i.e. zero);
s = the standard deviation of the difference values;
n = the number of pairs of measurements.

It may be noted that by mathematical manipulation the equation above can then be converted into an alternative form that may be more easily solved in practice:

$$t_{observed} = \frac{\Sigma d}{\sqrt{\dfrac{n\Sigma d^2 - (\Sigma d)^2}{n-1}}}$$

where: Σd = the sum of the differences between each pair of measurements.

The degrees of freedom for the test are $n - 1$ where **n is the number of data pairs**, i.e. the number of difference values. The observed value for t is then compared with the critical t value that would be expected under the null hypothesis, which can be obtained from the Student's t table (Appendix: Table A2) at the required level of probability α and the appropriate degrees of freedom. If the observed value is greater than or equal to the critical value, then the null hypothesis can be rejected and a significant difference between the paired samples can be claimed at the level of probability employed.

If a paired t-test is to be conducted as a two-tailed test, then the sign of the test statistic $t_{observed}$ can be ignored when making the comparison with the critical t value. If appropriate, the paired t-test can be performed as a one-tailed test, in which case the calculation procedure is exactly the same as that described above but the sign of $t_{observed}$ must then be taken into account when comparing $t_{observed}$ against one-tailed critical t values read from Student's t table. Example 6.7 illustrates the calculation of the paired t-test.

Example 6.7. A paired *t*-test of significance for testing the difference between two non-independent samples.

A test to determine whether there is a change with time in the respiration rate of a sample of stored apples.

The respiratory rates of each of ten labelled apples were measured when placed into storage in September and the respiratory rate of the same ten apples remeasured 4 weeks later in October. The results were:

Apple number	Resp. rate in Sept. (μmol O_2/g/h)	Resp. rate in Oct. (μmol O_2/g/h)	Difference (*d*)	d^2
1	10.3	12.2	+1.9	3.61
2	11.4	12.1	+0.7	0.49
3	10.9	13.1	+2.2	4.84
4	12.0	11.9	−0.1	0.01
5	10.0	12.0	+2.0	4.00
6	11.9	12.9	+1.0	1.00
7	12.2	11.4	−0.8	0.64
8	12.3	12.1	−0.2	0.04
9	11.7	13.5	+1.8	3.24
10	12.0	12.3	+0.3	0.09
$n = 10$	$\bar{x}_1 = 11.47$	$\bar{x}_2 = 12.35$	$\Sigma d = 8.8$	$\Sigma d^2 = 17.96$

$$(\Sigma d)^2 = 77.44$$

Mean of differences, $\bar{d} = 0.88$
Standard deviation of differences, $s = 1.065$

Null hypothesis: the mean respiration rate of the stored apples in September and in October is the same (H_0: $\mu_1 = \mu_2$).

$$t_{observed} = \frac{\bar{d}}{s/\sqrt{n}} = \frac{0.88}{1.065/\sqrt{10}} = \mathbf{2.613}$$

Alternative calculation :

$$t_{observed} = \frac{\Sigma d}{\sqrt{\dfrac{n\Sigma d^2 - (\Sigma d)^2}{n-1}}} = \frac{8.8}{\sqrt{\dfrac{10 \times 17.96 - 77.44}{10 - 1}}} = 2.613$$

From Student's *t* table the critical *t* value at $\alpha = 0.05$ and DF = 9 is 2.262. Since $t_{observed}$ is greater than 2.262 the null hypothesis is rejected.

Conclusion: there is a significant difference in the respiration rate of the apples between August and September ($P < 0.05$).

6.8 The Design and Analysis of Two-sample Experiments: Some Final Comments

This chapter has shown that while the design of experiments involving the comparison of effects caused by two treatments, or the comparison of a single treatment against a control, are relatively straightforward there are a number of important considerations that need to be addressed. The chosen method of data analysis will depend upon a number of design criteria and assumptions about the collected data. In particular, it needs to be recognized whether the samples are extracted from normally distributed populations and whether the population variances can be assumed to be statistically equal. If neither of these conditions is met, then an alternative non-parametric approach is required (see Chapter 10). In terms of design the major question will be whether the samples are independent or non-independent, that is, whether the two samples of data are derived from different sets of individuals or from the same set of individuals. At the risk of seeming to repeat the message from earlier chapters, it is extremely important that these questions are answered, as far as possible, prior to the performance of the experiment rather than at its conclusion when it may then be too late to ensure that a valid method of data analysis is available.

One of the more difficult questions concerning the design of two-sample experiments arises when on occasion a genuine choice is available between using an independent design or a paired-sample design. Consider again the scenario in Example 6.7. The aim of the experiment was to determine whether the respiration rate of apples significantly altered over time when stored in a particular storeroom. In order to inspect this, the investigator measured the respiration rate of ten labelled apples at the beginning of September and then remeasured the same ten apples at the beginning of October. Consequently the data samples were not independent and a paired-sample t-test was employed to inspect for a significant change in respiration rate. However, at the time of the second measurement the investigator could have randomly selected a second sample of apples and then analysed the difference in respiration rate using an independent sample Student's t-test. While both approaches are valid, there is a strong argument that where possible the paired sample approach is to be preferred as the paired sample t-test is more powerful and has a better chance of identifying small differences as being significant. The reason for this is fairly clear. When two independent samples are employed each sample will contain a certain amount of random within-sample variation that will not necessarily be similar and, since both sample variances are components of the calculation of the test statistic, their effect on the analysis will aggregate. In the case of paired samples, because the two sets of data arise from the same set of individuals, the within-sample variation should vary little between the samples and will cancel out when the differences between data pairs are calculated. Consequently the paired-sample t-test will be more sensitive to small differences and will be better able to detect significant differences when sample sizes are small than an independent sample t-test. On the other hand, if a biased sample had been inadvertently selected, then the effect of this would be compounded in the case of a paired-sample design because the same sample is used to provide both sets of data to be compared. In general terms what is being argued here is that, when applied to the same experimental design, an independent sample t-test would have the greater chance of producing a Type 1 statistical error (rejecting the null hypothesis when it is true) but a lower chance of producing a Type 2 statistical error (accepting the null hypothesis when it is false) than a paired-sample t-test. The investigator will therefore need to decide which is the most appropriate approach based on the nature of the experiment being conducted and the research question being asked.

Although fairly common, two-sample experimental designs are relatively simple and provide only a limited amount of information from one experiment. In agriculture and horticulture research, especially in the case of field trials, it is rather more usual to apply more than two treatments simultaneously, giving rise to multiple-sample experimental designs. It may appear that such designs could be analysed by performing a series of two-sample significance tests, for example, in the case of a three-sample design, then sample A might be compared with sample B, sample A with sample C and finally sample B with C. Even if the researcher had the stamina to undertake all the two-sample significance tests necessary – and the number required will increase exponentially as the number of samples increases – it can be shown very easily that such an approach is entirely invalid. The analysis of multiple-sample experiments requires a different approach and the statistical techniques required will be the subject of forthcoming chapters.

7 Analysis of Multiple-sample Experiments

Analyse: examine minutely the constitution of …

Analysis: resolution into simpler elements by analysing.

(Concise Oxford English Dictionary)

- Introduction to multiple-sample analysis.
- The principles of analysis of variance (ANOVA) tests.
- Calculation of one-way ANOVA.
- Types of ANOVA model.
- Multiple comparison tests following ANOVA.

7.1 Introduction to Multiple-sample Analysis

One of the most common experimental designs employed in agriculture and horticulture sciences, especially in field trials, is the **factorial design** in which a treatment factor is applied at a series of different levels of application. For example, the yield of a crop might be determined when a fertilizer is applied at a series of concentrations or the milk yield of cows might be compared among a number of different breeds. In such cases, where data are collected from three or more samples, a method of **multiple-sample analysis** is required to inspect for significant differences between all possible combinations of the sample means.

It might be considered that a relatively simple approach to multiple-sample analysis would be to carry out a series of two-sample comparison tests, such as the Student's t-test, between every possible combination of sample pairs in turn. For example, if there were three sample means to be examined, $\bar{x}_A$, $\bar{x}_B$ and $\bar{x}_C$, then one might test $\bar{x}_A$ against $\bar{x}_B$, then $\bar{x}_A$ against $\bar{x}_C$, and finally $\bar{x}_B$ against $\bar{x}_C$. Obviously this would become an increasingly long-winded exercise as the number of samples increased. Irrespective of the clumsiness of such an approach, it can, however, be shown that this approach is in fact statistically invalid. It has been discussed previously that whenever a statistical significance test is performed there will always be some probability of coming to the wrong conclusion, e.g. when a 5% significance test is carried out there must be, by definition, a 5% probability of concluding there is a significant difference when in reality there is not one present (i.e. a Type 1 error). Consequently, if a series of comparison tests are used to analyse a number of sample means then the probability of making a Type I error aggregates with each test undertaken. It can be shown mathematically that if a set of six sample means are analysed by a series of Student's t-tests the probability of making at least one Type 1 error is over 50%. A method of analysis is required, therefore, in which every possible combination of sample means are compared simultaneously but without the probability of making Type I errors accumulating with each comparison.

If the samples are selected fully randomly from normally distributed populations, and if the populations can be assumed to have equal variances, then the major parametric statistical technique available for multiple sample analysis is **analysis of variance** (commonly referred to by the acronym 'ANOVA'). This chapter will present a detailed explanation of the use of this technique in the analysis of single treatment factor experiments. If the data do not conform to the required assumptions for analysis by a parametric test and the researcher does not wish to apply a data transformation, then an alternative non-parametric technique will be required, the most frequently employed being the **Kruskal–Wallis test**, which is discussed in Chapter 10.

7.2 The Basic Principles of Analysis of Variance (ANOVA)

7.2.1 Partitioning the variation between treatment effects and random causes

Consider the situation where a series of samples is extracted from populations that are exposed to different experimental treatments and a particular variable is measured for all items in each of the samples. The question might be asked: why are the values observed in such an experiment not all exactly the same? There are two basic reasons for this: first, there are differences arising due to the different experimental treatments applied; and, second, there are differences between the measured items due to natural random variation. To analyse the data and come to a conclusion as to the cause of the observed results, what we need to do is ascertain which of these factors, i.e. the treatment or random effects, is the major source of variation within our particular experiment. ANOVA achieves this by providing a mathematical technique for separating out and estimating the variance due to the treatment factors imposed upon the subjects within the experiment and the variance due to random variation. The ratio between these variances then provides a test statistic which has a known probability distribution and allows the probability of the data supporting the null hypothesis to be determined, thereby leading to a conclusion as to whether or not significant treatment effects are present. Much of the ANOVA technique was originally developed by the British statistician Sir R.A. Fisher and the test statistic calculated by the ANOVA procedure is denoted F in acknowledgement of his contribution.

Although the technique is called 'analysis of variance' the null hypothesis for an ANOVA test is that there are no differences between the population means. It may be asked, therefore, how does an analysis of sample variances yield information concerning the differences between population means? To begin with it can be assumed that within a sample, where all replicate items are exposed to exactly the same experimental treatment, the only possible cause of variation must be natural random variation. As long as the amount of natural variation does not itself differ significantly between samples, then if the variance of measurements occurring within samples is combined and an average determined this should yield a good estimate of the total variation in the experiment that is due to random causes. Put in statistical terms, when samples are homogeneous, the pooled within-sample variance, denoted s_p^2, will provide a valid estimate of the overall population variance, σ^2. (For the moment do not worry about how the statistics are calculated; we will come to that in due course.)

Instead of taking a pooled within-sample variance, an alternative approach would be to inspect the variance of the sample means ($s_{\bar{x}}^2$) a quantity that we can call the 'between-sample' variance. If the samples are drawn from the same statistical population (i.e. the null hypothesis is correct), then both the between-sample variance, $s_{\bar{x}}^2$, and the within-sample variance, s_p^2, will be estimates of the same population variance, σ^2, and the ratio of one to the other will be close to 1. On the other hand, the samples may have been drawn from populations that have been exposed to different treatments that cause significant effects. In this case the variance of the means of the individual samples, $s_{\bar{x}}^2$, will now include all the treatment effects that exist in the experiment. The presence of treatment effects will therefore raise the ratio of the between-sample variance to the within-sample variance to above 1. Therefore the greater the ratio of the between-sample variance to the within-sample variance the greater is the proportion of the total variation in the experiment that can be attributed to a treatment effect.

The principles of ANOVA are illustrated in Fig. 7.1, where box plot diagrams are used to represent four samples that in one scenario show significant differences and in a second scenario show no significant differences. Under each scenario each equivalent sample has the same mean value; however, in the first scenario where differences are significant, the samples have low within-sample variances, giving rise to a high F-ratio. In the second scenario, where differences are not significant, the samples have relatively high within-sample variances, giving rise to a low F-ratio.

7.2.2 Determining the test statistic F

The basic calculation procedure begins by determining the total sum of squares in the experiment as a measure of the total variability in the data. The total sum of squares is then partitioned according to the sources of variation which exist, i.e. the variation between sample means due to treatments and the variation between replicates within samples due to random causes (natural as well as experimental). These partitioned sums of squares are then divided by their respective degrees of freedom to obtain the variances $s_{\bar{x}}^2$ and s_p^2 respectively. In ANOVA terminology, these variance values are called the **mean squares**. The ratio of the between-sample mean square value, $s_{\bar{x}}^2$, to the within-sample mean square value, s_p^2, then yields the test statistic denoted F. The greater the value the F test statistic is above 1 the greater is the ratio of treatment effects to random effects and therefore the greater the probability of a significant difference being present between the sample means.

7.2.3 The probability distribution of F

In order to determine the probability of obtaining the observed value of the test statistic F in circumstances where the null hypothesis is true, its value needs to be referred to the theoretical probability distribution of F. The F distribution is based on the assumptions that the variable under inspection is normally distributed, that the

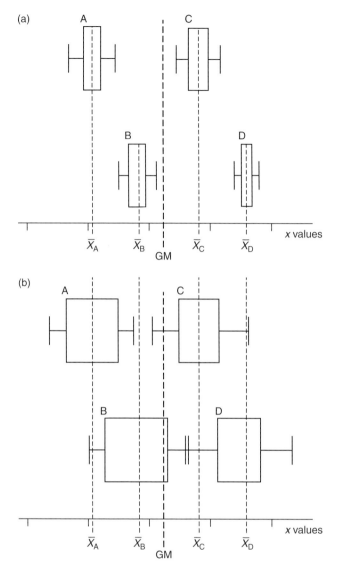

Fig. 7.1. The relationship between within-sample variability, F-ratio and presence of significant differences in a one-way ANOVA. The box plots show the distribution of four samples (A–D) around a grand mean (GM) in two scenarios in which there are (a) significant differences and (b) no significant differences present between sample means. In scenario (a) there are low within-sample variances compared with the between-sample variance, which gives rise to a large F-ratio, which infers the presence of significant differences. In scenario (b) there are high within-sample variances compared with the between-sample variance, which gives rise to a low F-ratio, which infers no significant differences present. (Note that an explanation of box plots was given in section 2.5.)

variances of two samples are being compared and the samples are drawn from two independent populations that have the same population variance. If all possible pairs of samples are selected randomly, then the smallest possible value of the ratio of their variances will be zero (which can only occur where the denominator sample variance is zero), while the largest value possible is infinity. Since, however, it is assumed that the populations have equal variances, then the most likely scenario is that the two samples selected will have similar variances and thus the value of F will be close to 1. Consequently the peak of the F distribution lies near to 1 while values approaching zero on the one hand and infinity on the other become increasing less likely, thereby producing two tails to the distribution. The precise shape of the distribution will depend on the size, i.e. the degrees of freedom, of the two samples. Just as the normal distribution and Student's t distribution can be calculated, the theoretical probability distribution of F for pairs of different sized samples can be mathematically determined (although the exact details of this complex calculation need not worry us here). Two examples of the F distribution for sample pairs of respective degrees of freedom of 30 and 10 and for degrees of freedom of 5 and 10 are shown in Fig. 7.2 and demonstrate that F can only take positive values and the distribution tends to be positively skewed.

The F probability distribution allows the critical values of F to be determined that cut off a given proportion of the values in the tails. Since in analysis of variance significant differences can only occur when the observed value of F is greater than 1, then only the right-hand upper tail of the F distribution needs to be considered. Consequently, *ANOVA is always performed as a one-tailed test of significance*. The critical one-tailed F values that cut off 5% ($\alpha = 0.05$) and 1% ($\alpha = 0.01$) of the right-hand tail can be obtained from statistical tables which are given in the Appendix (Tables A5.1 and A5.2). By comparing the calculated value of F with the appropriate critical values, the probability that the observed sample data comply with the null hypothesis (H_0) being true can be ascertained. Thus, if the critical probability α for the test is set at 5% and the calculated value for F is then larger than the critical value read from the 5% F probability table, then the probability that the observed data support H_0 must be <0.05 and a significant treatment effect can therefore be inferred.

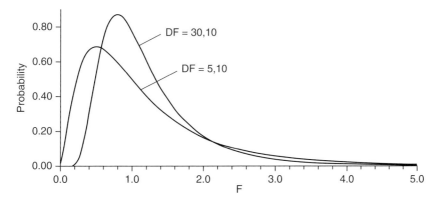

Fig. 7.2. The F probability distribution where the numerator and denominator degrees of freedom are (a) 30 and 10 and (b) 5 and 10 respectively.

7.2.4 The assumptions made by ANOVA

As with all parametric statistical tests the mathematics of ANOVA relies on a number of assumptions regarding the samples. These are:

1. Samples are independent. Each treatment sample must be represented by a set of different subjects. (This is exactly the same assumption as that made by the Student's *t*-test for comparing two independent samples.) If multiple measurements are made on the same subjects, this leads to a different type of experimental design referred to as a **repeated measures** or **within-subjects** design (this is the equivalent to a paired-sample design when comparing two samples) and requires a different approach to the analysis. This will be described in Chapter 9.

2. Samples are extracted from populations which are normally distributed. ANOVA is a parametric technique relying on the theory of the normal distribution. If the data are not normally distributed because, for example, they are skewed, they are in the form of percentages or proportions or they adhere to a different type of data distribution altogether, then the data should either be transformed so that they then adhere to a normal distribution or an alternative non-parametric method to ANOVA should be employed.

3. Sample variances are homogeneous. As explained above, the *F* probability distribution assumes that the samples are drawn from populations that have the same variance. The population variance σ^2 cannot of course be known but the within-sample variances are pooled in order to estimate σ^2. This approach can only be valid as long as the sample variances themselves do not differ significantly. (Note that the same assumption is also required by Student's *t*-test for much the same reason.) If the assumption of variance homogeneity does not hold, then either the data will need to be transformed or a non-parametric approach must be adopted.

7.3 Calculation of One-way Analysis of Variance

The standard procedure for the calculation of ANOVA involves the partitioning of the total sum of squares according to the effects of the treatment factor and random variation. The between-sample (treatment) variance and the within-sample (random effects) variance are then derived from these sums of squares. This procedure can be conveniently described in four main steps.

Step 1: Tabulation of data and determination of basic parameters

The data need to be carefully tabulated and the values of a series of basic statistics determined. The typical format and symbols used for this are shown in Table 7.1.

Step 2: Determination of the total sum of squares (SS) and the partitioned SS with respect to each source of variation

First recall that the formula for sum of squares is:

$$SS = \Sigma(x - \bar{x})^2 = \Sigma x^2 - \frac{(\Sigma x)^2}{n}$$

Table 7.1. Format of a data table and required definitions for performance of a one-way analysis of variance test.

Samples (treatments)	Observed values	Mean	x^2
A	x_1		x_1^2
	x_2		x_2^2
	x_3		x_3^2
	x_n		x_n^2
	Σx_s	$\overline{x}_A$	
B	x_1		x_1^2
	x_2		x_2^2
	x_3		x_3^2
	x_n		x_n^2
	Σx_s	$\overline{x}_B$	
C	x_1		x_1^2
	x_2		x_2^2
	x_3		x_3^2
	x_n		x_n^2
	Σx_s	$\overline{x}_C$	
	Σx		Σx^2

Where: Σx_s = sample total; Σx = grand total (often abbreviated GT); $\overline{x}$ = sample mean; k = no. of samples (i.e. treatment groups); n = no. of observations per sample; N = total no. of observations in whole experiment (= $k \times n$).

The $(\Sigma x)^2/n$ term in this equation is conveniently referred to as the 'correction term' (CT) and this may be determined initially:

$$CT = \frac{(\Sigma x)^2}{n}$$

The sum of squares calculations can then proceed as follows:

Total sum of squares. This measures the total variation present in the experiment and is given by:

$$SS_{total} = \Sigma x^2 - CT$$

Between-sample sum of squares. This is a measure of the variation in the experiment that is attributable to the treatment factor and is given by the sum of squares of the sample means ($\overline{x}$) around the grand mean (GM):

$$SS_{between-samples} = \Sigma(\overline{x} - GM)^2$$

In practice, it is obtained by summing the $(\Sigma x)^2/n$ value for each sample and subtracting the correction term, i.e.:

$$SS_{\text{between-samples}} = \left[\frac{(\Sigma x_A)^2}{n_A} + \frac{(\Sigma x_B)^2}{n_B} + \frac{(\Sigma x_C)^2}{n_C} + \dots \right] - CT$$

where: Σx_A, Σx_B, Σx_C ... are the totals for samples A, B, C etc.;
n_A, n_B, n_C ... are the number of observations in samples A, B, C, etc.

Within-samples sum of squares. This is a measure of the variation in the experiment that is attributable to random or error effects and is given by the combined sum of squares of all the individual samples. The sum of squares of a sample is given by the product of the sample variance and the sample DF; therefore the total within-sample SS can be calculated as:

$$SS_{\text{within-samples}} = s_A^2(n_A - 1) + s_B^2(n_B - 1) + s_C^2(n_C - 1) + \dots$$

where: s_A^2, s_B^2, s_C^2 ... are the variances of samples A, B, C, etc.

Since, however, the treatment SS and the within-sample SS must add up to give the total SS, the within-sample SS can be obtained more simply by subtraction, i.e.:

$$SS_{\text{within-samples}} = SS_{\text{total}} - SS_{\text{between-treatment}}$$

Degrees of freedom. The degrees of freedom (DF) for each source of variation are given by:

Total DF $= N - 1$ (total number of observations less 1)
Between-sample DF $= k - 1$ (total number of treatment samples less 1)
Within-sample DF $= N - k$ (total number of observations less the total number of treatment samples)

Note again that all DF values are additive; therefore, the within-sample DF is also equivalent to the total DF minus the between-sample DF.

Step 3: Determine the values for the mean squares and the F test statistic and formulate the analysis of variance table

The variance for each source of variation is now determined, although in ANOVA terminology this is referred to as the **mean square (MS)**. The mean square is calculated by dividing each SS value by its respective degrees of freedom:

Between-sample MS $= SS_{\text{between-sample}}/DF_{\text{between-sample}}$
Within-sample MS $= SS_{\text{within-sample}}/DF_{\text{within-sample}}$

Note that, because the between-sample MS represents the treatment effect and the within-sample MS is the remaining variability once the treatment effect has been accounted for, it is normal to refer to these quantities as the 'treatment MS' and 'residual MS' respectively. The terms 'treatment' and 'residual' will be commonly used to replace the terms 'between-samples and 'within-samples' from here on in this text.

Finally the test statistic F is calculated by the ratio of the treatment (i.e. between-sample) mean square to the residual (i.e. within-sample) mean square:

$$F\text{-ratio} = \frac{\text{treatment MS}}{\text{residual MS}}$$

The results of the analysis of variance are presented in the ANOVA table, which has the following standard format:

Table 7.2. ANOVA table standard format.

Source of variation	Sum of squares (SS)	Degrees of freedom (DF)	Mean Square (MS)	F-ratio
Treatment (between-sample)	$SS_{treatment}$	$DF_{treatment}$	$\dfrac{SS_{treatment}}{DF_{treatment}}$	$\dfrac{MS_{treatment}}{MS_{residual}}$
Residual (within-sample)	$SS_{residual}$	$DF_{residual}$	$\dfrac{SS_{residual}}{DF_{residual}}$	
Total	SS_{total}	DF_{total}		

Step 4: Comparison of the observed F-ratio with the critical F value and drawing a conclusion

For a 5% significance ANOVA test, the observed F-ratio is then compared with the one-tailed critical F value that would be obtained if the probability of the null hypothesis being correct was exactly 5%. In order to do this, the 5% one-tailed F table has to be consulted at both the treatment degrees of freedom (denoted by V_1) and the residual degrees of freedom (denoted by V_2). In the normal presentation of these tables, as in Table A5.1 in the Appendix, the treatment DF (V_1) is read along the rows and the residual DF (V_2) is read down the columns. If the observed F-ratio is greater than or equal to the 5% critical value, then the null hypothesis can be rejected at this level of probability and it may be concluded that a significant difference is present between the differently treated populations. If a significant difference is found at a critical level of probability of 5% ($\alpha = 0.05$), the difference may then be tested at higher levels of significance by reducing α. It is usual to check for greater significance at a critical probability of 0.01, and the appropriate F probability table for doing this is also provided in the Appendix (Table A5.2).

The full calculation of a one-way ANOVA is shown in Example 7.1.

Example 7.1. One-way analysis of variance.

A one-way ANOVA to determine whether different levels of irrigation affect the growth of *Prunus* trees

A multiple-sample experiment was performed to determine the effect of four different levels of irrigation on the growth rate of young newly transplanted *Prunus* trees. Growth rate was assessed by measuring the girth increase (1 m above the ground) that occurred over the following growing season. The data were as follows:

Irrigation level (% field capacity)	Girth increase (cm) Replicate 1	2	3	Sample total $\Sigma x_{treatment}$	Sample mean $\overline{x}$
25	4.98	5.76	4.41	15.15	5.05
50	6.06	5.76	6.09	17.91	5.97
75	7.59	6.12	7.02	20.73	6.91
100	4.08	3.45	3.48	11.01	3.67

$k = 4$ $n = 3$ $N = 12$ $\Sigma x = 64.80$
$(\Sigma x)^2 = 4199.04$
$\Sigma x^2 = 369.4176$

Null hypothesis: the mean girth of *Prunus* trees treated with the four given levels of irrigation is the same ($H_0: \mu_1 = \mu_2 = \mu_3 = \mu_4$).

Correction term (CT) $= (\Sigma x)^2/N$ $= 349.92$

Sum of squares:

Total SS $= \Sigma x^2 - CT = 369.4176 - 349.92$ $= 19.4976$

with DF $= N - 1 = 12 - 1 = 11$

Between-sample (treatment) SS $= \left[\dfrac{\left(\Sigma x_A\right)^2}{n_A} + \dfrac{\left(\Sigma x_B\right)^2}{n_B} + \dfrac{\left(\Sigma x_C\right)^2}{n_C} + ... \right] - CT$

$= \left[\dfrac{15.15^2}{3} + \dfrac{17.91^2}{3} + \dfrac{20.73^2}{3} + \dfrac{11.01^2}{3} \right] - 349.92 = 17.1612$

with DF $= k - 1 = 4 - 1 = 3$

Within-sample (residual) SS $= SS_{total} - SS_{treatment} = 19.4976 - 17.1612 = 2.3364$

with DF $=$ (total DF $-$ treatment DF) $= 11 - 3 = 8$

continued

Example 7.1. Continued.

ANOVA Table.

Source of variation	SS	DF	MS	F-ratio	F_{crit} $\alpha = 0.05$	$\alpha = 0.01$	Sig.
Treatment	17.1612	3 (V_1)	5.7204	19.59	4.07	7.59	$P < 0.01$
Residual	2.3364	8 (V_2)	0.2920				
Total	19.4976	11					

Conclusion: the observed F-ratio is greater than the 1% critical F value; therefore the null hypothesis is rejected and it is concluded that there is a significant effect of irrigation treatment on the girth of *Prunus* trees ($P < 0.01$).

7.4 Types of ANOVA Model

The discussion so far has assumed that the application levels of the treatment factor have been set by the investigator. In a factorial experiment of this sort, where the level of each factor is fixed rather than being random, the ANOVA performed is termed a **'fixed effects'** or **Model I ANOVA**. For example, the one-way factorial experiment described in Example 7.1 involves the growth of trees in response to application of four specific levels of irrigation. In this case the conclusions reached cannot be extrapolated beyond these fixed levels of irrigation to cover other treatments not included in the experiment and, if the experiment were to be repeated to check the results, the same levels of irrigation would be used again.

In some cases, however, the levels of the treatment factors are not fixed by the investigator but are random. For example, the investigator may want to study whether differences in annual rainfall affect the growth rate of trees. In this case differences in growth between trees grown at randomly selected locations each with a different annual rainfall are tested for significance. The rainfall in each location at the time of the investigation is, however, random and cannot be set at a predetermined level by the investigator. In such cases we are only using a random selection of all the possible levels of the factor and, if the experiment were to be repeated, different levels of the factor would be employed. Furthermore, at the end of the analysis we would normally wish to make inferences about all possible levels of the treatment factor, not just for the level employed. In this case the type of ANOVA to be employed is termed a **'random effects'** or **Model II ANOVA**.

In single-factor experiments, the ANOVA model employed makes no practical difference to the way the ANOVA is mathematically performed, although it will affect the way the null hypothesis and final conclusions are expressed. In the fixed-effects irrigation experiment the null hypothesis is that there is no difference in tree growth between the levels of irrigation applied and the conclusion is that a significant difference is present between these fixed treatment levels. In a random-effects experiment the null hypothesis would be rather more generalized. In the investigation of rainfall on tree growth, for example, the null hypothesis would state that rainfall had no effect on tree growth and the conclusion would subsequently state that the effect of rainfall per se was either significant or not significant. In multiple-factor experiments, specifying the ANOVA model becomes very important as the model affects the way that the F-ratio test statistic is finally determined. Furthermore, where two or more treatment factors are present, the possibility arises that one factor may be fixed while a second factor is random. For example, the effect of different annual rainfall (i.e. a random factor) on tree growth could be tested in the same experiment on different varieties of the tree (i.e. a fixed factor). Multiple-factorial experiments containing both fixed and random factors are then analysed by a 'mixed effects' or Model III ANOVA. These multiple-factor ANOVA models will be discussed further in the following chapter.

7.5 Comparison between Samples: Multiple Comparison Tests

The beauty of analysis of variance is that it detects whether or not a significant difference exists between all possible combinations of population means from within the factorial experiment. Analysis of variance does not, however, reveal the exact location of any significant differences present. Thus, in Example 7.1, it was determined that irrigation has caused a significant effect but we do not precisely know between which levels of irrigation significant differences occur. Following a fixed effects ANOVA, a so-called **multiple comparison test** can be conducted to ascertain between which treatment means significant differences exist. Of course, if the treatment levels are random (i.e. a model II ANOVA has been applied), the question of differences between specific treatment levels does not arise.

A number of possible multiple comparison tests is available. Arguably the simplest is the calculation of the **least significance difference (LSD)**, which is essentially a type of Student's t-test. This test, however, does not take account of the number of sample means present in the experiment and therefore has a relatively high risk of producing Type 1 errors. For this reason its use must be carefully planned in advance of the experiment. More widely accepted as a means of blanket testing of differences between means following an experiment are the **Tukey HSD test** and the similar **Student–Newman–Keuls test**. Both these tests are based on the distribution of a statistic termed q, the critical value of which varies with the total number of sample means that are to be compared, and are less prone to Type I errors than the LSD test, but conversely more likely to produce a Type II error. Another useful test is the **Dunnet test**, which compares a single control sample mean with all other sample means in the experiment.

Before proceeding further, a very important rule of use should be clearly noted. Multiple comparison tests should not be performed if the original ANOVA test has failed to produce a significant conclusion. The ANOVA test should be considered overriding in this regard, thus *if the ANOVA yields a non-significant F-ratio for a particular treatment factor then any further analysis of the differences between the treatment means is totally invalid.*

The basis of most multiple comparison tests is the statistic termed the **standard error of the difference** between two sample means. Before describing the determination of multiple comparison tests, we will recall the definition of this statistic.

7.5.1 The standard error of the difference

The standard error of the difference (SED) was first encountered during the discussion of the z-test and Student's t-test for comparing two sample means. It should be recalled that the SED for two means is given by the square root of the sum of the sample variances divided by their respective sample sizes n, i.e.:

$$\text{SED} = \sqrt{\frac{s_1^2}{n_1} + \frac{s_2^2}{n_2}}$$

Since it was an original assumption of ANOVA that all sample variances are equal, then the already calculated value for residual variance (denoted by $\text{MS}_{\text{residual}}$) can be used as a measure of the variance of all the samples in the experiment. The equation above thus simplifies to:

$$\text{SED} = \sqrt{\text{MS}_{\text{residual}}\left(\frac{1}{n_1} + \frac{1}{n_2}\right)}$$

Further, if the sizes (n) of the samples being compared are equal then the above equation further simplifies to:

$$\text{SED} = \sqrt{\frac{2 \times \text{MS}_{\text{residual}}}{n}}$$

The SED gives a measure of the random error in measuring the difference between two means against which the observed difference can be compared. It is used in the following multiple comparison tests.

7.5.2 Least significant difference test

The least significant difference (LSD) is the smallest numerical difference that must be present between two sample means within a multiple-sample experiment for those two means to be considered significantly different at a defined level of probability. In effect the LSD is a confidence limit interval based on Student's t distribution and is determined by the product of Student's t value and the standard error of the difference (SED) between the means. Assuming that the samples have equal variance and have similar size, then:

$$\text{LSD} = t \times \text{SED}$$

where: t = Student's t value read from Student's tables at the required level of probability (usually $\alpha = 0.05$) and at the residual degrees of freedom;
SED = standard error of the difference (calculated as described above).

Having obtained the value of the LSD, any pair of samples that show a numerical difference between their means that is equal to or greater than the LSD may then be concluded to have been drawn from two populations that are significantly different at the level of probability for which the LSD was determined.

The major problem with the LSD test is that, being a type of Student's *t*-test, it compounds the probability of making a Type I error every time it is employed to compare two means within the experiment. Therefore, in order to reduce the incidence of Type I errors *LSD tests should only be used for making a limited number of comparisons between means and the means to be compared should be preselected before the experiment is conducted*. The temptation to use the LSD to test between all possible pairs of means or for making any unplanned comparisons after the experiment has been conducted should be firmly resisted.

The use of a LSD test is illustrated in Example 7.2.

Example 7.2. Calculation of least significant difference following a one-way ANOVA.

Data from the ANOVA of the effect of irrigation on stem girth increase in *Prunus* trees shown in Example 7.1.

$MS_{residual}$ (from the ANOVA table) $= 0.292$ (Residual DF $= 8$)
t (from Student's *t* table; $\alpha = 0.05$, residual DF $= 8$) $= 2.306$
n (for all samples) $= 3$

$$LSD_{(\alpha = 0.05)} = t \times \sqrt{\frac{2 \times MS_{residual}}{n}} = 2.306 \times \sqrt{\frac{2 \times 0.292}{3}} = 1.018$$

Conclusion: where the observed difference between a preselected pair of treatment sample means is ≥ 1.018, then that difference is significant ($P < 0.05$).

7.5.3 Tukey honest significant difference test

The Tukey honest significance difference (HSD) test differs from the LSD test in that, irrespective of how many pairs of sample means are compared, the probability of making a Type I error is maintained at a constant level and does not aggregate as more comparisons are made. For this reason the Tukey HSD test can be used to make any number of sample mean comparisons after the experiment has been completed and which have not been planned for in advance. On the other hand, the Tukey HSD test is less powerful than the LSD test and does run the risk that small but true differences may fail to be detected.

The Tukey honest significant difference (HSD) test employs a data distribution known as the **Studentized range**, from which a critical value of a statistic referred to as *q* is obtained. (Note here that, in consulting the Studentized range, the critical value of *q* varies with the number of samples (*k*) in the experiment. It is in this way that the Tukey test takes into account the number of potential comparisons that can be made and maintains a constant probability of making a Type I error). The critical *q* value is then used in the calculation of the **minimum critical difference (MCD)**,

---

which is the smallest numerical difference that must be present between a pair of sample means for a significant difference between the two samples to be claimed.

The first step in the procedure is to determine what we might call a 'Tukey estimate' of the standard error for the differences between the sample means (which will be symbolized as $TukeySED$). This is obtained by dividing the residual variance ($MS_{residual}$) by 2, multiplying by the sum of the reciprocal of the sample sizes and taking the square root. Where the sample sizes are equal this simplifies as shown below:

For equal sample size:

$$TukeySED = \sqrt{\frac{MS_{residual}}{n}}$$

For unequal sample size:

$$TukeySED = \sqrt{\frac{MS_{residual}}{2} \times \left(\frac{1}{n_a} + \frac{1}{n_b}\right)}$$

where: n_a, n_b, = size of the samples chosen for comparison.

Note that the formula for determining the $TukeySED$, unlike the standard formula for SED, does not involve multiplying the residual variance ($MS_{residual}$) by 2. The reason for this is simply that the Studentized range from which the Tukey q statistic is obtained, and which is then used to calculate the minimum critical difference, builds in the assumption that the SED is too small by a factor of $\sqrt{2}$.

The critical value for the test statistic (q_{crit}) is obtained by consulting the Studentized range (Table A7 in the Appendix) at the selected critical level of probability (α), the residual degrees of freedom (V) and also the number of samples present in the experiment (k). The minimum critical difference (MCD) between means is then determined by multiplying the $TukeySED$ value by q_{crit}:

$$MCD = q_{crit} \times TukeySED$$

If the difference between a pair of selected sample means is equal to or greater than the MCD, then the populations from which the samples were extracted can be concluded to be significantly different at the level of probability at which the test was performed. In presenting the final results of a Tukey test, a useful procedure is to formulate a grid showing the calculated difference between every possible pair of sample means. In this way it is very easy to compare the observed differences between means with the MCD and thus to distinguish those treatments which are significantly different. This is illustrated in Example 7.3.

7.5.4 Student–Newman–Keuls multiple range test

The Student–Newman–Keuls (SNK) test can be considered as a compromise between the LSD and Tukey HSD tests; it is slightly more powerful than the Tukey HSD test, although on most occasions the two tests will produce the same conclusion. If, however, the difference between two means is close to the critical difference for significance, there is a greater likelihood of detecting it as a significant effect using the SNK test than the Tukey test. It may be argued, however, that the Tukey HSD test has less chance of producing Type I errors and is, therefore, the safer option, although there is not universal agreement upon this.

Example 7.3. Calculation of a Tukey HSD test following a one-way ANOVA.

Data from the ANOVA of the effect of irrigation on stem girth increase in *Prunus* trees shown in Example 7.1.

$MS_{residual}$ (from the ANOVA Table = 0.292 (DF = 8)

$$n = 3$$

$$TukeySED = \sqrt{\frac{MS_{residual}}{n}} = 0.312$$

Critical q value from table (α = 0.05, residual DF = 8, k = 4) = 4.529
MCD between means = 4.529 × 0.312   = 1.41

Conclusion: where the observed difference between any pair of sample means is >1.41, then that difference is significant ($P < 0.05$).

Table of differences between sample means
(Significant differences detected by Tukey HSD test are denoted by an asterisk)

	Difference between sample means			
Irrigation level (%)	100 ($\bar{x}$ = 3.67)	75 ($\bar{x}$ = 6.91)	50 ($\bar{x}$ = 5.97)	25 ($\bar{x}$ = 5.05)
25 ($\bar{x}$ = 5.05)	1.38	1.86*	0.92	
50 ($\bar{x}$ = 5.97)	2.30*	0.94		
75 ($\bar{x}$ = 6.91)	3.24*			
100 ($\bar{x}$ = 3.67)				

To undertake the SNK test, the sample means need first to be ordered according to their magnitude, i.e. they are ranked. A standard error of the difference between means (SED) is estimated by exactly the same formula as in the Tukey HSD test. A critical value for the test statistic q is then obtained from the Studentized range (Appendix: Table A7), which is consulted at the required critical probability (α), the residual degrees of freedom (V) and at the number of ranked sample means (p) that lie in the range defined by the two sample means to be compared. (Therefore p replaces k that was used for a Tukey HSD test.) For example, if the difference between the second largest sample mean (rank = 2) and the fifth largest sample mean (rank = 5) was to be tested, then the value of p would be 4. The SED is then multiplied by q_{crit} to obtain a critical minimum difference that must separate the two selected sample means for a significant difference between them to be claimed. A minimum critical difference (MCD) is first obtained for a comparison between the smallest mean and the largest mean, and then for the next smallest mean and the largest mean and so on. It may be noted that for two adjacent means the Student–Newman–Keuls MCD will in fact be the same as the LSD value. Once all

MCD values have been determined, the difference between every pair of means can be inspected and a significant difference between means can be claimed where the numerical difference between the means is equal to or greater than the the appropriate MCD. Of course, once a pair of means have been shown not to be significantly different, then the differences between any pair of means that are ranked closer together must also be non-significant and do not need to be tested.

The calculation of the Student–Newman–Keuls test is shown in Example 7.4. If the outcome of the SNK test in Example 7.4 is compared with that of the Tukey

---

Example 7.4. Calculation of a Student–Newman–Keuls multiple range test following a one-way ANOVA.

Data from the ANOVA of the effect of irrigation on stem girth increase in *Prunus* trees shown in Example 7.1.

$MS_{residual}$ (from the ANOVA Table) = 0.292 (DF = 8)

$$n = 3$$

$$Tukey\text{SED} = \sqrt{\frac{MS_{residual}}{n}} = 0.312$$

Treatment samples are ranked by their means:

Irrigation treatment	Mean	Rank
25%	5.05	2
50%	5.97	3
75%	6.91	4
100%	3.76	1

Determination of the minimum critical difference (MCD) for all paired comparisons in ranking order

Sample comparison		Difference between means $(\overline{x}_1 - \overline{x}_2)$	p	$q_{(p;\ DFresid,\ \alpha\ =\ 0.05)}$	MCD	Significance
Rank	Irrigation treatment					
1 v 4	100% v 75%	3.24	4	4.529	1.413	$P < 0.05$
2 v 4	25% v 75%	1.86	3	4.041	1.261	$P < 0.05$
3 v 4	50% v 75%	0.94	2	3.261	1.017	n.s.
1 v 3	100% v 50%	2.30	3	4.041	1.261	$P < 0.05$
2 v 3	25% v 50%	0.92	2	3.261	1.017	n.s.
1 v 2	100% v 25%	1.38	2	3.261	1.017	$P < 0.05$

Conclusion: where the observed difference between any pair of sample means $(\overline{x}_1 - \overline{x}_2)$ is ≥ MCD value for those means, then that difference is significant $(P < 0.05)$.

HSD test for the same data shown in Example 7.3, it will be noted that for the comparison between 25% and 100% irrigation the SNK test indicated a significant difference while the Tukey test did not. This illustrates the point that the SNK test is relatively more powerful than the Tukey test at detecting small differences as being significant.

7.5.5 Dunnet test

The Dunnet test is used when the investigator, rather than being interested in all possible comparisons between sample means in the experiment, is really only interested in the difference between one specified sample mean, identified as the 'control', and the other sample means. In this case a slightly different probability distribution is employed from that used for the Tukey test and is referred to as the q' distribution. The standard error of the difference, SED, is estimated by:

for equal sample size:

for unequal sample size:

$$SED = \sqrt{\frac{2 \times MS_{residual}}{n}}$$

$$SED = \sqrt{MS_{residual}\left(\frac{1}{n_a} + \frac{1}{n_{control}}\right)}$$

where: $MS_{residual}$ = residual MS from the ANOVA
n = size of samples to be compared

A critical level of the test statistic q' is obtained from the q' probability distribution at the required critical probability α, the residual DF and k samples (see Table A8.1 in the Appendix). A minimum critical difference (MCD) for testing the significance of the difference between each sample mean and the control mean is then determined by:

$$MCD = q'_{crit} \times SED$$

Since it is possible to specify whether the control mean should be larger or smaller than the compared sample mean, the Dunnet test can be performed as a one-tailed test if required. A separate table of one-tailed critical q' values is provided in Table A8.2 in the Appendix.

7.6 Testing the Assumption that Multiple Sample Variances are Homogeneous

As previously explained, the analysis of variance technique depends on the assumption that the samples are extracted from populations that have homogeneous variances. If there is any doubt that this assumption is valid, then it should be examined before proceeding with the ANOVA. This is, however, easier said than done. One of the more common tests available for inspecting homogeneity of sample variances is

Example 7.5. Calculation of a Dunnet test following a one-way ANOVA.

Data from the ANOVA of the effect of irrigation on stem girth increase in *Prunus* trees shown in Example 7.1.

Assume that the 100% irrigation treatments has been previously identified as the control against which all other treatments are to be compared.

$MS_{residual}$ (from the ANOVA Table) = 0.292 (DF = 8)

$$n = 3$$

$$SED = \sqrt{\frac{2 \times MS_{residual}}{n}} = 0.441$$

Determination of the minimum critical difference (MCD) for all paired comparisons between the control and treatments.

Irrigation treatment comparison	Difference between means $(\bar{x}_1 - \bar{x}_2)$	SED	$q'_{(k; DFresid, \alpha = 0.05)}$	MCD	Significance
25% v 100%	**3.24**	0.441	2.88	**1.27**	$P < 0.05$
50% v 100%	**1.86**	0.441	2.88	**1.27**	$P < 0.05$
75% v 100%	**0.94**	0.441	2.88	**1.27**	n.s.

Conclusion: where the observed difference between any treatment mean and the control mean $\geq$ 1.27, then that difference is significant ($P < 0.05$).

Bartlett's test. Unfortunately this test is both complicated to calculate and very sensitive to small deviations of the data from a normal distribution. Although this test is made available in a number of statistics software packages, it is not generally recommended for the purpose of checking population variance homogeneity before proceeding with ANOVA and will not be discussed further here. (The interested reader will find further discussion of Bartlett's test in alternative texts, e.g. Zar, 2007) A simpler test for equal variances is the **variance ratio F_{max} test**, which is similar to the test described earlier in the context of Student's *t*-test (see section 6.4) and which examines the difference between the largest sample variance in the experiment and the smallest. Clearly if the difference between the largest and smallest variances proves to be insignificant then no significant differences can exist between any of the other intermediate sample variances.

The null hypothesis for the test is that there are no differences in variances of the samples. The test statistic F_{max} is determined by:

$$F_{max} = \frac{s^2_{max}}{s^2_{min}}$$

where: S^2_{max} = largest sample variance, S^2_{min} = smallest sample variance.

Unlike the test for checking the homogeneity of variances prior to Student's t-test, in this case we have to account for the fact that there are more than two samples in the experiment as a whole. Consequently the table of critical values of F_{max} (Table A6 in the Appendix) needs to be consulted at the number of samples, k, present in addition to the degrees of freedom of the samples being compared. Where the sample sizes are different, the DF of the smaller sample are employed. The observed value is then compared with the critical value read from the table; if $F_{max_{observed}} < F_{max_{critical}}$, then samples may be concluded to have been extracted from populations with equal variances; however, if $F_{max_{observed}} \geq F_{max_{critical}}$, then the null hypothesis must be rejected and a difference between the population variances must be assumed to exist.

In theory, if variances are shown to differ significantly, then ANOVA becomes invalid and either a data transformation or an alternative non-parametric technique should be employed. In practice, however, ANOVA is actually a very robust test and, as long as sample sizes are reasonably large, it can satisfactorily cope with data which do not show complete homogeneity of variance, or indeed small departures from normality.

7.7 The Power of an ANOVA Multiple-sample Test and Determination of Appropriate Sample Size

The power of a significance test, as previously explained (see section 5.4), is the probability of a test correctly rejecting the null hypothesis when it really is not true. The advantage of determing the power of an ANOVA test, particularly if this can be undertaken in advance of the experiment being undertaken, is that it can provide a warning that a certain experimental design is unlikely to yield conclusive outcomes. For example, were the power of a particular ANOVA test shown to be less than 0.25, this would indicate a greater than 75% chance of making a Type II error. This would strongly suggest a need to increase sample size and/or try to improve homogeneity of sample variances before performing the experiment. Unfortunately, determing the power of an ANOVA test is rather a complex process and depends on the value of the so-called **non-centrality parameter** (ϕ), which itself is based on the distribution of the F-ratio when the null hypothesis is **false** (the so-called **non-central F distribution**). Calculation of ϕ for a particular experiment depends on being able to make sensible estimates of the population variance σ^2 and each of the treatment population means μ_i which are usually estimated from the residual MS and sample means observed from previous similar experiments using the same material. Having determined ϕ, the value is then referred to the theoretical distributions of ϕ (usually displayed in graphical rather than tabular form) at the appropriate treatment and residual DF values, and at the required critical probability α, from which the power of the test can be read. Since most statistical software packages will determine the power of ANOVA tests, the complex calculations involved and distribution plots of ϕ are not illustrated here.

In section 6.6 we saw how it was particularly useful to be able to express the power of a Student's t-test in terms of its ability to identify a given numerical difference between two sample means as being significant and further, to determine the smallest sample size required to enable a test to detect the presence of a significant difference. Similarly, the smallest sample size required (assuming n is constant) can be determined that would allow a difference between any pair of means of a given value to be detected as significant by an ANOVA operating with a given level of power. To determine this we again need an estimate of σ^2 in order to calculate a value for the non-centrality parameter (ϕ). Then, by a reiterative process in which different values of n are tested against the required minimum difference that we wish to detect, the smallest value of n can be identified.

Again, the full details of the calculation will not be presented here but the process can be illustrated using the irrigation trial described in Example 7.1. As commercial growers, we might only be interested if we can observe at least a 1 cm difference in girth of trees in response to irrigation treatment. If we stipulate that our ANOVA test should have at least an 80% probability of detecting a significant difference (i.e a power of 0.8), what we then need to know is the smallest sample size that would be required to allow a difference (δ) of 1 cm between any pair of sample means to be detectable as significant at the 5% level ($P \leq 0.05$). We might start by guessing that a sample size of ten should suffice (giving a residual DF of $k \times [10 - 1] = 36$). Then given that there are four treatment samples (giving a treatment DF of $k - 1 = 3$) and assuming from previous experience that s^2 is 0.3, the non-centrality parameter (ϕ) is determined by:

$$\phi = \sqrt{\frac{n\delta^2}{2ks^2}} = \sqrt{\frac{10 \times 1^2}{2 \times 4 \times 0.3}} = 2.04$$

By consulting the distribution of ϕ (at DF = 3 and 36; α = 0.05) this gives a power of ≈ 0.91.

By a similar process a sample size n of eight confers a power of ≈ 0.82 to the test while a sample size n of seven confers a power of only ≈ 0.76. Thus the smallest sample size that provides at least an 80% probability of being able to detect a difference of as little as 1.0 cm as significant at a P level of ≤ 0.05 is eight.

The statistical procedures involved here are rather complex and most researchers today will employ computer software to undertake these. The reader who requires further detail, including graphical representation of the ϕ distributions, is referred to Zar (2009).

8 Analysis of Multiple-factorial Experiments

Factor: circumstance, fact or influence contributing to a result.

(Concise Oxford English Dictionary)

- Introduction to multiple-factor experimental designs.
- The concept of treatment interaction.
- Principles of multiple-factor analysis of variance (ANOVA) and types of model.
- Calculation of two-way ANOVA with and without replication.
- The general linear model approach to ANOVA.
- Multiple comparison testing following two-way ANOVA.
- Three-way ANOVA.
- Handling unequal sample replication and invalid assumptions.

8.1   Introduction to Multiple-factorial Experiments

In the previous chapter we examined how multiple samples arising from single-factor experiments can be analysed for significant differences. Frequently, however, more than one treatment factor will be investigated simultaneously in the same experiment, each factor being applied at a number of different levels. For example, the effect of a fertilizer might be tested at a range of concentrations on a crop grown in a range of different soil types. In this experiement there are two factors investigated simultane-ously, namely fertilizer and soil, and it is then referred to as a **two-way factorial** design. This type of experimental design is illustrated in Fig. 8.1. In the same way a **three-way factorial** design can be used to inspect the effects of three factors; for exam-ple, the effect of fertilizer could be examined on different varieties of a crop grown in different soil types all in the same experiment. In practice, three-way factorial designs are relatively uncommon but in this age of computer power they can be analysed fairly readily. Although theoretically possible to perform, factorial designs above three-way require so many resources and demand such high levels of replication in order to enable their statistical analysis that they are more or less unknown.

Another type of multiple-factor experiment that is encountered in agriculture and horticulture, particularly in trial work, is the 2^k **design**. This is an experiment in which there are k treatment factors present, where k may actually be quite a large number, but each factor is investigated at just two levels of application. For example, the combined effect of three diet supplements in an animal feed on the rate of growth could be investigated by simply adding or deleting each supplement from the diet. This would give a total of 2^3 treatments each being replicated at an

   © C. Ireland 2010. *Experimental Statistics for Agriculture and Horticulture* (C. Ireland)

'Extragrow' fertilizer
Fertilizer concentration (kg/ha)

SOIL TYPE	5	10	50	100
Clay loam	Sample 1 Replicate 1 Replicate 2 Replicate 3 Replicate n	Sample 2 Replicate 1 Replicate 2 Replicate 3 Replicate n	Sample 3 Replicate 1 Replicate 2 Replicate 3 Replicate n	Sample 4 Replicate 1 Replicate 2 Replicate 3 Replicate n
Boulder clay	Sample 5 Replicate 1 Replicate 2 Replicate 3 Replicate n	Sample 6 Replicate 1 Replicate 2 Replicate 3 Replicate n	Sample 7 Replicate 1 Replicate 2 Replicate 3 Replicate n	Sample 8 Replicate 1 Replicate 2 Replicate 3 Replicate n
Stony clay	Sample 9 Replicate 1 Replicate 2 Replicate 3 Replicate n	Sample 10 Replicate 1 Replicate 2 Replicate 3 Replicate n	Sample 11 Replicate 1 Replicate 2 Replicate 3 Replicate n	Sample 12 Replicate 1 Replicate 2 Replicate 3 Replicate n

Fig. 8.1. The design of a typical two-way factorial experiment: the effect of fertilizer on crops grown in different soil types. In this experiment a fertilizer is applied at four different concentrations on crops grown in three different soil types, producing a 4×3 two-way factorial experimental design. There are 12 samples in total ($k = 12$) with n replicate observations made per sample.

appropriate level. Such designs can be fully randomized but are more frequently conducted as randomized block experiments, the analysis of which will be considered in the next chapter.

The most important difference between single-factor and two (or more)-factor experiments is that in the latter case, in addition to the effect of the individual treatments, the possibility of interaction between treatment factors arises as a potential major effect that influences the results and requires analysis. The nature of treatment interaction and the different ways that it may arise within an experiment therefore need some initial explanation.

8.2　The Concept of Treatment Interaction

When two or more treatment factors are applied simultaneously in a multiple-factorial experiment, the possibility for detecting **interaction** between the treatment factors is evoked. By the term 'interaction' we mean the effect that one treatment may have in modifying the response of the subjects to another simultaneous treatment. For example, in a tissue culture experiment the required optimum concentration of a certain plant growth hormone required for maximum growth rate was investigated with different concentrations of sugar present in the growth media. It was found that inclusion of a certain level of the hormone zeatin on its own produced a 10% increase in growth rate and a certain level of sucrose on its own produced a similar 10% increase. When, however, zeatin and sucrose were included in combination an increase in growth rate of 25% occurred. The hormone and the sugar together produced an effect that was greater than the sum of their individual effects. This type of response is termed a positive **synergistic interaction**. Sometimes, of course, a negative interaction might occur in which the response to a combined treatment is decreased compared with the sum of the effects of the individual treatments. Some of the different types of interaction that may occur within a two-factor experiment are illustrated in Fig. 8.2.

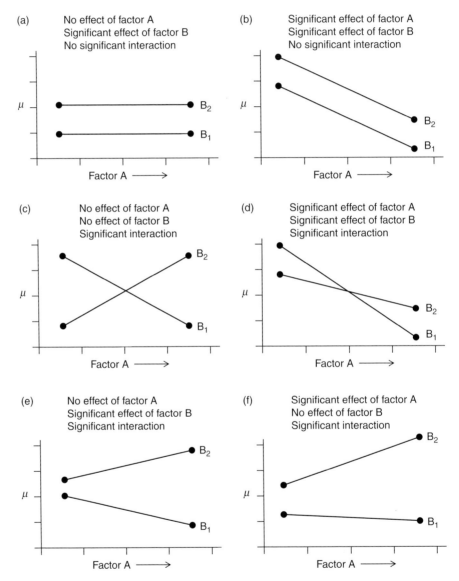

Fig. 8.2. An illustration of the different types of factor interaction that may occur within a two-factor experimental design. The effect of an increasing level of factor A under two levels of factor B on a population mean μ under six different scenarios.

Chapter 8

It is important to understand that the existence of a significant interaction is independent of the existence of any other significant effects. For example, in a particular glasshouse trial, the yield of cucumbers showed no significant response to an increase in the level of supplementary lighting and similarly showed no significant response to an increase in the CO_2 concentration in the glasshouse atmosphere. However, when both the level of lighting and CO_2 concentration were increased together a significant yield increase then occurred. This represents an example of a positive interaction between two treatments but one in which neither treatment was capable of producing a significant effect by itself (as in Fig. 8.2c).

Interaction between treatments is very important in experimental biology. Very often the effects of the individual treatments will be well known beforehand and it is the question of whether the treatment factors interact in producing a response that then becomes the primary concern of the investigator and the main reason for performing a multiple-factorial experiment.

8.3 Two-way Analysis of Variance

8.3.1 The general principles of two-way ANOVA

Two-way factorial experiments, such as that illustrated in Fig. 8.1, involve the simultaneous application of two treatment factors, each at more than one level of application. Such experiments require an analysis that is able to separate the combined treatment effect into its components and thereby allow the significance of effects of each of the two treatment factors, e.g. soil and fertilizer, and their interaction to be determined independently. If the samples are drawn from normally distributed populations, then the statistical test that achieves this is a parametric **two-way analysis of variance** (two-way ANOVA).

In one-way ANOVA, the total variance present in an experiment is partitioned between the single treatment and the random sources of variation (see Chapter 7). Two-way ANOVA is a logical extension of this procedure; however, each treatment factor is now considered as a separate source of variation. Thus the total variation in the data set, described by the total sum of squares, is partitioned between each of the two treatment factors and the random (or 'residual') variation. In addition, as long as there is replication present within samples, the interaction between treatments may be analysed as a further source of variation. In a two-way factorial experiment (with replication), there are, therefore, three separate null hypotheses to be tested:

$H_{0\,(1)}$ = there is no effect of treatment A
$H_{0\,(2)}$ = there is no effect of treatment B
$H_{0\,(3)}$ = there is no effect of treatment A×B interaction

The two-way ANOVA subsequently produces three separate values of the test statistic F, i.e. an F-ratio for each of the two treatment factors and an F-ratio for their interaction, which are used to independently test the three null hypotheses.

8.3.2 Assumptions and limitations of two-way ANOVA

Two-way ANOVA depends on the same principles as one-way ANOVA and, therefore, also depends on the same assumptions that are required by one-way ANOVA and were discussed in the previous chapter. These are that:

1. The samples are independent.
2. The samples are drawn from normally distributed populations.
3. The samples are drawn from populations that have equal variances.

In the case of two-way factorial experiments, a distinction needs to be made between the cases when there is replication and when there is no replication within samples. Clearly a one-way factorial experiment without any replication is untenable since the sample means would be based on single observations and no within-sample variance would exist. It is possible, however, to conduct a two-way factorial experiment without replication so that the single values for the different levels of one treatment factor are employed as the replicates for the other treatment factor. However, when no replication is present, interaction is not then detectable and the sources of variation are simply the two treatment factors and the residual variation. The reason for this is fairly obvious. Without replication, no within-group variances can be calculated and therefore the response to a treatment cannot be compared under different application levels of a second treatment. Thus, *treatment interaction only becomes a detectable source of variation when sample size is ≥ 3.*

8.3.3 Standard calculation procedure for performing two-way ANOVA with replication for a fully randomized experimental design

Due to the number of summations required it becomes extremely important that the data for a two-way ANOVA are tabulated in a clear fashion. A typical format for a two-way factorial data table for a fully randomized design and the definition of a number of required parameters is shown in Table 8.1. It should also be noted that there is a very important practical limitation that applies to the calculation of two-way ANOVA by the standard partitioning of sum of squares method, namely that *all samples need to be of equal size.*

The calculation of the ANOVA commences with the calculation of the total sum of squares (SS_{total}) as a measure of the total variation in the experiment. The total between-sample sum of squares ($SS_{between-samples}$), often termed the 'main effects' SS, is then determined. This item contains the total variation due to the two treatment factors plus the variation due to the treatment interaction. The sums of squares due to each of the treatment factors (SS_A and SS_B) are then determined independently. The variation due to interaction ($SS_{A \times B}$) can then be calculated simply as the difference between the sum of the two individual treatment SS and the between-sample SS. Finally, the random variation in the experiment, represented by the residual SS, is the sum of the product of the sample variances and their respective degrees of

Table 8.1. Typical format for the data table for a two-way factorial experiment with replication.

Treatment B Level	Treatment A Level 1	2	3	n	Treatment B totals
1	x_1 x_2 x_3 x_n	x_1 x_2 x_3 x_n	x_1 x_2 x_3 x_n	x_1 x_2 x_3 x_n	
	$\Sigma_{1,1}$	$\Sigma_{1,2}$	$\Sigma_{1,3}$	$\Sigma_{1,n}$	$T_{B,1}$
2	x_1 x_2 x_3 x_n	x_1 x_2 x_3 x_n	x_1 x_2 x_3 x_n	x_1 x_2 x_3 x_n	
	$\Sigma_{2,1}$	$\Sigma_{2,2}$	$\Sigma_{2,3}$	$\Sigma_{2,n}$	$T_{B,2}$
3	x_1 x_2 x_3 x_n	x_1 x_2 x_3 x_n	x_1 x_2 x_3 x_n	x_1 x_2 x_3 x_n	
	$\Sigma_{3,1}$	$\Sigma_{3,2}$	$\Sigma_{3,3}$	$\Sigma_{3,n}$	$T_{B,3}$
n	x_1 x_2 x_3 x_n	x_1 x_2 x_3 x_n	x_1 x_2 x_3 x_n	x_1 x_2 x_3 x_n	
	$\Sigma_{n,1}$	$\Sigma_{n,2}$	$\Sigma_{n,3}$	$\Sigma_{n,n}$	$T_{B,n}$
Treatment A totals	$T_{A,1}$	$T_{A,2}$	$T_{A,3}$	$T_{A,n}$	Σx (Grand Total)

Where: $\Sigma_{n,n}$ = sample total for level n of treatment factors A and B; $T_{A,n}$ = total for level n of treatment factor A; $T_{B,n}$ = total for level n of treatment factor B; Σx = grand total (often abbreviated GT; the grand mean is abbreviated GM).

freedom, i.e. $\Sigma(s^2 \times n-1)$. This is, however, more simply determined as the difference between the total SS and the between-sample SS. The degrees of freedom associated with each of these partitioned SS values must also be identified.

The following parameters are defined:

k = total number of treatment samples
n = no. of replicate measurements per sample
N = total no. of measurements in whole experiment (= $k \times n$)
N_A = number of levels of application of treatment A
N_B = number of levels of application of treatment B
n_a = number of observations made for each level of treatment A (= $n \times N_B$)
n_b = number of observations made for each level of treatment B (= $n \times N_A$)

The formal calculations involved in the two-way ANOVA procedure are as follows:

Step 1: Calculation of the *correction term* (CT)

$$CT = \frac{(\Sigma x)^2}{N}$$

Step 2: Calculation of the *total* sum of squares

$$SS_{total} = \Sigma (x - \bar{x})^2$$

In practice, this is calculated by: $\Sigma x^2 - CT$

Step 3: Calculation of the *between-samples* sum of squares

$$SS_{between\text{-}samples} = \Sigma (\bar{x} - GM)^2$$

In practice this is calculated by:

$$SS_{between\text{-}samples} = \left[\frac{(\Sigma x_{1,1})^2}{n} + \frac{(\Sigma x_{1,2})^2}{n} + \frac{(\Sigma x_{1,3})^2}{n} + ... + \frac{(\Sigma x_{N_A N_B})^2}{n} \right] - CT$$

where: $\Sigma x_{i,j}$ is the sample total for the *i*th application level of the first factor and the *j*th application level of the second factor.

Step 4: Calculation of the *individual treatment* sum of squares

This is the sum of the squared deviations of the overall mean for each treatment ($\bar{T}$) and the grand mean:

$$SS_{treatment} = \Sigma (\bar{T} - GM)^2$$

In practice the sums of squares for each treatment A and B are calculated by:

$$SS_A = \left(\frac{\Sigma T_A^2}{n_A} \right) - CT \quad \text{and} \quad SS_B = \left(\frac{\Sigma T_A^2}{n_B} \right) - CT$$

where, ΣT_A^2 = sum of the squares of treatment factor A totals
ΣT_B^2 = sum of the squares of treatment factor B totals
n_A = no. of observations made for each level of treatment A
n_B = no. of observations made for each level of treatment B

(It may be noted here that the number of items per sample, n, is employed as a constant throughout the calculation of the different SS terms. This explains the necessity of all samples being of equal size.)

Step 5: Calculation of the treatment *interaction* sum of squares

Since the total between-samples SS includes both the individual treatment effects and the interaction between the two treatments, the interaction SS ($SS_{A \times B}$) can be calculated by subtraction:

$$SS_{A \times B} = SS_{between-samples} - (SS_A + SS_B)$$

Step 6: Calculation of the *residual* sum of squares

$$SS_{residual} = s_{1,1}^2(n-1) + s_{1,2}^2(n-1) + s_{1,3}^2(n-1) + \ldots\ldots s_{N_A N_B}^2(n-1)$$

where: $s_{i,j}^2$ is the variance for the sample representing the ith application level of the first factor and the jth application level of the second factor.

However, the residual SS can be more easily obtained by simple subtraction:

$$SS_{residual} = SS_{total} - SS_{between-samples}$$

Step 7: Determination of the *degrees of freedom* (DF) for each source of variation

The degrees of freedom (DF) for each sum of squares item are calculated as in one-way ANOVA except that the DF for the interaction between the two treatments needs also to be determined. The interaction DF is simply the product of the two treatment DF values.

DF total	$= N - 1$	(total no. of observations less 1)
DF treatment A	$= N_A - 1$	(total no. of levels of treatment A less 1)
DF treatment B	$= N_B - 1$	(total no. of levels of treatment B less 1)
DF interaction A × B	$= (N_A - 1 \times N_B - 1)$	(product of the treatment DF values)
DF residual	$= k \times (n - 1)$	(this is equivalent to the total DF minus both the treatment and interaction DF)

Step 8: Determination of the *mean square* (variance) for each source of variation

Each partitioned sum of squares item is now converted to a mean square (or variance) by dividing by the degrees of freedom:

MS treatment A	$= SS_{treatment\ A} / DF_{treatment\ A}$
MS treatment B	$= SS_{treatment\ B} / DF_{treatment\ B}$
MS interaction A × B	$= SS_{interaction\ A \times B} / DF_{interaction}$
MS residual	$= SS_{residual} / DF_{residual}$

Step 9: Calculation of the *F*-ratio and presentation of the results in an ANOVA Table

The method of determination of the *F*-ratio test statistic depends on the ANOVA model being applied (as previously noted in section 7.4).

1. Fixed effects Model I ANOVA. In the most common model in which the level of all treatment factors is fixed by the investigator (Model I ANOVA), the null hypothesis states there is no difference between the fixed levels at which a treatment has been applied, e.g. between 0, 5, 10 and 50 kg/ha fertilizer applications. The null hypothesis for a treatment A applied at fixed levels can thus be stated in the form: $\mu_{A_1} = \mu_{A_2} = \dots \mu_{A_N}$, where N is the number of levels at which the treatment was applied. If the null hypothesis were to be true, then the variance due to the treatments and random error effects would be equal. Therefore the *F* test statistic for each treatment source of variation and the interaction is given by the ratio of each treatment mean square to the residual mean square. This provides a measure of the variation due to the treatment relative to that due to random effects. Thus:

$$F\text{-ratio treatment A} = MS_{\text{treatment A}}/MS_{\text{residual}}$$

$$F\text{-ratio treatment B} = MS_{\text{treatment B}}/MS_{\text{residual}}$$

$$F\text{-ratio interaction A} \times \text{B} = MS_{\text{interaction A×B}}/MS_{\text{residual}}$$

2. Random effects Model II ANOVA. Where a treatment is applied at randomized levels, the null hypothesis is that there is no effect of the treatment at any possible level of application, e.g. the annual rainfall has no effect on mildew infection in a particular crop. In statistical terms, this is equivalent to the null hypothesis stating that the random treatment factor produces no extra variance in the experiment. In the relatively uncommon case where the two-way factorial design consists of both treatment factors being applied at random levels (i.e. a Model II ANOVA), under the null hypothesis the variance due to the treatments and the interaction will therefore be equal (i.e. H_0: $\sigma_A^2 = \sigma_B^2 = \sigma_{A×B}^2$). The *F* test statistic for a randomized treatment is, therefore, determined as the ratio of the treatment MS to the interaction MS while the *F* value for the interaction remains the ratio of the interaction MS to the residual MS. Thus:

$$F\text{-ratio (all treatments)} = MS_{\text{treatment}}/MS_{\text{interaction}}$$

$$F\text{-ratio (interaction)} = MS_{\text{interaction}}/MS_{\text{residual}}$$

3. Mixed effects Model III ANOVA. Where one factor is applied at a series of fixed levels but the second factor at random levels, then a Model III ANOVA is employed. In this case, under the null hypotheses of no treatment effects the variance of the fixed effects treatment would be equal to the variance due to the interaction while the variance of the randomized treatment will be equal to the residual variance. The *F*-ratios are therefore determined by:

$$F\text{-ratio (fixed effects treatment)} = MS_{\text{treatment}}/MS_{\text{interaction}}$$

$$F\text{-ratio (random effects treatment)} = MS_{\text{treatment}}/MS_{\text{residual}}$$

$$F\text{-ratio (interaction)} = MS_{\text{interaction}}/MS_{\text{residual}}$$

Table 8.2. Two-way ANOVA table standard format.

Source of variation	Sum of squares (SS)	Degrees of freedom (DF)	Mean square (MS)	F-ratio Fixed effects Model I ANOVA	F-ratio Random effects Model II ANOVA	F-ratio Mixed effects Model III ANOVA[a]
Treatment factor A	$SS_{treatment A}$	$DF_{factor A}$	$\dfrac{SS_{treatment A}}{DF_{factorA}}$	$\dfrac{MS_{treatment A}}{MS_{residual}}$	$\dfrac{MS_{treatment A}}{MS_{A \times B}}$	$\dfrac{MS_{treatment A}}{MS_{A \times B}}$
Treatment factor B	$SS_{treatment B}$	$DF_{factor B}$	$\dfrac{SS_{treatment B}}{DF_{factor B}}$	$\dfrac{MS_{treatment B}}{MS_{residual}}$	$\dfrac{MS_{treatment B}}{MS_{A \times B}}$	$\dfrac{MS_{treatment B}}{MS_{residual}}$
A × B interaction	$SS_{A \times B}$	$DF_{A \times B}$	$\dfrac{SS_{A \times B}}{DF_{A \times B}}$	$\dfrac{MS_{A \times B}}{MS_{residual}}$	$\dfrac{MS_{A \times B}}{MS_{residual}}$	$\dfrac{MS_{A \times B}}{MS_{residual}}$
Residual (within-samples)	$SS_{residual}$	$DF_{residual}$	$\dfrac{SS_{residual}}{DF_{residual}}$			
Total	SS_{total}	DF_{total}				

[a]Mixed effects model assuming factor A levels are fixed and factor B levels are random.

The final stages in the calculation of two-way ANOVA are presented in the two-way ANOVA table (Table 8.2).

Step 10: Determine the probability of significance

In order to determine the probability of statistical significance, the calculated F-ratio for both treatments and the interaction are compared in turn with the one-tailed critical value of F at the desired level of probability read from the table of F values (see Tables A5.1 and A5.2 in the Appendix). The F probability table is consulted at both the degrees of freedom of the mean square used as the numerator when determining the F-ratio (V_1) and the degrees of freedom associated with the denominator mean square (V_2). Therefore, for a fixed effects (Model I) ANOVA, V_1 will be the DF of the treatment or interaction and V_2 will be the residual DF. In a random or mixed effects ANOVA (Model II or Model III), V_2 will become the interaction DF for testing some of the treatments (see above). If the observed F-ratio for a particular treatment or the interaction is equal to or greater than the critical F value, then the null hypothesis may be rejected and a significant effect of that treatment or interaction may be claimed at the level of probability at which the test was performed.

A fully worked example of a two-way ANOVA for a typical fixed effects model is shown in Example 8.1.

Example 8.1. A two-way analysis of variance with replication (Model I: fixed effects ANOVA).

The effect of irrigation treatment in conjunction with plant density on growth of young birch trees.

In a two-way factorial experiment different levels of irrigation were applied to birch saplings grown at a series of plant densities. The girth of the trees was subsequently determined.

Plant density (no./m²)		Stem girth (cm)					
		Irrigation treatment field capacity					
		0% FC	33% FC	66% FC	100% FC	*Total*	*Mean*
1		15	16	21	21		
		13	20	19	19		
		17	19	23	14		
		11	15	22	17		
		12	21	26	18		
	Total	68.0	91.0	111.0	89.0	359	
	Mean	13.6	18.2	22.2	17.8		17.95
	± SE	1.1	1.2	1.2	1.2		
2		16	18	22	24		
		19	20	26	18		
		12	21	20	18		
		14	21	21	22		
		14	21	18	17		
	Total	75.0	101.0	107.0	99.0	382.0	
	Mean	15.0	20.2	21.4	19.8		19.10
	± SE	1.2	0.6	1.3	1.4		
5		10	13	16	15		
		9	10	19	15		
		13	14	20	15		
		10	11	23	22		
		9	15	16	16		
	Total	51.0	63.0	94.0	83.0	291	
	Mean	10.2	12.6	18.8	16.6		14.55
	± SE	0.7	0.9	1.3	1.4		
Total		194	255	312	271	1032	
Mean		12.93	17.00	20.80	18.07		

$n = 5$, $N_{irr} = 4$, $N_{den} = 3$, $k = 12$, $n_{irr} = 15$, $n_{den} = 20$, $N = 60$

Null hypotheses: there is no effect of the irrigation treatments on the girth of *Prunus* trees; there is no effect of the plant density treatments on the girth of *Prunus* trees; there is no irrigation × plant density interaction effect on the girth of *Prunus* trees.

The analysis
$\Sigma x^2 = 18814$, $\quad \Sigma T^2 = 92518$, $\quad \Sigma T_{irr}^2 = 273446$, $\quad \Sigma T_{den}^2 = 359486$
$CT = (\Sigma x)^2/N = 1032^2/60 = 17750.40$

continued

Example 8.1. Continued.

Calculation of sum of squares

$$SS_{total} = \Sigma x^2 - CT = 18814 - 17750.40 = 1063.60$$

$$\text{with DF} = N - 1 = 35$$

$$SS_{between\text{-}samples} = \left(\frac{\Sigma T^2}{n}\right) - CT = (92518 / 5) - 17750.40 = 753.20$$

$$SS_{irrigation} = \left(\frac{\Sigma T_{irr}^2}{n_{irr}}\right) - CT = (273446 / 15) - 17750.40 = 479.33$$

$$\text{with DF} = N_{irr} - 1 = 3$$

$$SS_{density} = \left(\frac{\Sigma T_{den}^2}{n_{den}}\right) - CT = (359486 / 20) - 17750.40 = 223.90$$

$$\text{with DF} = N_{den} - 1 = 2$$

$$SS_{irrigation \times density} = SS_{between\text{-}samples} - (SS_{irrigation} + SS_{density})$$
$$= 753.20 - (479.33 + 223.90) = 49.97$$
$$\text{with DF} = (N_{irr} - 1) \times (N_{den} - 1) = 6$$

$$SS_{residual} = SS_{total} - SS_{between\text{-}samples} = 1063.60 - 753.20 = 310.40$$
$$\text{with DF} - k \times (n - 1) = 48$$

ANOVA Table

Source of variation	SS	DF	MS	F-ratio	F_{crit} $\alpha = 0.05$	$\alpha = 0.01$	Sig.
Irrigation treatment	479.33	3 (V_1)	159.78	24.70	2.80	4.23	$P < 0.01$
Density treatment	223.90	2 (V_1)	111.95	17.30	3.19	5.09	$P < 0.01$
Irrigation × density	49.97	6 (V_1)	8.33	1.29	2.30	3.22	n.s.
Residual	310.40	48 (V_2)	6.47				
Total	1162.93	59					

continued

Example 8.1. Continued.

Both the irrigation treatment F-ratio and the density treatment F-ratio > 1% critical F values; therefore the null hypotheses concerning these treatments are rejected. The irrigation × density interaction F-ratio < 5% critical F value; therefore the null hypothesis concerning the interaction is accepted.

Conclusion: there is a significant irrigation effect present ($P < 0.01$); there is a significant plant density effect present ($P < 0.01$); there is no significant irrigation × density interaction effect present.

8.4 Two-way ANOVA Without Replication

It is, of course, unusual to conduct a two-way factorial experiment without including any replication; however, when there is limited experimental material available this may become necessary. Without replication treatment interaction cannot be detected, therefore the ANOVA is only concerned with three possible sources of variation, namely the two treatment factors, detected as the treatment variances, and natural random variation, detected as the residual variance. To allow both treatment variances to be determined, the totals for both treatments have to be calculated separately and a typical data table will therefore take a form similar to that shown in Table 8.3.

Table 8.3. Typical format for the data table for a two-way factorial experiment without replication.

Treatment factor B level	Treatment factor A level				Treatment B Totals
	1	2	3	n	
1	$x_{1,1}$	$x_{1,2}$	$x_{1,3}$	$x_{1,n}$	T_{B1}
2	$x_{2,1}$	$x_{2,2}$	$x_{2,3}$	$x_{2,n}$	T_{B2}
3	$x_{3,1}$	$x_{3,2}$	$x_{3,3}$	$x_{3,n}$	T_{B3}
n	$x_{n,1}$	$x_{n,2}$	$x_{n,3}$	$x_{n,n}$	T_{Bn}
Treatment A totals	$T_{A,1}$	$T_{A,2}$	$T_{A,3}$	$T_{A,n}$	Σx (Grand total)

The following terms are defined:

N_A = no. of levels of application of treatment A
N_B = no. of levels of application of treatment B
N = total no. of observations ($= N_A \times N_B$)

The calculation of the two-way ANOVA without replication can now proceed as follows:

Step 1: Calculation of the *correction term* (CT)

$$CT = \frac{(\Sigma x)^2}{N}$$

Step 2: Calculation of the *total* sum of squares

$$SS_{total} = \Sigma(x - \bar{x})^2 = \Sigma x^2 - CT$$

Step 3: Calculation of the treatment sum of squares

Where two treatment factors A and B are applied, the treatment sum of square values (SS_A and SS_B) are calculated by:

$$SS_A = \Sigma\left(\bar{T}_A - GM\right)^2 = \left(\frac{\Sigma T_A^2}{n_A}\right) - CT \text{ and } SS_B = \Sigma\left(\bar{T}_A - GM\right)^2 = \left(\frac{\Sigma T_B^2}{n_B}\right) - CT$$

where: ΣT_A^2 = sum of the squares of treatment A totals;
ΣT_B^2 = sum of the squares of treatment B totals;
n_A = no. of observations made for each level of treatment A (= N_B);
n_B = no. of observations made for each level of treatment B (= N_A).

Step 4: Calculation of the residual sum of squares (i.e. the *SS* due to random variation)

Since the treatment SS and the residual SS must sum to give the total SS, the residual SS can now be obtained by simple subtraction, i.e.:

$$SS_{residual} = SS_{total} - (SS_A + SS_B)$$

Step 5: Determination of the degrees of freedom (DF) for each source of variation

Total DF $= N - 1$ (total no. of observations less 1)

Treatment A DF $= N_A - 1$ (total no. of levels of treatment A less 1)

Treatment B DF $= N_B - 1$ (total no. of levels of treatment B less 1)

Residual DF $= (n_A - 1) \times (n_B - 1)$ (this is equivalent to the total DF minus both treatment DF)

Step 6: Calculation of the mean square (variance) for each source of variation

The mean square for each source of variation is calculated by dividing each SS value by its respective degrees of freedom:

Treatment A MS $= SS_{treatment\ A}/DF_{treatment\ A}$

Treatment B MS $= SS_{treatment\ B}/DF_{treatment\ B}$

Residual MS $= SS_{residual}/DF_{residual}$

Analysis of Multiple-factorial Experiments

Step 7: Calculation of the test statistic F and presentation of the results in an ANOVA table

For a normal fixed effects Model I ANOVA, the F-ratio test statistic is obtained for each of the treatment factors by division of the treatment mean square by the residual mean square. If a Model II or Model III ANOVA is employed, then F for each treatment factor is determined as explained in section 8.3 ('Standard calculation procedure').

The final stages in the calculation are presented in a two-factor ANOVA table:

Table 8.4. Two-way ANOVA (without replication) table standard format.

Source of variation	Sum of squares (SS)	Degrees of freedom (DF)	Mean square (MS)	F-ratio[a]
Treatment factor A	$SS_{treatment\ A}$	$N_A - 1$	$SS_{treatment\ A}/DF_{treatment\ A}$	$\dfrac{MS_{treatment\ A}}{MS_{residual}}$
Treatment factor B	$SS_{treatment\ B}$	$N_B - 1$	$SS_{treatment\ B}/DF_{treatment\ B}$	$\dfrac{MS_{treatment\ B}}{MS_{residual}}$
Residual (within-sample)	$SS_{residual}$	$(n_a - 1) \times (n_b - 1)$	$SS_{residual}/DF_{residual}$	
Total	SS_{total}	$N - 1$		

[a]Fixed effects Model I ANOVA assumed.

Step 8: Comparison with the critical value of F and drawing a conclusion

In order to assess the significance of each treatment independently, their respective F-ratio values are compared with the critical value of F obtained from tables at the desired level of probability and at the treatment degrees of freedom (V_1) and the residual degrees of freedom (V_2). If the observed F-ratio for either treatment factor is greater or equal to the critical F value, then the null hypothesis may be rejected and a significant effect for that treatment factor may be claimed.

8.5 The General Linear Model (GLM) Approach to Analysis of Variance

A more sophisticated approach to analysing multiple-sample experiments by ANOVA is offered by employing a general linear model (GLM) to describe the separate effect of the treatment factor and the random variation on the sample mean. When analysing relatively simple single-factor experiments, the adoption of the GLM approach has little advantage over the standard technique for calculating ANOVA. However, the GLM technique does have major advantages when analysing multiple-factor experiments, i.e., experiments that involve more than one treatment factor applied to the same subjects at the same time, and is increasingly becoming the basis of computer statistical software packages. Consequently, the concept of general linear modelling will now be introduced. It should be noted that the GLM technique is derived from the multiple regression models which describe the relationship between a measured variable, e.g crop yield, and a number of independent 'predictor' variables, e.g. fertilizer concentration, soil type, variety, etc. Therefore, if the reader is unfamiliar with the concept of linear regression analysis, it may help to read Chapter 12 first, which covers this subject, before returning to this section.

To begin with, it is necessary to understand what the term 'linear' means in the context of general linear models. The term linear does not imply that there is necessarily a strict linear relationship between the variable measured and the level of the treatment factor. In other words, a graph of the measured values of the variable plotted against the level of the treatment does not need to produce a straight line but could equally well produce a complex curve. Instead the term 'linear' implies that the value of the measured variable can be predicted by a linear equation which contains a number of components that are added together. These components include a quantitative expression of the treatment effect and a quantitative expression of the random error effect so that a linear effects model for a single factor ANOVA has the general form:

$$y = \mu + \alpha + \varepsilon$$

where: y = the measured value of a variable under investigation;
 μ = the population mean value;
 α = the quantitative effect on the variable caused by the level of treatment applied;
 ε = the quantitative effect on the variable caused by random (error) effects

Since μ by definition is constant across the whole experiment, it is the effects of the treatment (α) and the random error (ε) that operate together to determine each replicated y value in a sample; thus the F test statistic in the ANOVA becomes the ratio of the amount of systematic variation present in the treatment effects (α) to the amount of random variation (ε). Since μ, the set of α terms and the variance of the ε terms all describe the behaviour of the population, their values cannot be measured directly but can only be estimated from the sample data. Thus when ANOVA is calculated by the standard partitioning of sum of squares technique described in the previous section, μ is estimated by the grand mean, variation in α is estimated by the treatment mean square and variation in ε by the residual mean square. In order to utilize these estimates, it is then necessary to assume that the samples come from normally distributed populations that have equal variances.

Now consider an experiment in which two treatment factors are applied simultaneously, e.g. the effect on crop yield of fertilizer concentration and soil type. The single-factor linear model described above can be extended fairly simply to describe these further factors and their interaction:

$$y = \mu + \alpha + \beta + (\alpha \times \beta) + \varepsilon$$

where: y = the measured value of a variable under investigation;
   μ = the population mean of the variable;
   α = the quantitative effect on the variable caused by the level of treatment factor α;
   β = the quantitative effect on the variable caused by the level of treatment factor β;
   $(\alpha \times \beta)$ = the quantitative effect on the variable caused by the interaction of treatment factor α and treatment factor β;
   ε = the quantitative effect on the variable caused by random (error) effects.

In section 8.3, it was shown how, as long as the experimental design remains balanced (i.e. there is equal replication across samples), the standard ANOVA technique can be extended to analyse two (or more) factors. Where, however, samples are of unequal size, either planned or due to missing data, and the experimental design becomes complex, then GLM offers a more flexible and efficient approach. Furthermore, the general linear model also allows for the analysis of the simultaneous effect of both continuous and categorical treatment factors upon a responding variable. For example, the yield of a crop might be compared between a number of varieties (i.e. a fixed categorical factor) grown at locations with different rainfall (i.e. a random continuous factor).

The general linear model for a multiple-factor experiment is based on multiple linear regression models that take the general form:

$$Y = \beta_0 + \beta_1 X_1 + \beta_2 X_2 + \ldots + \beta_k X_k + \varepsilon$$

where: Y = the dependent response variable;
   X_i = the level of the ith treatment factor;
   β_0 = a constant referred to as the 'intercept' (equivalent to μ in the linear model above);
   β_i = a constant referred to as the regression coefficient associated with the ith independent treatment factor;
   k = the number of treatment factors;
   ε = a random error term associated with the measurement of y for a given set of treatment factors.

In the context of a multiple-sample experiment, β_1 to β_k represent the contributions that each treatment factor makes to the value of the measured variable Y. There are some major limitations to the multiple regression model, however, which restrict its use in analysing multiple-sample experiments, namely that the responding measured Y variable must be linearly dependent on the X variables and the independent X variables must not be correlated to each other. These restrictions can be overcome, however, by the use of matrix mathematics, which can transform the multiple regression model into the so-called **general linear model**. Solving ANOVA using a GLM approach remains subject to the same assumptions of adherence of the data to a normal distribution and equality of sample variances that are demanded by the standard ANOVA calculations; however, even these restrictions can be lifted by further complex mathematical treatment to form the so-called **generalized linear**

model. For further explanation of the use of general and generalized linear models in statistical data analysis, the reader is referred to Quinn and Keough (2002) and StatSoft (2006).

Since it is almost certain that anyone applying the GLM procedure to multiple-sample analysis will undertake this using a statistics software package, a worked example of the matrix calculations involved in calculating a GLM is not provided here (in any case this would take up a number of pages!). In the future it is almost certain that researchers will need to become increasingly familiar with the GLM approach and it has in fact been convincingly argued that running general linear models using computer packages will become, sooner rather than later, the expected approach for conducting analysis of variance.

8.6 Multiple Comparisons between Samples in a Two-way Factorial Experiment

ANOVA tests establish the existence of significant differences but they do not locate precisely between which treatment samples the differences occur. As in the case of single-factor analysis, two-factor analysis can be extended by undertaking a multiple comparison test in order to determine the location of significant differences. This can be achieved by a number of possible tests, the most usual being the **least significant difference (LSD) test**, the **Tukey HSD test** and the **Student–Newman–Keuls (SNK) test**. The calculation of these involves essentially the same procedure as that used when following one-way ANOVA (see Chapter 7, section 7.5) and is based on the standard error of the difference between means (SED). In the case of multiple-factorial experiments, the application of such tests does, however, have to take very careful account of the experimental design employed and the outcome of the ANOVA.

First, if one or other of the factors has been detected by ANOVA to cause a significant main treatment effect, then a multiple comparison test may be applied to the treatment means (rather then the sample means) in order to establish between which levels of the treatment significant differences are present. If, however, the ANOVA reveals significant interaction to be present, then the main treatment factors cannot be inspected independently in this way and the multiple comparison is aimed at seeking significant differences between the sample means across the main treatments.

Secondly, it is only useful to apply multiple comparison tests to treatments that were applied at fixed levels. For treatments inspected at random levels in a random effects (Model II) ANOVA or within a mixed effects (Model III) ANOVA, the only interest is in whether there is an effect of the main treatment overall. Whether a treatment produces a significant effect overall is shown by the ANOVA and making multiple comparisons between randomized levels of a treatment becomes inconsequential.

Thirdly it should be noted that a practical limitation to the computation of multiple comparison tests (e.g. LSD and Tukey HSD tests, etc.) following two-way ANOVA is that all samples must have equal size, i.e. n must be constant.

Finally it must be reiterated that it is completely invalid to use such tests unless the ANOVA has established the existence of significant effects in the first place.

8.6.1 LSD test to inspect for significant differences between treatment means following two-way ANOVA

An example of an LSD test following a two-way ANOVA is shown in Example 8.2. Remember that an LSD test should only be used to compare the means of a limited number of sample pairs within the experiment and these should be selected before the experimental data are analysed.

Example 8.2. Calculation of an LSD multiple comparison test for a two-factor analysis.

The effect of irrigation treatment in combination with plant density treatment on growth of young birch trees. Data from Example 8.1.

$MS_{residual}$ (residual mean square read from the ANOVA Table in Example 8.1)
$$= 6.47 \ (DF = 48)$$
t (from Student's t table; $\alpha = 0.05$; residual DF = 48) $\quad = 2.011$
n (no. of replicates per sample) $\quad = 5$

Since samples to be compared are of equal size LSD can be calculated by:

$$LSD_{(\alpha=0.05)} = t \times \sqrt{\frac{2 \times MS_{resid}}{n}} \quad = 2.011 \times \sqrt{\frac{2 \times 6.47}{5}} \quad = 3.24$$

Conclusion: where the observed difference between a **preselected** pair of sample means is ≥ 3.24, then the difference is significant ($P < 0.05$).

8.6.2 Tukey HSD test to inspect for significant differences between treatment means following two-way ANOVA

In contrast to the LSD test, the Tukey HSD test accounts for the number of samples in the experiment and potentially allows all pairs of treatment means to be compared. While the Tukey test is less likely to find a significant difference between a specific pair of sample means than the LSD test, it is less likely to produce Type 1 statistical errors and for this reason it may be preferred. Following two-way ANOVA, the Tukey HSD test may be performed to inspect the location of significant differences between levels of whichever treatments have been revealed to exert significant effects. For example, the two-way ANOVA in Example 8.1 indicated that an irrigation treatment produced a significant effect on birch stem girth. In this case the Tukey HSD test would be used to identify between which levels of irrigation significant differences were present. Since plant density also produced a significant effect, the procedure would be repeated to establish between which densities there were significant differences. The irrigation × density interaction was, however, non-significant and comparisons between samples should not therefore be further pursued. The estimation of the standard error of the differences (*Tukey*SED) is obtained in the same manner as for a one-factor analysis except that n becomes the total number of observations made for each level of the treatment being tested irrespective of the second factor. The critical Tukey q value is read from tables (Table A7 in the Appendix) for the number of treatment means being compared (N_t) and the residual degrees of freedom. The product of $q_{critical}$ and the *Tukey*SED gives the minimum critical difference (MCD), which can then be used to compare the means for the levels of the treatment factor in question. This is illustrated in Example 8.3.

Example 8.3. Calculation of a Tukey LSD multiple comparison test for a two-factor analysis.

The effect of irrigation treatment in combination with plant density treatment on growth of young birch trees. Data from Example 8.1.

(a) Test the differences between the means for irrigation treatments
$MS_{residual}$ (residual mean square from the ANOVA Table) $\qquad$ = 6.47 $\quad$ (DF = 48)
$n_{irrigation}$ (total no. of observations for each level of irrigation) $\quad$ = 15

$$TukeySED = \sqrt{\frac{MS_{resid}}{n_{irr}}} = \sqrt{\frac{6.47}{15}} \qquad = 0.66$$

Critical $q_{(\alpha = 0.05)}$ (read from tables at $k = 4$, residual DF = 48) $\quad$ = 3.77
Critical minimum difference = 3.77 × 0.66 $\qquad\qquad\qquad$ = 2.47

Conclusion: where the observed difference between any pair of irrigation treatment means is ≥2.47, then the difference is significant ($P < 0.05$).

Irrigation levels to be compared	Difference between means	Tukey critical minimum difference ($\alpha = 0.05$)	Significance
0%–33%	4.07	2.47	$P < 0.05$
0%–66%	7.87	2.47	$P < 0.05$
0%–100%	5.14	2.47	$P < 0.05$
33%–67%	3.80	2.47	$P < 0.05$
33%–100%	1.07	2.47	n.s.
67%–100%	2.73	2.47	$P < 0.05$

(b) Test the differences between the means for plant density treatments
$MS_{residual}$ (residual mean square from the ANOVA Table) $\qquad$ = 6.47 (DF = 48)
$n_{density}$ (total no. of observations for each level of density) $\qquad$ = 20

$$TukeySED = \sqrt{\frac{MS_{resid}}{n_{density}}} = \sqrt{\frac{6.47}{20}} \qquad = 0.57$$

Critical $q_{(\alpha = 0.05)}$ (read from tables at $k = 3$, residual DF = 48) $\quad$ = 3.42
Critical minimum difference = 3.42 × 0.62 $\qquad\qquad\qquad$ = 1.95

continued

Example 8.3. Continued.

Conclusion: where the observed difference between any pair of density treatment means is ≥2.12, then the difference is significant ($P < 0.05$).

Densities (no./m²) to be compared	Difference between means	Tukey critical minimum difference ($\alpha = 0.05$)	Significance
1–2	1.15	1.95	n.s.
1–5	3.40	1.95	$P < 0.05$
2–5	4.55	1.95	$P < 0.05$

In cases where the interaction has been shown to be significant, it is not appropriate to conduct a multiple comparison test between levels of each of the main treatment factors. Under these circumstances, however, all the sample means may be compared across the main treatments by calculating an SED as the square root of the residual mean square divided by the sample size, n.

8.7   Three-way Analysis of Variance

Three-way factorial experimental designs are much less commonly employed than one- and two-way designs because of their much greater demand on resources. Where a three-way factorial design is required, their analysis by ANOVA is in principle a logical extension of two-way ANOVA and today computers can readily handle the large number of calculations involved. Since, however, there is now an extensive array of possible interactions between the three treatment factors, the major problem faced by the investigator will generally be one of interpretation of the outcome rather than calculating the statistics. The number of interactions involved becomes clear when the linear model is developed to describe a three-way factorial design. The general linear model becomes:

$$y = \mu + \alpha + \beta + \gamma + (\alpha \times \beta) + (\alpha \times \gamma) + (\beta \times \gamma) + (\alpha \times \beta \times \gamma) + \varepsilon$$

where:

y = the measured value of a variable under investigation;

μ = the population mean of the variable;

α = the quantitative effect on the variable caused by the level of treatment factor α;

β = the quantitative effect on the variable caused by the level of treatment factor β;

γ = the quantitative effect on the variable caused by the level of treatment factor γ;

$\alpha \times \beta \times \gamma$ = the quantitative effect on the variable caused by the interaction of treatment factor α, treatment factor β and treatment factor γ;

ε = the quantitative effect on the variable caused by random (error) effects.

Consequently there is now an increase in the number of null hypotheses to be tested:

$H_{0\,(1)}$ = there is no effect of treatment A

$H_{0\,(2)}$ = there is no effect of treatment B

$H_{0\,(3)}$ = there is no effect of treatment C

$H_{0\,(4)}$ = there is no effect of treatment A×B interaction

$H_{0\,(5)}$ = there is no effect of treatment A×C interaction

$H_{0\,(6)}$ = there is no effect of treatment B×C interaction

$H_{0\,(7)}$ = there is no effect of treatment A×B×C interaction

The procedure for partitioning the sum of squares for a three-way ANOVA follows the same steps as for a two-way ANOVA, albeit there are now four rather than just a single interaction term to be determined. For a fixed effects Model I ANOVA, the *F* test statistic for each main treatment and for each interaction will be the ratio of the treatment or interaction MS to the residual MS. (In theory, random and mixed effects three-way ANOVA models are possible but determination of the appropriate *F*-ratios is then highly complex and, since these are so unlikely to be encountered in practice, they really need no further consideration here.) The final ANOVA table will thus take the form shown in Table 8.5.

Table 8.5. Three-way ANOVA table standard format.

Source of variation	Sum of squares (SS)	Degrees of freedom (DF)	Mean Square (MS)	*F*-ratio
Treatment factor A	$SS_{\text{treatment A}}$	$DF_{\text{treatment A}}$	$\dfrac{SS_{\text{treatment A}}}{DF_{\text{treatment A}}}$	$\dfrac{MS_{\text{treatment A}}}{MS_{\text{residual}}}$
Treatment factor B	$SS_{\text{treatment B}}$	$DF_{\text{treatment B}}$	$\dfrac{SS_{\text{treatment B}}}{DF_{\text{treatment B}}}$	$\dfrac{MS_{\text{treatment B}}}{MS_{\text{residual}}}$
Treatment factor C	$SS_{\text{treatment C}}$	$DF_{\text{treatment C}}$	$\dfrac{SS_{\text{treatment C}}}{DF_{\text{treatment C}}}$	$\dfrac{MS_{\text{treatment C}}}{MS_{\text{residual}}}$
A × B interaction	$SS_{\text{A×B}}$	$DF_{\text{A×B}}$	$\dfrac{SS_{\text{A×B}}}{DF_{\text{A×B}}}$	$\dfrac{MS_{\text{A×B}}}{MS_{\text{residual}}}$
A × C interaction	$SS_{\text{A×C}}$	$DF_{\text{A×C}}$	$\dfrac{SS_{\text{A×C}}}{DF_{\text{A×C}}}$	$\dfrac{MS_{\text{A×C}}}{MS_{\text{residual}}}$
B × C interaction	$SS_{\text{B×C}}$	$DF_{\text{B×C}}$	$\dfrac{SS_{\text{B×C}}}{DF_{\text{B×C}}}$	$\dfrac{MS_{\text{B×C}}}{MS_{\text{residual}}}$
A × B × C interaction	$SS_{\text{A×B×C}}$	$DF_{\text{A×B×C}}$	$\dfrac{SS_{\text{A×B×C}}}{DF_{\text{A×B×C}}}$	$\dfrac{MS_{\text{A×B×C}}}{MS_{\text{residual}}}$
Residual	SS_{residual}	DF_{residual}	$\dfrac{SS_{\text{residual}}}{DF_{\text{residual}}}$	
Total	SS_{total}	DF_{total}		

The F-ratios are then employed to test each null hypothesis independently in the normal way.

Since the modern researcher employing three-way ANOVA is most unlikely to undertake the analysis without the aid of computer statistics software, a fully worked example of the analysis is not provided here. Example 8.4 does, however, illustrate the use of a three-way ANOVA where the analysis has been generated by a computer program.

While the significance of the possible interaction effects within a three-way factorial design may be determined fairly straightforwardly, understanding the meaning of the interactions in the context of the experiment performed may be an entirely different matter. If, for example, three treatments A, B, C, were to show no interactions when considered in pairs but displayed a significant interaction when all three were combined – and this is an entirely possible outcome of a three-way ANOVA – then it becomes very problematic to place a meaningful interpretation upon this. A similarly difficult interpretation arises in Example 8.4, where a significant interaction is found between two treatments (irrigation and mulch) but when considered with a third factor (plant density) in a three-way interaction the effect is not significant. It is recommended, therefore, that the researcher should very carefully consider in advance the appropriateness of employing three-way factorial designs and the meanings that might be attached to the possible alternative outcomes. Unless it is absolutely necessary to resolve all the combinations of interactions between three factors simultaneously, it may often be more profitable in the long run to perform a series of two-factor trials to compare the factors in paired combinations.

Example 8.4. A three-way analysis of variance with replication (Model I: fixed effects ANOVA).

The effect of different types of mulch in combination with different irrigation levels and different plant densities on growth of young birch trees.

In a three-way factorial experiment, three different types of mulch and four different levels of irrigation were applied to birch saplings grown at three different plant densities. The girth of the trees was subsequently determined.

					Stem girth (cm)							
		No mulch				Bark mulch				Plastic mulch		
		Irrigation (% FC)				Irrigation (% FC)				Irrigation (% FC)		
Plant density	0	33	66	100	0	33	66	100	0	33	66	100
1	15	16	21	21	18	18	23	21	20	19	20	21
	13	20	19	19	17	20	26	18	16	23	22	18
	17	19	23	14	18	27	27	16	16	26	25	9
	11	15	22	17	12	15	24	17	14	19	16	17
	12	21	26	18	20	24	28	20	22	20	26	18
$\bar{x}$	13.60	18.20	22.20	17.80	17.00	20.80	25.60	18.40	17.60	21.40	21.80	16.60

continued

Example 8.4. Continued.

2	16	18	22	24	15	20	25	22	15	23	18	18
	19	20	26	18	22	23	26	15	22	18	21	14
	12	21	20	18	14	27	19	13	18	27	23	13
	14	21	21	22	15	24	27	20	19	26	20	20
	14	21	18	17	13	21	20	16	24	26	19	15
$\bar{x}$	15.00	20.20	21.40	19.80	15.80	23.00	23.40	17.20	19.60	24.00	20.20	16.00
5	10	13	16	15	10	17	19	14	11	19	16	14
	9	10	19	15	14	17	27	16	20	17	22	18
	13	14	20	15	23	24	24	12	22	20	21	11
	10	11	23	22	18	15	28	12	21	17	16	10
	9	15	16	16	18	16	22	12	23	17	20	12
$\bar{x}$	10.20	12.60	18.80	16.60	16.60	17.80	24.00	13.20	19.40	18.00	19.00	13.00

Table of treatment means.

Irrigation (% FC)	**0**	**33**	**66**	
	16.06	19.56	21.82	
Density (no./m²)	**1**	**2**	**5**	
	19.25	19.63	16.60	
Mulch	**None**	**Bark**	**Plastic**	
	17.20	19.40	18.88	
Density (no./m²) × Irrigation (% FC)	**0**	**33**	**66**	**100**
1	16.07	20.13	23.20	17.60
2	16.80	22.40	21.67	17.67
5	15.40	16.13	20.60	14.27
Density (no./m²) × Mulch	**None**	**Bark**	**Plastic**	
1	17.95	20.45	19.35	
2	19.10	19.85	19.95	
5	14.55	17.90	17.35	
Irrigation (% FC) × Mulch	**None**	**Bark**	**Plastic**	
0	12.93	16.47	18.87	
33	17.00	20.53	21.13	
66	20.80	24.33	20.33	
100	18.07	16.27	15.20	
Density (no./m²) × Irrigation (% FC) × Mulch	Means given in data table			

ANOVA Table.

Source of variation	SS	DF	MS	F-ratio	P
Irrigation	986.417	3	328.806	33.23	< 0.001
Density	327.411	2	163.706	16.55	< 0.001
Mulch	158.811	2	79.406	8.03	< 0.001
Irrigation × density	153.967	6	25.661	2.59	0.020
Irrigation × mulch	464.567	6	77.428	7.83	< 0.001
Density × mulch	41.722	4	10.431	1.05	0.382

continued

Example 8.4. Continued.

Source of variation	SS	DF	MS	F-ratio	P
Irrigation × density × mulch	69.300	12	5.775	0.58	0.853
Residual	1424.800	144	9.894		
Total	3626.994	179			

Least significant differences of means (5% level).

Irrigation	1.311
Density	1.135
Mulch	1.135
Irrigation × density	2.270
Irrigation × mulch	2.270
Density × mulch	1.966
Irrigation × density × mulch	3.932

Conclusion: there is a significant difference between mulch treatments ($P < 0.001$); there is a significant difference between irrigation treatments ($P < 0.001$); there is a significant difference between density treatments ($P < 0.001$); there is a significant irrigation × mulch interaction effect ($P < 0.001$); all other interactions are non-significant.

8.8 Missing Data and Unequal Sample Replication

As previously stated a full multiple-factor ANOVA can only be performed satisfactorily when there are an equal number of replicates in each treatment sample. Unfortunately, as all experimental investigators know, data are frequently lost during the course of an experiment. Unless some compensation for the missing data can be made, it will not then be possible to employ ANOVA to analyse the experimental results.

One possible solution is to randomly remove items from other treatment samples in the analysis until the number of items in each sample is the same. As long as data items are independent of each other and items are removed absolutely randomly, this is probably the most valid procedure available. However, it will be successful only if the replication level is sufficiently high to permit a decrease in the sample degrees of freedom and of course it does involve discarding what otherwise might have been useful data.

When a single or a small number of items are lost during the course of an experiment, an obvious approach is to replace missing data items by a value equal to the mean of the remaining items in the sample. The problem here is that in doing this the variance of the sample will be reduced and this will therefore bias the outcome of the ANOVA test. Such an approach must therefore be considered only when the replication level is high and, to compensate for the increase in bias, the residual degrees of freedom and the appropriate treatment degrees of freedom must both be reduced by the number of missing items that have been replaced. Obviously the original data table must clearly show where data items have been replaced by means. Overall this is not a method to be recommended.

The general linear modelling (GLM) approach to ANOVA does, however, provide a technique for coping with missing data within a multiple-factorial designed experiment. The concept of applying a GLM approach to analysis of variance was described in section 8.5. The matrix mathematics involved are complex but are readily handled by the more sophisticated statistical computer packages available today.

Occasionally the problem is not that data are lost during the course of the experiment but that there is insufficient material available at the outset to permit the same number of replicates in every treatment sample. The only solution to this problem is through use of special types of experimental design. Experimental designs are possible in which the experimental material is proportioned so that, while sample sizes may not be equal, the variation in sample sizes is systematic. Specialized ANOVA techniques are then available to analyse such experiments. The next chapter will discuss the design of multiple-factorial experiments in the context of performing analysis of variance and will describe some of these more specialized designs.

8.9 Analysing Multiple-factor Experiments when the Assumptions Required by ANOVA are Invalid

If data samples cannot be assumed to adhere to a normal distribution and/or they do not possess statistically equal variances, then parametric ANOVA cannot, in theory, be employed. (Strictly speaking, it is the populations from which the samples are drawn that must conform to these assumptions but the populations can only be assessed by using the sample statistics as population estimates.) In Chapter 10 it will be explained that in the case of simple multiple-sample designs the non-parametric Kruskal–Wallis test provides a non-parametric alternative to one-way ANOVA but unfortunately there is no extension of this test, or any other non-parametric test, that satisfactorily handles multiple-factor designs where treatment interaction is involved. If interactions are not considered important, then the factorial experiment can be treated as a randomized multiple-sample design and the Kruskal–Wallis test applied to simply determine if significant differences exist between different combinations of treatments. Another possible approach is to perform a **rank transformation** of the data values, that is, to rank all the values in the entire factorial experiment, and then perform a standard two-way ANOVA on the ranked values. Rank transformation differs from those types of transformation that depend on applying a constant mathematical function (e.g. $\log_{10}$, arcsin, etc.) in that there is no clear quantitative relationship between the magnitude of the original values and their rank values. For this reason, while rank transformation may be successful in allowing significant main treatment effects to be identified, it does not allow the safe identification of interactions. Furthermore, rank transformation will not necessarily guard against unequal sample variances, since if the sample variances are not homogeneous then the ranks may also be non-homogeneous. Therefore, a test of the equality of sample variances is necessary for rank-transformed data prior to ANOVA just as it is for untransformed data. It is not surprising, therefore, that this is another controversial area in statistics and not all statisticians agree on the validity or usefulness of performing ANOVA on rank-transformed data.

In conclusion, if the data from a factorial experiment do not adhere to the assumptions of ANOVA but it is required to analyse for interactions in addition to main treatment effects, then the soundest approach is to perform a mathematical data transformation to convert the data to a close approximation of the normal distribution, as discussed previously (see Chapter 3; section 3.3.4). Two-way ANOVA is then performed on the transformed values. Even if transformation fails to produce perfect normality or homogeneity of sample variances, the ANOVA test is still fairly robust and is still able to produce a reliable outcome as long as discrepancies from these assumptions are not too large.

Design and Analysis of More Complex Factorial Experiments

Design: contrive, plan, make preliminary sketch of, draw plan of, construct the groundwork or plot of ...

(Concise Oxford English Dictionary)

- Introduction to the design of complex multiple-factor experiments.
- Randomized block experimental designs and their analysis.
- Latin square experimental designs and their analysis.
- Split-plot experimental designs and their analysis.
- Nested (or hierarchical) experimental designs and their analysis.
- Repeated-measures experimental designs and their analysis.

9.1 Introduction to the Design of Factorial Experiments

We have considered previously the use of samples and the importance of using valid sample selection procedures when designing experiments. It is obvious that if a sample is to accurately represent the population from which it is drawn it must be both as large and as unbiased as possible. When statistically analysing factorial experiments the ideal situation is where samples are selected totally randomly. Frequently, however, complete randomization in sample selection is either not valid or not possible. This is especially likely in field experiments where experimental conditions are not uniform. Under such circumstances a more systematic approach to experimental design and sample selection is required that deliberately distributes the effects of non-uniformity evenly across the experiment. The major way this is achieved is to employ so-called **blocking** procedures and to modify the mode of statistical analysis to account for the curtailing of randomization that block designed experiments involve. In this chapter the design and analysis of **randomized block experiments** will be considered in some detail. Once the design and analysis of randomized block designs are understood, more systematic designs such as **split-plot**, **nested** (or hierarchical) and **Latin square** designs can be developed that in turn handle more complex experimental situations.

A further assumption that has been made in the previous discussion of factorial experiments is that the samples are independent, that is, that each sample of data arises from a different set of measured subjects. This may not be the case, however, when either there is a limited amount of experimental material, requiring multiple measurements to be made on the same material, or the nature of the experimental design actually demands repeat measurements to be made. For example, we may wish

---

to examine the effect on milk yield in a single sample of cows with changes in diet or the change in respiration rate over time of a sample of fruit in storage. This type of experiment is referred to as a **repeated-measures design** and requires a modified approach to the subsequent data analysis.

9.2 Randomized Block Designs

Full randomization in treatment allocation and sample selection is not valid when a systematic bias or error runs through the experimental material. For example, consider the plan of a farm field shown in Fig. 9.1 in which a crop trial is to take place involving a yield comparison between four crop varieties. The plan shows not only the location of plants in the field but some external landscape features that may be expected to impinge on the trial site. Clearly each variety should not be sown in single plots in each quarter of the field since the external factors would then not affect each variety equally and true varietal differences would not be discernible. If, however, the varieties were sown in replicate plots that were located totally randomly across the field (by using, for example, a random number function on a computer to generate the plot positions) there would be nothing to prevent the majority of the plants of one of the varieties all occurring at one end of the field while the majority of another variety all occurred at the opposite end. The impact of the shading by the tall hedge, the drainage effect of the slope and the waterlogging effect of the stream would not then be equal on each variety and it would become likely that these effects would overshadow the effect of any true genotypic differences in yield. Furthermore, there may be other factors present that may not be so obviously apparent, such as patches in the field where the soil compaction changes, and again we need to ensure as far as possible that these affect all varieties equally. In such cases it is clear that the most valid design is one in which the replicate plots of each variety are systematically spread out across the field so that there is a reasonable expectation that the external factors affect each variety equally.

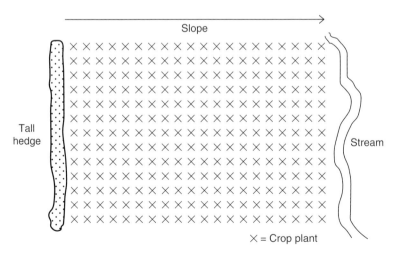

Fig. 9.1. Plan of an experimental field site which is likely to show systematic variation between replicates.

Treatments can be dispersed evenly across an experimental site by a process called 'blocking'. In a blocked design the experimental material is segregated into replicated plots, properly called 'blocks', and within each block the treatments are assigned to subjects totally randomly, thereby retaining a randomized component. However, by replicating the blocks a number of times right across the experiment it is ensured that all treatments are more or less equally influenced by any non-uniformity in the experimental conditions. Most commonly, each treatment is represented an equal number of times in each block and this is then termed a **randomized complete block (RCB)** design. Occasionally not all treatments may be represented in every block (although there will normally be equal replication of each treatment in the experiment as a whole), thus giving rise to an **incomplete block design**. This usually arises when there is insufficient experimental material available to allow for a complete block design.

As an example of a randomized complete block design, we will return to the crop variety trial described above. If there are four varieties, A, B, C and D, these may be conveniently sown in four plots within each block. Each plot within each block will give rise to a single result, although this result might be the mean of a number of sample measurements. For example, the yield of each of five plants per plot may be determined and the result for each plot then expressed as the mean yield per plant. The plots will be allocated within each block in a totally random way, e.g.:

C	D
B	A

In many cases it may be desirable to have replication of treatments within a block. For example, in a comparative growth trial of sheep breeds, blocks were represented by pens within a grazing field, each pen held four sheep of each of four breeds A, B, C, D, and the growth rate of each individual sheep was subsequently determined. Clearly the individual sheep within the pens are not located in a fixed position, as are plants in a crop trial, and on the basis that the sheep move around the pens in a random manner the pens may be considered as randomized blocks, e.g.:

B	A	A	D
D	B	C	B
C	A	D	C
C	B	D	A

Once the design of the individual blocks has been resolved, the blocks can then be distributed evenly across the experimental site. In this way the systematic external factors would potentially affect all the treatments equally. An example of a randomized complete block design is shown in Fig. 9.2.

9.2.1 The size, number and position of blocks

In an RCB design there is no requirement that the blocks themselves are located randomly within the experimental site; indeed, doing so would be likely to reduce their usefulness. Block designs only operate efficiently if the conditions within each block

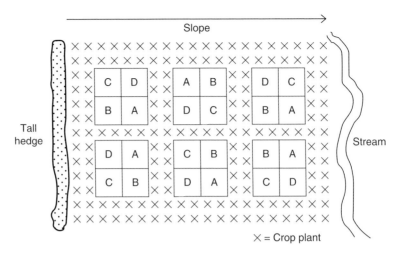

Fig. 9.2. Plan of an experimental crop field trial using a randomized complete block design. There are four main treatments A, B, C, D, which are allocated randomly to plots within six blocks. The blocks are then distributed across the field so that the effect of the hedge, stream and slope can be assumed to be equal on each of the four treatments.

are relatively uniform and the natural variation within blocks is smaller than that between blocks. The size and location of the blocks will be selected to achieve this and this therefore depends on some prior recognition by the investigator of the nature of the variation within the site. If there is a marked gradient in variation across the site of the experiment, then the blocks will need to be kept fairly small but can be dispersed regularly across the site. On the other hand, if it is recognized that variation in conditions is patchy across the site, then blocks will need to be deliberately located within homogeneous regions so that the within-block variation is as low as possible.

In deciding on the number of blocks to be included, a compromise needs to be reached between having too few blocks, rendering them inefficient in distributing the systematic error equally among treatments, and having more blocks than is necessary, leading to an inefficient use of resources. One important criterion is that when the design is statistically analysed by ANOVA there are sufficient residual degrees of freedom to enable the analysis to operate effectively. In an RCB design with one measurement made for each treatment in each block the normal recommendation is that there should be at least 12 residual degrees of freedom (i.e. $DF_{residual} \geq 12$). The residual degrees of freedom in this case are given by the product of the treatment DF and the block DF (i.e. $DF_{residual} = N_{treatment} - 1 \times N_{blocks} - 1$), so, for example, where there are four treatments the required number of blocks would be at least five.

9.2.2 Statistical analysis of randomized complete block designs by ANOVA

Randomized complete block (RCB) designed experiments can be analysed by the analysis of variance (ANOVA) technique, assuming that the samples are extracted from normally distributed populations and sample variances are homogeneous. The aim of the ANOVA is to separate the effects of external factors from the effects of the treatment factors. The test employed may be considered as a multiple-factorial ANOVA in which the blocks are considered as a source of variation. In the ANOVA the total sum of squares is therefore partitioned into the main treatment effects, the block effects and residual random effects. Since the blocks contain equal numbers of replicates of each treatment, it can be assumed that differences between blocks are due to external factors alone. A significant difference between blocks would therefore indicate the presence of significant non-uniformity and thereby justify the use of the randomized block design. Whether or not any interaction exists between blocks and treatment is of no interest and the block × interaction is not determined separately but is contained within the residual variation. The general linear model describing a randomized complete block design with a single treatment factor is therefore:

$$y = \mu + \alpha + blk + \varepsilon$$

where: y = the measured value of a variable under investigation;
μ = the population mean of the variable;
α = the quantitative effect on the variable caused by the level of the treatment factor α;
blk = the quantitative effect on the variable caused by blocking of treatments;
ε = the quantitative effect on the variable caused by random residual (error) effects.

To test the treatment effects the test statistic F is the ratio of the treatment variance to the residual variance, i.e.:

$$F\text{-ratio} = \frac{s^2_{treatment}}{s^2_{residual}}$$

The final ANOVA table for a randomized complete block design therefore takes the form shown in Table 9.1.

Table 9.1. RCB design ANOVA table standard format.

Source of variation	Sum of squares (SS)	Degrees of freedom (DF)	Mean Square (MS)	F-ratio
Block effect	SS_{block}	$N_{block} - 1$	SS_{block}/DF_{block}	$\dfrac{MS_{block}}{MS_{residual}}$
Treatment	$SS_{treatment}$	$N_{treatment} - 1$	$SS_{treatment}/DF_{treatment}$	$\dfrac{MS_{treatment}}{MS_{residual}}$
Residual	$SS_{residual}$	$(N_{blocks} - 1) \times (N_{treatment} - 1)$ (or by subtraction)	$SS_{residual}/DF_{residual}$	
Total	SS_{total}	$N - 1$		

Where N = total no. of measurements in whole experiment; $N_{treatment}$ = number of levels of application of treatment; N_{block} = number of blocks.

An illustration of the calculation of an ANOVA for a randomized complete block designed experiment is shown in Example 9.1, which is based on the experimental design illustrated in Fig. 9.2. Four potato varieties have been grown in a field arranged into six blocks, the varieties being randomly allocated to four plots within each block. At harvest, the individual tuber yield of a sample of individual plants from each plot was recorded and the mean yield (kg/plant) determined for each plot. It is required to test the null hypothesis that there are no yield differences between varieties.

First, the total sums of squares and associated degrees of freedom are determined and the total SS then partitioned between the block effects, the main treatment (i.e. variety) effects and the residual random effects. Any interaction between blocks and treatment is ignored and remains within the residual SS. The mean squares are then derived (MS = SS/DF) and presented in the normal way in an ANOVA table. The F-ratio test statistic is given by the $MS_{treatment}/MS_{residual}$ and the probability of significance subsequently determined by comparison of the F-ratio

---

Example 9.1. Analysis of variance of a randomized complete block experiment.

Field trial of yield in four varieties of potato.

The design: in order to control variation in the external conditions and in the field, a randomized complete block design was used. Six blocks located across the field were each divided into four plots. The four potato varieties, A, B, C, D, were allocated randomly to the plots (see Fig. 9.2).

The data

	Mean potato yield (kg/plant)					
	Variety plots				Block total	
Blocks	A	B	C	D	(T_{block})	Block mean
1	1.30	1.50	1.98	1.18	5.96	1.488
2	1.35	1.50	2.03	1.13	6.01	1.500
3	2.08	2.00	2.18	1.34	7.60	1.906
4	1.95	2.10	2.40	2.20	8.65	2.163
5	1.83	2.20	2.33	1.95	8.31	2.075
6	1.90	1.73	2.10	1.63	7.36	1.838
Variety total $(T_{treatment})$	10.41	11.03	13.02	9.43	$\Sigma x = 43.89$	
Variety mean	1.733	1.838	2.167	1.575		

No. observations (N) = 24; no. blocks (N_{block}) = 6; no. variety treatments $(N_{treatment})$ = 4

Total no. of measurements per treatment level $(n_{treatment})$ = 6

Total no. of measurements per block (n_{block}) = 4

Null hypothesis: potato yield is the same in the four variety of potatoes investigated.

continued

Example 9.1. Continued.

The analysis:

$\Sigma x^2 = 83.5001;$ $\Sigma T_{treatment}^2 = 488.4743;$ $\Sigma T_{block}^2 = 327.4499;$

$CT = (\Sigma x)^2/N = 80.2638$

The calculation procedure is now similar to that of a two-way ANOVA, treating the blocks as a source of variation; however, the block × treatment interaction is not calculated but is contained within the residual variation.

Sum of squares:

Total	$= \Sigma x^2 - CT$	$=$	$83.5001 - 80.2638$	$= 3.2363$
			with DF $= N - 1 = 23$	
Between blocks	$= \Sigma T_{block}^2/n_{block} - CT$	$=$	$(327.4499)/4 - 80.2638 = 1.5986$	
			with DF $= N_{block} - 1 = 5$	
Between treatment $= \Sigma T_{treat}^2/n_{treatment} - CT$		$=$	$(488.4743)/6 - 80.2638 = 1.1485$	
			with DF $= N_{var} - 1 = 3$	
Residual	$= SS_{total} - (SS_{treatment} + SS_{block})$			$= 0.4892$
		with DF $= DF_{total} - (DF_{treat} + DF_{block}) = 15$		

ANOVA table.

Source of variation	SS	DF	MS	F-ratio	F_{crit} $\alpha = 0.05$	$\alpha = 0.01$
Blocks	1.5986	5	0.3197	9.81	2.90	4.56
Treatment (variety)	1.1485	3	0.3828	11.74	3.29	5.42
Residual	0.4892	15	0.0326			
Total	3.2363	23				

Conclusion: the variety treatment F-ratio is greater than the critical F value required for rejection of the null hypothesis and it is concluded that there is a significant difference in yield between varieties ($P < 0.01$).

There is a significant block effect ($P < 0.01$), indicating that non-uniformity in external conditions had an additional significant effect on potato yield in the field trial.

The analysis can be continued by an appropriate multiple comparison test. In this case an LSD test is performed to identify between which varieties a significant difference occurs.

$$SED = \sqrt{\frac{2 \times MS_{resid}}{n}} = \sqrt{\frac{2 \times 0.0326}{6}} = 0.104$$

$LSD_{(P = 5\%)}$ for comparison of selected pairs of variety means $= t_{(DF = 15; P = 0.05)} \times SED$

$= 1.753 \times 0.104$

$= 0.182$

Conclusion of LSD test: any pair of variety means selected for comparison that numerically differ by ≥ 0.182 are significantly different ($P < 0.05$).

with critical values given in tables. If of interest an *F*-ratio may also be determined for the block effects but it is not essential to do this. In this example there is in fact a significant difference between blocks, demonstrating the existence of systematic variation running through the trial site. By using an RCB design in conjunction with ANOVA, the variation due to the non-uniform external factors has been accounted for and thereby has allowed a valid conclusion to be drawn regarding the effect of the main treatment.

It may be noted that, where an ANOVA test of a randomized complete block design leads to the conclusion that there is no difference between blocks, this indicates that there is no significant systematic variation in the data attributable to external factors. This suggests the possibility of removing the blocks from the analysis and summing all the treatment replicates across the experiment, thereby allowing the data to be reanalysed using a simple ANOVA for a completely randomized design. This would have the effect of increasing the probability of obtaining a significant treatment effect and might change the conclusion to the experiment if the treatment *F*-ratio was close to the critical value for rejecting the null hypothesis. However, just because there are no significant differences between blocks this does not mean that external influences that affect the data are definitely absent. Given that a block design was selected in the first place to guard against the possibility of such influences, then the function of the blocks has not been removed just because there are no statistically significant block effects. Most statisticians would therefore argue against recombining data when block effects are non-significant but there is not universal agreement upon this (see e.g. Clarke and Kempson (1997) for further discussion of this issue).

9.2.3 Multiple-factor randomized complete block designs

Randomized complete block designs can be extended to cater for two treatment factors although the analysis of such experiments does become more complex, particularly if it is required to examine the interaction between the two factors. In the potato crop trial discussed above (illustrated in Fig. 9.2), in addition to comparing the yields between different varieties it might also be required to test the effect of a particular fertilizer spray. In this case each block would contain an equal number of replicates of each combination of potato variety and fertilizer spray. If, for example, there are four varieties, A, B, C and D, and two fertilizer regimes, plus fertilizer (+f) and zero fertilizer (−f), these might be arranged randomly into plots within an individual block as shown below:

B −f	C +f
D +f	A −f
B +f	C − f
A +f	D −f

A number of blocks will subsequently be dispersed across the field.

The general linear model for this design is:

$$y = \mu + \alpha + \beta + (\alpha \times \beta) + blk + \varepsilon$$

where: y = the measured value of a variable under investigation;
μ = the population mean of the variable;
α = the quantitative effect on the variable caused by the level of treatment factor α;
β = the quantitative effect on the variable caused by the level of treatment factor β;
$\alpha \times \beta$ = the quantitative effect on the variable caused by the interaction of treatment factor α and treatment factor β;
blk = the quantitative effect on the variable caused by blocking;
ε = the quantitative effect on the variable caused by random (error) effects.

The ANOVA will therefore involve the partitioning of the total sum of squares between five sources of variation, namely blocks, treatment factor A, treatment factor B, treatment A×B interaction and the random residual variation. The main calculations required to achieve this are as follows:

$$SS_{total} = \Sigma x^2 - CT \qquad\qquad SS_{block} = \Sigma B^2 / n_{block} - CT$$

$$SS_{treatment} = \Sigma T^2 / n_{treatment} - CT$$

where:

ΣB^2 = sum of the squared totals of each block;
ΣT^2 = sum of the squared totals of each treatment group;
n_{block} = total no. of measurements per block;
$n_{treatment}$ = total no. of measurements per treatment group.

The treatment SS comprises three components:

$$SS_{treatment\ A} = \Sigma T_A^2 / n_A - CT \qquad SS_{treatment\ B} = \Sigma T_B^2 / n_B - CT$$

$$SS_{A \times B\ interaction} = SS_{treatment} - (SS_{treatment\ A} + SS_{treatment\ B})$$

where: ΣT_k^2 = sum of the squared totals of each level of treatment k;
n_k = total no. of measurements per level of treatment k.

Finally the residual SS is determined:

$$SS_{residual} = SS_{total} - (SS_{block} + SS_{treatment})$$

Note that the interactions between the blocks and each of the treatment factors is of no interest to the experimenter and are not calculated separately but remain within the residual variance.

The ANOVA table is then formatted as shown in Table 9.2.

Table 9.2. RCB multiple factor ANOVA table standard format.

Source of variation	SS	DF	MS	F-ratio
Blocks	SS_{block}	$N_{block} - 1$	SS/DF	$\dfrac{MS_{blocks}}{MS_{resid}}$
Treatment A	$SS_{treatment\ A}$	$N_A - 1$	SS/DF	$\dfrac{MS_{treat\ A}}{MS_{resid}}$
Treatment B	$SS_{treatment\ B}$	$N_B - 1$	SS/DF	$\dfrac{MS_{treat\ B}}{MS_{resid}}$
A×B interaction	$SS_{A×B\ interaction}$	$(N_A - 1) \times (N_B - 1)$	SS/DF	$\dfrac{MS_{A \times B}}{MS_{resid}}$
Residual	$SS_{residual}$	$(N_A \times N_B) - 1 \times (N_{block} - 1)$ (or by subtraction)	SS/DF	
Total	SS_{total}	$N - 1$		

The full calculation of a two-factor randomized complete block design is illustrated in Example 9.2 In theory it would be quite possible to design a three-factor randomized complete block experiment based on exactly the same principles, although in practice such designs are so demanding of resources that they are very rarely used.

Example 9.2. Analysis of variance of a multiple-factor randomized complete block design experiment.

Field trial of the effect of a treatment fertilizer spray on yield in four varieties of potato.

The design: six blocks located across the field were each divided into eight plots. Each of four potato varieties (A, B, C, D) plus and minus a treatment fertilizer spray, were allocated randomly to the plots in each block.

The data:

									Mean potato yield (kg/plant)	
					Variety plots					
	− Fertilizer spray				+ Fertilizer spray				Block total	Block
Blocks	A	B	C	D	A	B	C	D	(T_{block})	mean
1	1.30	1.50	1.98	1.18	1.48	1.61	1.90	1.66	5.96	1.488
2	1.35	1.50	2.03	1.13	1.82	1.38	2.10	1.45	6.01	1.500
3	2.08	2.00	2.18	1.34	2.40	1.96	1.94	1.89	7.60	1.906

continued

Example 9.2. Continued.

Blocks	A	B	C	D	A	B	C	D	T_{block}	Block Mean
4	1.95	2.10	2.40	2.20	2.10	2.17	2.65	2.31	8.65	2.163
5	1.83	2.20	2.33	1.95	2.10	2.28	2.16	2.25	8.31	2.075
6	1.90	1.73	2.10	1.63	2.07	1.91	2.18	2.00	7.36	1.838
Treatment total ($T_{treatment}$)	10.41	11.03	13.02	9.43	11.97	11.31	12.93	11.56	Σx= 91.66	
Treatment mean	1.74	1.84	2.17	1.57	2.00	1.89	2.16	1.93		

Variety	A	B	C	D
total (T_{var})	22.38	22.34	25.95	20.99
mean	1.87	1.86	2.16	1.75

Fertilizer	− fertilizer	+ fertilizer
total (T_{fert})	43.89	47.77
mean	1.83	1.99

No. observations (N) = 48; No. blocks (N_{block}) = 6;
No. fertilizer treatments (N_{fert}) = 2; No. variety treatments (N_{var}) = 4.

Total no. of measurements per treatment ($n_{treatment}$) = 6
Total no. of measurements per block (n_{block}) = 8
Total no. of measurements per fertilizer (n_{fert}) = 24
Total no. of measurements per variety (n_{var}) = 12

Null hypotheses: there is no difference in potato yield between the different fertilizer treatments; there is no difference in potato yield between the different varieties; there is no difference in potato yield due to fertilizer × variety interaction.

The analysis:
Σx^2 = 180.8142 ΣT_{block}^2 = 1424.1286
$\Sigma T_{treatment}^2$ = 1060.4898 ΣT_{fert}^2 = 4208.3050 ΣT_{var}^2 = 2113.9226
$CT = (\Sigma x)^2/N$ = 175.0324

The calculation procedure is now similar to that of a three-way ANOVA, treating the blocks as a source of variation; however, the interaction terms involving blocks are not calculated but are contained within the residual variation.

Sum of squares (SS):

Total	$= \Sigma x^2 - CT$	$= 180.8142 - 175.0324$	=
		5.7818	
		with DF = $N - 1$	= 47
Between treatment	$= \Sigma T_{treatment}^2/n_{treatment} - CT$	= (1060.4898)/6 − 175.0324	=
		1.7159	
Between blocks	$= \Sigma T_{block}^2/n_{block} - CT$	= (1424.1286)/8 − 175.0324	=
		2.9837	
		with DF = $N_{block} - 1$	= 5

continued

Example 9.2. Continued.

Between fertilizer $= \Sigma T_{fert}^2/n_{fert} - CT$ $= (4208.3050)/24 - 175.0324$
$= 0.3136$
with DF $= N_{fert} - 1 = 1$

Between variety $= \Sigma T_{var}^2/n_{var} - CT$ $= (2113.9226)/12 - 175.0324$
$= 1.1278$
with DF $= N_{var} - 1 = 3$

Fert × var interaction $= SS_{treatment} - (SS_{fert} + SS_{var})$ $= 1.7159 - (0.3136 + 1.1278)$
$= 0.2745$
with DF $= DF_{fert} \times DF_{var} = 3$

Residual $= SS_{total} - (SS_{treatment} + SS_{block}) = 5.7818 - (1.7159 + 2.9837)$
$= 1.0822$
with DF $= (N_A \times N_B) - 1$
$\times (N_{block} - 1) = 35$

ANOVA table

Source of variation	SS	DF	MS	F-ratio	F_{crit} $\alpha = 0.05$	$\alpha = 0.01$
Blocks	2.9837	5	0.5967	19.30	2.49	3.70
Fertilizer	0.3136	1	0.3136	10.14	4.12	7.42
Variety	1.1278	3	0.3759	12.16	2.87	4.40
Fert × var interaction	0.2745	3	0.0915	2.96	2.87	4.40
Residual	1.0822	35	0.0309			
Total	5.7818	47				

Conclusion: both the fertilizer and the variety treatment F-ratio values are greater than the critical F values required for rejection of the null hypothesis and it is concluded that there is a significant effect of the fertilizer treatment and variety on yield ($P < 0.01$). There is also a significant fertilizer × variety interaction effect on yield ($P < 0.05$).

There is a significant block effect ($P < 0.01$), indicating that non-uniformity in external conditions had an additional significant effect on potato yield in this field trial.

Since the interaction is significant, the analysis could be continued by performance of an appropriate multiple comparison test. Simple observation of the data, however, would clearly suggest that the interaction present is primarily due to variety C, which, unlike the other varieties, fails to respond positively to the additional fertilizer spray treatment.

9.3 Incomplete Block Designs

Occasionally, due to a lack of resources (experimental material, manpower or time) or because it is not possible to have blocks that are physically large enough to accommodate all the planned treatments, it is not possible to employ a complete block design in which every treatment is represented in every block. Consider, for example, a field trial designed to assess the impact of five different irrigation treatments on tree growth. Due to the limited area and uniformity of the land available and the large spaces occupied by the trees, it is not possible to form blocks that hold more than three trees. Therefore each block will be lacking in two of the treatments. The essential problem in designing this trial is how to avoid confounding the true effect of the treatments with the effects that arise from differences between the blocks. If the treatments were allocated to blocks randomly, then it is quite possible that one or more pairs of the treatments never occurred together in the same block. If a difference in response were to be subsequently found between any such pair of treatments, it would not be certain whether this was due to the treatments or due to a difference in the blocks containing the treatments. This is illustrated in Fig. 9.3(a). Another similar situation arises when, due to the nature of the measurement being made, it is not possible for all items in an experiment to be measured on each occasion. For example, in a tree irrigation trial it might be necessary to measure the rate of transpiration; however, due to the time taken to make such a measurement, it might not be possible to measure transpiration in every tree in any one day, so that a number of days are required to collect one complete set of measurements. Clearly, differences in weather between the days are then likely to confound the results. Each day of measurement would then need to be considered as an incomplete block.

9.3.1 The design of incomplete block experiments

To prevent the occurrence of confounding in an incomplete block design, it is necessary to allocate the treatments to blocks such that every possible pair of treatments occurs together in the same block an equal number of times. This is referred to as a **balanced incomplete block design**. To achieve such a design, a very careful consideration of the method for the allocation of treatments to the blocks is required. The design of a balanced incomplete block experiment is illustrated in Fig. 9.3(b).

The minimum number of blocks required to achieve a balanced incomplete block design can be determined from the following formula:

$$\text{blocks} = \frac{t!}{k! \times (t-k)!}$$

where: t = no. of treatments;
k = no. of experimental units per block.

The total number of replicates necessary for each treatment across the whole experiment, $n_{\text{treatment}}$, is then given by:

$$n_{\text{treatment}} = \frac{\text{blocks} \times k}{t}$$

Therefore in the example illustrated in Fig. 9.3(b), as there are five treatments ($t = 5$) and each block contains three trees ($r = 3$), the minimum number of blocks required is:

$$\frac{5!}{3!(5-3)!} = \frac{120}{12} = 10$$

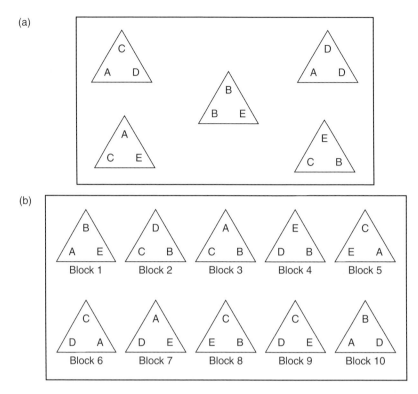

(a)

(b)

Block 1 Block 2 Block 3 Block 4 Block 5

Block 6 Block 7 Block 8 Block 9 Block 10

Fig 9.3. Design of an incomplete block experiment. Five irrigation treatments, A, B, C, D, E, are to be applied to trees in an incomplete block design with three trees per block and with a minimum of three replicates per treatment. **(a) Unbalanced incomplete design.** Treatment allocated to blocks randomly: requires a minimum of five blocks. *In this case the treatment and block effects are confounded:* by random chance, replicates of treatments A and B, treatments B and D and treatments D and E never occur together in the same block. Therefore any apparent difference in response to these pairs of treatments may be due to a difference between the blocks rather than to a difference between the treatments. **(b) Balanced incomplete design.** Treatments allocated to blocks to ensure that each pair of treatments occurs together in same block an equal number of times (i.e. three): requires six replicates per treatment and a minimum of ten blocks. *In this case treatment and block effects are separable:* because all pairs of treatments occur together in the same block an equal number of times (i.e. three), and also occur in different blocks an equal number of times (i.e. three), the effects of treatments can be separated from the effects of the blocking.

This would allow for $(10 \times 3)/5 = 6$ replicates per treatment. If more replicates are required, then the number of blocks would need to be doubled to 20, which would allow for 12 replicates per treatment.

9.3.2 Analysis of balanced incomplete block designs

As long as the design is completely balanced, the analysis of an incomplete block design can be achieved by ANOVA, following the same procedures employed for randomized complete block designs described in section 9.2. The total sum of squares (SS) is partitioned between the SS for blocks, the SS for treatment and the random residual SS and the mean square (MS) values determined by SS/DF in the normal way. The ratio of the treatment MS to the residual MS then provides the F-ratio for determining the significance of the treatment effects independently of the block effects.

A major complication arises, however, when the difference between selected pairs of treatment means are required to be inspected by a multiple comparison test following ANOVA. The problem is that each treatment mean arises from a different combination of blocks and they are not, therefore, strictly comparable. This is illustrated in Fig. 9.3(b) where irrigation treatment A does not occur in blocks 2, 4, 8 while treatment B is present; conversely, treatment A is present in blocks 5, 6, 7 while treatment B is absent. Therefore any major difference in the means of these two sets of blocks would confound the comparison of the A and B treatment means. In other words, some portion of the difference in treatment means will be due to the effect of blocks rather than the treatments themselves. This can be overcome by determining an estimate for the missing values in the blocks and then recalculating the treatment means. For example, in Fig. 9.3(b) estimates for the missing treatment C and D values would be determined for block 1, estimates for treatment A and E values determined for block 2, etc. The new treatment means, based on the actual values together with the estimates of the missing values, are termed the **least squares means** and the subsequent determination of the standard error of difference (SED) is then based on the least squares means. The calculation of the estimates of the missing value is a complex reiterative process that strives to cause as little change as possible to the individual block variances and therefore minimizes any change to the overall residual variance. A full description of the techniques involved is beyond the scope of this chapter and the reader is referred to alternative texts for description of the complex procedure involved (e.g. Cochran and Cox, 1992; Clarke and Kempson, 1997; Mead et al., 2003; Zar, 2007). It may also be noted that more advanced statistical computer packages such as GenStat are able to handle the full analysis of incomplete block designs.

9.4 Latin Square Designs

Latin square experimental designs are employed in factorial experiments where two independent external factors are specifically identified that each produce variation that systematically runs through the experiment. Very often the factors operate to produce gradients in variation that run in different directions. For example, within a glasshouse there may be a gradient of shade across the crop rows while a temperature gradient may be running in parallel with the rows. Latin square designs are particularly common in animal experimentation where the animals are segregated according to some shared characteristics. In an animal growth experiment, for example, the animals available for a particular experiment may come from different litters and there may be a number of different enclosures in which the animals are kept. A Latin square design is also used to overcome the problem of being unable to make all measurements at the same time. For example, if it was possible to measure only one replicate for each treatment on each day, then the day of measurement could become a factor within the Latin square design.

The purpose of the Latin square design is to distribute the effects of two well-defined sources of variation across all treatments in a systematic manner. To achieve this, either the experimental site or the experimental material is segregated into plots such that each plot has an equal number of units in each 'row' and each 'column' of a square grid. The treatments are then assigned to plots so that each treatment is represented once in each row and once in each column and, therefore, the number of treatments must be equal to both the number of rows and the number of columns. (The terms 'columns' and 'rows' do not need to literally refer to a field site. For example, in the animal growth experiment cited above, the litters and enclosures would be referred to as the 'row' and 'column' factors respectively.) If, for example, there were four treatments A, B, C, D, possible arrangements that fulfil this requirement are:

A B C D		B D C A		C D B A		D A B C	
B C D A	or	C A B D	or	D C A B	or	A D C B	etc.
C D A B		D B A C		A B D C		B C D A	
D A B C		A C D B		B A C D		C B A D	

In designing the overall experiment, a random element is introduced by randomizing the order of the rows and columns within the plots and allocating the treatments randomly to the labels A, B, C, D, etc.

9.4.1 Analysis of Latin square designs

In analysing Latin square designs, the effects of the two identified external factors are recognized specifically and referred to as the 'row effect' and the 'column effect'. The general linear model for the design thus becomes:

$$y = \mu + \alpha + \delta + \gamma + \varepsilon$$

where: y = the measured value of a variable under investigation;
μ = the population mean of the variable;
α = the quantitative effect on the variable caused by the level of the treatment factor α;
δ = the quantitative effect on the variable caused by row variation;
γ = the quantitative effect on the variable caused by column variation;
ε = the quantitative effect on the variable caused by random residual (error) effects.

In the calculation of the ANOVA, the total sums of squares (SS) is partitioned between the row effect, column effect, treatment effect and residual effect. There is assumed to be no interaction of the treatment with either the row factor or the column factor; indeed, if there were to be interaction present, then the Latin square design would become invalid. The final ANOVA table for a Latin square design takes the form shown in Table 9.2.

The calculation of an ANOVA for a Latin square designed experiment is illustrated in Example 9.3.

Table 9.3. Latin square design ANOVA table standard format.

Source of variation	Sum of squares (SS)	Degrees of freedom (DF)	Mean Square (MS)	F-ratio
Row effect	SS_{rows}	$N_{rows} - 1$	SS_{rows}/DF_{rows}	$\dfrac{MS_{rows}}{MS_{residual}}$
Column effect	$SS_{columns}$	$N_{columns} - 1$	$SS_{columns}/DF_{columns}$	$\dfrac{MS_{columns}}{MS_{residual}}$
Treatment	$SS_{treatment}$	$N_t - 1$	$SS_{treatment}/DF_{treatment}$	$\dfrac{MS_{treatment}}{MS_{residual}}$
Residual	$SS_{residual}$	$(N_t - 1) \times (N_t - 2)$ (or by subtraction)	$SS_{residual}/DF_{residual}$	
Total	SS_{total}	$N - 1$		

Where N = total no. of measurements in whole experiment; N_t = number of levels of application of the treatment factor (= $N_{rows} = N_{columns}$).

Example 9.3. Design and analysis of a Latin square designed experiment.

The growth rate of weaned lambs in response to diet supplements. Five different diet supplements (S1, S2, S3, S4, S5) were fed to each of five lambs kept in grazing pens over a 6-month period and the growth rate determined as the mean daily increase in body weight.

The design: since there may be natural similarities between lambs from the same breeding line but differences in conditions may occur between the pens in which lambs were kept, a Latin square design was employed in order to distribute these effects evenly among the treatments. In the experiment, each pen held five lambs, one from each of five breeding lines, and one of each of the five supplements was fed to one each of the lambs. Across all pens it was ensured that each of the five supplements was fed to a lamb from each of the five breeding lines. The full design is shown in the table below.

Allocation of five diet supplements (S1–S5) to lambs in a Latin square designed experiment.

Breeding line	Pen 1	Pen 2	Pen 3	Pen 4	Pen 5
A	S5	S2	S1	S4	S3
B	S3	S5	S2	S1	S4
C	S2	S1	S4	S3	S5
D	S4	S3	S5	S2	S1
E	S1	S4	S3	S5	S2

continued

Example 9.3. Continued.

The data:

Breeding line	Pen 1 Diet	Pen 2 Diet	Pen 3 Diet	Pen 4 Diet	Pen 5 Diet	Breeding line totals
			Mean daily increase in body weight (g)			
A	(S5) 34.4	(S2) 26.8	(S1) 24.2	(S4) 25.6	(S3) 30.2	141.2
B	(S3) 39.4	(S5) 41.4	(S2) 33.3	(S1) 35.0	(S4) 35.3	184.4
C	(S2) 37.9	(S1) 35.7	(S4) 34.2	(S3) 36.8	(S5) 40.5	185.1
D	(S4) 37.5	(S3) 42.6	(S5) 43.8	(S2) 35.6	(S1) 38.2	197.7
E	(S1) 42.5	(S4) 39.5	(S3) 43.7	(S5) 46.8	(S2) 40.4	212.9
Pen totals	191.7	186.0	179.2	179.8	184.6	

	S1	S2	S3	S4	S5
			Diet totals and means		
Diet total	175.6	174.0	192.7	172.1	206.9
Diet mean	35.12	34.80	38.54	34.42	41.38

No. of observations (N) = 25; no. of treatments (N_t) = no. of columns = no. of rows = 5

Null hypothesis: the growth rate of lambs fed with the five different diet supplements is the same.

The analysis:

Σx^2 = 34740.61; $\Sigma T_{\text{column}}^2$ = 169862.70; ΣT_{rows}^2 = 172614.50;

$\Sigma T_{\text{diets}}^2$ = 170670.70

Correction term (CT)   = $(\Sigma x)^2/N$ = 33951.75

Sum of squares:

Total	= Σx^2 − CT	= 34740.61 − 33951.75	= 788.86
		with DF = N − 1	= 24
Between columns	= $\Sigma T_{\text{column}}^2 / N_t$ − CT =	169862.70/5 − 33951.75	= 20.79
		with DF = N_t − 1	= 4
Between rows	= $\Sigma T_{\text{rows}}^2 / N_t$ − CT =	172614.50/5 − 33951.75	= 571.15
		with DF = N_{rows} − 1	= 4
Between treatment =	$\Sigma T_{\text{diets}}^2/ N_t$ − CT =	170670.70/5 − 33951.75	= 182.39
		with DF = N_{columns} − 1	= 4
Residual	= $SS_{\text{total}} - (SS_{\text{coluums}} + SS_{\text{rows}} + SS_{\text{diets}})$		= 14.53
		with DF (by subtraction)	= 12

continued

Example 9.3. Continued.

ANOVA table.

Source of variation	SS	DF	MS	F-ratio	F_{crit} $\alpha = 0.05$	F_{crit} $\alpha = 0.01$
Columns	20.79	4	5.20	4.30	3.26	5.41
Rows	571.15	4	142.79	118.01	3.26	5.41
Treatment (diets)	182.39	4	45.60	37.69	3.26	5.41
Residual	14.53	12	1.21			
Total	788.86	24				

Multiple comparison test:

$$SED = \sqrt{\frac{2 \times MS_{resid}}{n}} = \sqrt{\frac{2 \times 1.21}{5}} = 0.70$$

$LSD_{(P=5\%)}$ for comparison of selected diet means $= t_{(DF=12; P=0.05)} \times SED$
$$= 1.782 \times 0.70 = 1.25$$

Conclusion: the diet treatment F-ratio is greater than the critical value required for rejection of the null hypothesis and it is concluded that there is a significant difference in growth rate between diet treatments ($P < 0.01$).

There is also both a significant column ($P < 0.05$) and row effect ($P < 0.01$), indicating that there was variation in growth rate due to the different breeding lines and the different pens.

The LSD test indicates that any pair of diet means selected for comparison that numerically differ by ≥ 1.25 are significantly different ($P < 0.05$).

The major practical problem with the use of Latin square designs is securing sufficient residual degrees of freedom to enable a valid analysis. If the number of treatments is three, then the $DF_{residual}$ is only two while with four treatments $DF_{residual}$ is still only six. Only with five treatments does $DF_{residual}$ reach the normally recommended minimum value of 12 for performance of a valid ANOVA, by which time the total number of plots required will be 25 (i.e. N_t^2). Depending on the nature of the experiment the need for 25 plots may challenge the resources available.

9.5 Split-plot Designs

In many two-factorial experiments, one of the treatment factors to be applied, because of practical limitations, may not be easily replicated. Consider, for example, a field investigation into the interaction between soil type and fertilizer treatment

---

on crop growth. In the field it is unlikely to be possible to establish crops in replicated small plots each with different soil types. Instead, variation in soil type is only likely to occur between relatively large fields. Each field of a different soil type then represents a main treatment plot and within these main field plots crops subjected to different fertilizer regimes are grown in replicated subplots. Experimental designs of this sort, in which main treatment plots are subdivided into smaller replicated subplots that accommodate a second treatment factor, are called **split-plot designs**.

Commonly, randomized blocks are incorporated into split-plot designs so that each replicate block is divided into main plots (one for each level of the first treatment factor) and the main plots subdivided into subplots that contain the second treatment factor. An example of such a design involving the investigation of day-length extension and variety on yield of a strawberry crop grown in polytunnels is illustrated in Fig. 9.4.

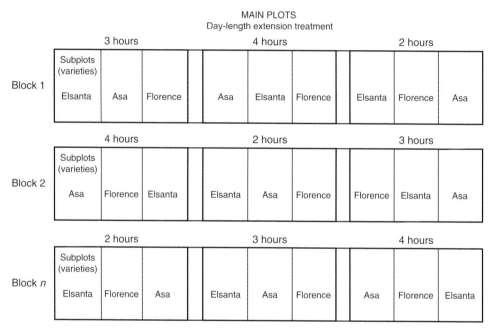

Fig. 9.4. Design of a split-plot experiment. An investigation into the effect of day-length extension treatment and variety on yield of strawberries. Three day-length extension treatments (2 hours, 3 hours, 4 hours) were applied to main plots housed within replicated blocks. Three varieties were tested, cvs Elsanta, Asa and Florence, which were grown in subplots within the main plots. The main features of this design are: (i) the blocks are each divided into three main plots; (ii) the day-length extension treatments are allocated randomly to the main plots; (iii) the main treatment plots are each divided into three subplots; (iv) the varieties are allocated randomly to the subplots; (v) the day-length extension treatments are therefore replicated only between the blocks. The variety treatments are replicated between the blocks and between subplots and are therefore analysed with a greater amount of precision than the day-length extension treatments.

The main problem with split-plot designs is that the two treatment factors are not equally replicated. The treatment applied to the main plots will be substantially less well replicated than the treatment applied to the subplots and consequently the two treatments will not be investigated with the same degree of precision. Where possible the treatment that is the less important of the two should therefore be applied to the main plots; however, practical limitations may prevent this. If it is essential that the two factors are considered with equal weighting, then a split-plot design cannot be used.

9.5.1 Analysis of split-plot designs

The analysis of split plots, although rather complex, is a logical extension to the two-way ANOVA technique employed for randomized blocks. The analysis is performed in two parts, first the analysis of the main plot treatment, followed by the analysis of the subplot treatment.

Analysis of main plot treatment

To analyse the effect of the treatment assigned to the main plots (treatment A), the subplots are treated as homogeneous replicates. The main plot treatment can then be analysed as if it were housed within a randomized block design. The ANOVA partitions the total main plot sum of squares (SS_{MP}) between a sum of squares for blocks (SS_{block}), a sum of squares for the main plot treatment (SS_{treatA}) and a sum of squares for the main plot residual error ($SS_{MPresidual}$). The mean squares (MS) for these items are determined in the normal way by dividing the SS values by their respective degrees of freedom. The ratio of the main plot MS to the main plot residual MS gives an *F*-ratio that allows the significance of the effect of the main plot treatment to be assessed.

Analysis of subplot treatment

To analyse the treatment assigned to the subplots (treatment B), the total sum of squares concerned with the subplots ($SS_{subplot}$) is partitioned between a sum of squares for the subplot treatment ($SS_{treat\ B}$), a sum of squares for the interaction between the subplot and main plot treatments ($SS_{A \times B}$), and a residual term that represents subplot error ($SS_{SPresidual}$). This subsequently produces a mean square (MS) for each of the items. The ratio of the subplot MS to the subplot residual MS gives an *F*-ratio that allows the significance of the effect of the subplot treatment to be assessed. Similarly, the ratio of the interaction MS to the subplot residual MS gives a value of *F* that allows the significance of the interaction to be assessed.

The ANOVA table for the analysis of a blocked, split-plot two-factor design is shown in Table 9.4.

Table 9.4. Split plot design ANOVA table standard format.

Source of variation	SS	DF	MS	F-ratio
Main plot analysis				
Blocks	SS_{block}	$N_{block} - 1$	SS/DF	$\dfrac{MS_{blocks}}{MS_{MP\ residual}}$
Main plot treatment A	$SS_{treatment\ A}$	$N_A - 1$	SS/DF	$\dfrac{MS_{treatment\ A}}{MS_{MP\ residual}}$
Main plot residual	$SS_{MP\ residual}$	$(N_A - 1) \times (N_{block} - 1)$ (or by subtraction)	SS/DF	
Main plot total	$SS_{main\ plot\ total}$	$(N_{block} \times N_A) - 1$		
Subplot analysis				
Subplot treatment B	$SS_{treatment\ B}$	$N_B - 1$	SS/DF	$\dfrac{MS_{treatment\ B}}{MS_{SP\ residual}}$
A × B interaction	$SS_{A \times B\ interaction}$	$(N_A - 1) \times (N_B - 1)$	SS/DF	$\dfrac{MS_{A \times B}}{MS_{SP\ residual}}$
Subplot residual	$SS_{SP\ residual}$	$N_A \times (N_B - 1) \times (N_{block} - 1)$ (or by subtraction)	SS/DF	
Subplot total	$SS_{SP\ total}$	$N_A \times N_{block} \times (N_B - 1)$		
Total	SS_{total}	$N - 1$		

Where N = total no. of measurements in whole experiment; N_{block} = number of blocks; N_A = number of levels of treatment factor A; N_B = number of levels of treatment factor B.

The exact procedure for the calculation is explained in the context of Example 9.4, which is based on the design described in Fig. 9.4.

Example 9.4. Design and analysis of a split-plot designed experiment.

The yield of three different varieties of glasshouse-grown strawberries in response to three day-length extension treatments.

The design: as it was not practical to apply day-length treatments to small plots, the experiment was performed as a blocked split-plot experiment. Four beds of strawberries representing replicate blocks were divided into three main plots and each main plot was split into three subplots. The day-length treatments were applied to the main plots, and the variety treatments were applied to subplots. The design is illustrated in Fig. 9.4.

continued

Example 9.4. Continued.

The data: the total fresh weight fruit yield (expressed as kg/m^2) was determined within each subplot.

Fruit yield (kg/m^2)										
Main plots: Day-length extension	2 hours			4 hours			6 hours			
Subplots: Variety treatment	Asa	Elsanta	Florence	Asa	Elsanta	Florence	Asa	Elsanta	Florence	Block mean
Blocks										
1	1.5	2.8	2.1	1.8	2.7	2.3	1.2	3.0	1.7	2.12
2	1.3	1.9	1.6	1.2	3.2	2.1	1.0	2.9	2.0	1.91
3	0.8	2.8	1.3	1.1	3.1	1.7	1.0	3.3	1.8	1.88
4	1.3	2.2	1.6	1.2	3.0	1.9	1.0	3.4	1.7	1.92
Treatment total	4.9	9.7	6.6	5.3	12.0	8.0	4.2	12.6	7.2	
Treatment mean	1.20	2.50	1.67	1.37	3.00	2.03	1.07	3.07	1.83	
Main plot mean		1.79			2.13			1.99		

No. of observations (N) = 36; grand total (GT) = 70.5
No. of main plots (N_{MP}) = 3; No. subplots ($N_{subplot}$) = 36; No. blocks (N_{block}) = 4

Null hypotheses: there is no effect of day-length extension on fruit yield in strawberries; there is no effect of variety on fruit yield in strawberries; there is no daylength × variety interaction effect on fruit yield in strawberries.

The analysis: the correction term (CT) and the total sum of squares are obtained in the normal manner:

$$CT = GT^2/N \qquad = 70.5^2/36 \qquad\qquad = 138.0625$$
$$SS_{total} = \Sigma x^2 - CT \qquad = 158.6100 - 138.0625 \qquad = 20.5475$$
$$\text{with DF} = N - 1 = 35$$

The ANOVA is now performed in two parts, a main plot analysis to establish the significance of the main plot factor (i.e. day-length extension) and a subplot analysis to establish the significance of the subplot factor (i.e. variety) and the treatment interaction (i.e. day-length extension × variety).

continued

Example 9.4. Continued.

Main plot analysis: the main plot analysis is based on the main plot totals as follows:

| Block | Day-length extension | | | |
	2 hour	4 hour	6 hour	Total
1	6.4	6.8	5.9	19.1
2	4.8	6.5	5.9	17.2
3	4.9	5.9	6.1	16.9
4	5.1	6.1	6.1	17.3
Total	21.2	25.3	24.0	70.5

A total sum of squares for the main plots ($SS_{main plot}$) is derived from the main plot totals and is then partitioned between the SS for the blocks (SS_{block}), the main plot treatment ($SS_{day length}$) and the main plot residual ($SS_{MPresidual}$). Note that all the main plot sum of square values need to be adjusted by division to account for the number of values from which they are derived.

Main plot sums of squares:

$$SS_{main plot} = \frac{\sum T_{main plot}^2}{3} - CT \qquad = \frac{418.57}{3} - 138.0625 \qquad = 1.4608$$

$$\text{with DF} = N_{MP} - 1 = 11$$

$$SS_{block} = \frac{\sum T_{blocks}^2}{9} - CT \qquad = \frac{1245.55}{9} - 138.0625 \qquad = 0.3319$$

$$\text{with DF} = N_{block} - 1 = 3$$

$$SS_{day length} = \frac{\sum T_{day length}^2}{12} - CT \qquad = \frac{1665.53}{12} - 138.0625 \qquad = 0.7317$$

$$\text{with DF} = N_{day length} - 1 = 2$$

$$SS_{MPresidual} = SS_{main plot} - (SS_{block} + SS_{day length}) = 1.4608 - (0.3319 + 0.7317) = 0.3972$$

$$\text{with DF} = (N_{day length} - 1) \times (N_{block} - 1) = 6$$

Sub-plot analysis: the subplot analysis is based on the total of the subplot replicates as follows:

| Variety | Day-length extension | | | |
	2 hours	4 hours	6 hours	Total
Asa	4.9	5.3	4.2	14.4
Elsanta	9.7	12.0	12.6	34.3
Florence	6.6	8.0	7.2	21.8
Total	21.2	25.3	24.0	70.5

continued

Chapter 9

Example 9.4. Continued.

The total subplot sum of squares ($SS_{subplot}$) is obtained by subtracting the main plot SS from the total SS. $SS_{subplot}$ is then partitioned into the SS for the variety treatment (SS_{var}), the treatment interaction ($SS_{day\ length \times var}$) and the subplot residual ($SS_{SPresidual}$). To achieve this the total treatment SS ($SS_{treatment}$) is required and is derived from the nine replicate subplot totals given in the above table. SS_{var} is similarly derived from the variety treatment totals in the above table. Note that both $SS_{treatment}$ and SS_{var} need to be adjusted by division to account for the number of subplots on which they are based. The interaction between the main plot and sub-plot treatment ($SS_{day\ length \times var}$) is then determined by the difference between $SS_{treatment}$ and SS_{var}. Finally the subplot residual SS is obtained by subtraction of the variety treatment SS and the interaction SS from the total subplot SS.

Subplot sums of squares:

$$SS_{treatment} = \frac{\sum T_{treatment}^{2}}{n_{treat}} - CT \qquad = \frac{625.99}{4} - 138.0625 \qquad = 18.435$$

$$\text{with DF} = (N_{day\ length} \times N_{var}) - 1 = 8$$

$$SS_{subplot} = SS_{total} - SS_{main\ plot} \qquad = 20.5475 - 1.4608 \qquad = 19$$

$$\text{with DF} = N_{day\ length} \times N_{block} \times (N_{var} - 1) = 24$$

$$SS_{var} = \frac{\sum T_{var}^{2}}{n_{var}} - CT \qquad = \frac{1859.09}{12} - 138.0625 \qquad = 16.8617$$

$$\text{with DF} = N_{var} - 1 = 2$$

$$SS_{day\ length \times var} = SS_{treatment} - (SS_{day\ length} + SS_{var}) = 18.435 - (0.7317 + 16.8617)$$
$$= 0.8416$$

$$\text{with DF} = (N_{day\ length} - 1) \times (N_{var} - 1) = 4$$

$$SS_{SPresidual} = SS_{subplot} - (SS_{var} + SS_{day\ length \times var}) = 19.0867 - (16.8617 + 0.8416)$$
$$= 1.3834$$

$$\text{with DF} = N_{day\ length} \times (N_{var} - 1) \times (N_{block} - 1) = 18$$

ANOVA table.

Source of analysis	SS	DF	MS	F-ratio	F_{crit} $\alpha = 0.05$	$\alpha = 0.01$	Sig.
Main plot analysis:							
Blocks	0.3319	3	0.1106	1.67	4.76	9.78	n.s.
Main plot treatment (day length)	0.7317	2	0.3659	5.53	5.14	10.92	$P < 0.05$
Main plot residual	0.3972	6	0.0662				
Main plot total	1.4608	11					

continued

Example 9.4. Continued.

Source of analysis	SS	DF	MS	F-ratio	F_{crit} $\alpha = 0.05$	$\alpha = 0.01$	Sig.
Subplot analysis:							
Subplot treatment (variety)	16.8617	2	8.4309	109.63	3.55	6.01	$P < 0.01$
A × B interaction	0.8416	4	0.2104	2.74	2.93	4.58	n.s.
Subplot residual	1.3834	18	0.0769				
Subplot total	19.0867	24					
Total	20.5475	35					

Conclusion: the F-ratio for the day-length extension treatment (main plots) > critical F value at $\alpha = 0.05$; there is a significant difference in yield of the strawberries between the day-length extension treatments ($P < 0.05$).

The F-ratio for the variety treatment (subplots) > critical value at $\alpha = 0.01$; there is a significant difference in yield between the different varieties of the strawberries ($P < 0.01$).

The F-ratio for the day-length extension × variety interaction < critical F value at $\alpha = 0.05$; there is no significant interaction between the day-length extension and variety treatment factors.

Note also that the F-ratio for the block is smaller than the critical F value at $\alpha = 0.05$; it is concluded that there is no significant difference between the blocks.

9.6 Nested or Hierarchical Designs

In all of the factorial designs that have been discussed so far every individual replicate in the experiment is of equal status and contributes a single measurement. For example, in an experiment to examine the effect of irrigation treatments on the water status of trees, a given number of leaves may be collected randomly from differently irrigated plots of trees and their water content measured. A set of sample means is then determined and, if the differences are large enough, we may be able to conclude that irrigation significantly affects the leaf water status. We will not, however, be able to say anything about the trees from which the leaves are collected. Alternatively, we could randomly select a given number of trees within each plot and then subsample a given number of leaves from each of these trees. There is now the possibility for inspecting the variation between trees within the plots as well as the differences between the treatment plots. This type of design is termed a **nested** or **hierarchical** design because one factor, in this case the different trees, becomes 'nested' within another factor, in this case irrigation level.

The key feature of a nested design that distinguishes it from a normal factorial design is that the categories of the nested factor within each level of the main factor are different. In the example described here, different trees give rise to the leaf samples

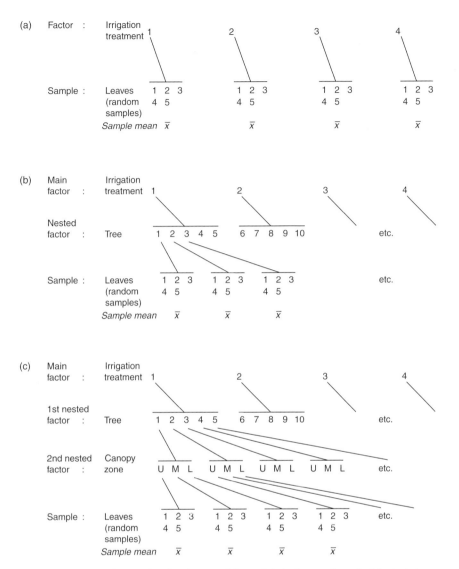

Fig. 9.5. Comparison of a randomized normal factorial design and nested factorial experimental designs. Experiment to investigate the effect of irrigation on leaf water status in trees using (a) fully randomized design; (b) single-factor nested design; (c) two-nested factor design.

within each of the main irrigation treatments. This essential difference between a normal factorial design and a nested design is illustrated in Fig. 9.5. Nested designs can be applied to more complex multiple factor experiments. For example, in the tree irrigation experiment it might be of interest to determine to what extent the position in the canopy affects the leaf water content. In this case the tree canopy within each irrigation plot may be divided into an upper, middle and lower zone. A given number of replicate leaves may then have been collected from each of these canopy zones within each tree. Again this is illustrated in Fig. 9.5.

9.6.1 Analysis of nested designs

In the statistical analysis of a nested design, it is necessary to test a null hypothesis concerning the main factor effect and, in addition, a null hypothesis concerning each of the nested factors. The ANOVA needs, therefore, to consider and account for the variation between main groups (i.e. main factor) and subgroups (i.e. nested factors) as well as the variation within samples (i.e. residual effects). Thus, in the tree irrigation experiment, we would need to determine the sum of squares (SS) and mean square (MS) for the irrigation treatment, for the trees nested within the irrigation treatments and for the leaves within samples. The calculation procedure to achieve this is shown in Example 9.5.

Example 9.5. Analysis of variance of a nested design.

The effect of irrigation on leaf water status of trees.

The design: four irrigation treatments (A, B, C, D) were applied to a sample of trees. Leaf water content was determined in subgroups of five randomly selected leaves from each of five trees nested within each irrigation treatment. (The design of this experiment is shown in Fig. 9.5b.) The table shows the total and mean leaf water content for each subsample of leaves and for each irrigation treatment.

The data:

Irrigation treatment		Leaf water content (g/g leaf fresh weight) (Total and mean of subgroups comprising five leaves from each tree)					Irrigation total	Irrigation mean
				Tree				
A		1	2	3	4	5		
	Total	4.04	3.77	4.25	4.13	4.21	20.40	0.816
	Mean	0.808	0.754	0.85	0.826	0.842		
				Tree				
B		6	7	8	9	10		
	Total	4.24	3.94	4.37	3.92	4.26	20.73	0.830
	Mean	0.848	0.788	0.874	0.784	0.852		
				Tree				
C		11	12	13	14	15		
	Total	3.54	3.25	3.47	3.54	3.68	17.48	0.699
	Mean	0.708	0.65	0.694	0.708	0.736		
				Tree				
D		16	17	18	19	20		
	Total	3.27	3.25	3.29	3.14	3.39	16.34	0.654
	Mean	0.654	0.65	0.658	0.628	0.678		

$$\Sigma x = 74.95$$
$$\Sigma x^2 = 56.9513$$

continued

Example 9.5. Continued.

N = total no. of leaves = 100; n = no. leaves per subgroup = 5
$N_{irrigation}$ = no. of irrigation treatments = 4; $n_{irrigation}$ = no. leaves per irrig. treatment = 25
$N_{subgroup}$ = no. of subgroups (trees) = 20

Null hypothesis: the water content of leaves from different trees treated with different amounts of irrigation is the same.

The analysis: the correction term (CT) and the total sum of squares are now determined in the normal manner:

CT = $(\Sigma x)^2 / N$ = $74.59^2/100$ = 56.1750
SS_{total} = $\Sigma x^2 - CT$ = 56.9513 − 56.1750 = 0.7763
                with DF = $N - 1$ = 99

The sum of squares for all subgroups, which contains the main factor (irrigation) effect and the nested factor (trees) effect, is determined:

$SS_{subgroups}$ = $\Sigma T_{subgroup}^2 / n - CT$ = $(4.04^2 + 3.77^2 + ... + 3.39^2)/5 - 56.1570$ = 0.6510
                with DF = $N_{subgroup} - 1$ = 19

The sum of squares for the main factor is determined by:

$SS_{irrigation}$ = $\Sigma T_{irrigation}^2 / n_{irrigation} - CT$ = $(20.40^2 + 20.73^2 + 17.48^2 + 16.34^2) / 25 - 56.1570$
                                = 0.5625
                with DF = $N_{irrigation} - 1$ = 3

The sum of squares for the nested factor and the residual effects can then be found by subtraction:

SS_{trees} = $SS_{subgroups} - SS_{irrigation}$ = 0.6510 − 0.5625 = 0.0885
                with DF = $DF_{subgroups} - DF_{irrigation}$ = 16
$SS_{residual}$ = $SS_{total} - SS_{subgroups}$ = 0.7763 − 0.6510 = 0.1252
                with DF = $DF_{total} - DF_{subgroups}$ = 80

In the ANOVA the mean square (MS) at each nested level estimates the variability for that level plus all levels below. F-ratios are, therefore, determined by comparing the MS for each factor with that of the factor immediately below it in the nested hierarchy.

ANOVA table.

Source of variation	SS	DF	MS	F-ratio	F_{crit} $\alpha = 0.05$	$\alpha = 0.01$	Sig.
All sub-groups	0.6510	19					
Main factor (irrigation)	0.5625	3	0.1875	$\dfrac{0.1875}{0.0055} = 34.01$	3.24	5.29	$P < 0.01$

continued

Example 9.5. Continued.

Nested factor (trees)	0.0885	16	0.0055	$\dfrac{0.0055}{0.0016} = 3.44$	1.77	2.24	$P < 0.01$
Residual	0.1252	80	0.0016				
Total	0.7763	99					

Conclusion: there is a significant statistical difference between irrigation treatments ($P < 0.01$) and also between replicate trees within treatments ($P < 0.01$). Therefore, while a significant effect of the irrigation treatments has been detected, the test also indicates that the trees are not uniform and respond significantly differently to each other.

In the ANOVA the mean square (MS) at each nested level estimates the variability for that level plus all levels below it in the hierarchy. The F-ratios are, therefore, determined by comparing the MS for each factor with that of the factor immediately below it, thus:

$$F\text{-ratio (main factor)} = \frac{\text{MS main factor}}{\text{MS nested factor}} \text{ and}$$

$$F\text{-ratio (nested factor)} = \frac{\text{MS nested factor}}{\text{MS residual}}$$

Therefore in the irrigation experiment illustrated in Fig. 9.5(b) the F values would be obtained by:

$$F\text{-ratio (main factor)} = \frac{\text{MS irrigation}}{\text{MS trees}} \text{ and}$$

$$F\text{-ratio (nested factor)} = \frac{\text{MS trees}}{\text{MS residual}}$$

9.6.2 Nesting within multiple-factorial experiments

In the nested designs illustrated in Fig. 9.5, there is a linear structure in which one factor is totally nested within another, i.e. trees within irrigation treatments, canopy zones within trees, leaves within canopy zones, so that, while the significance of each factor may be determined, interactions between the factors are not identifiable. Occasionally, however, nested factors are incorporated into two or more crossed factors, leading to very complex experimental designs. For example, in the irrigation experiment, it may have also been required to inspect the interactive effect of applying a mulch around the trees on the water status of leaves. Each irrigation treatment is therefore combined which each mulch treatment but within each combination there are nested a different set of trees and replicate leaves are sampled from each tree. This design is illustrated in Fig. 9.6. In order to inspect the relevant null hypotheses concerning the two treatment factors, their interaction and the nested factor, five different mean square values are computed; MS for factor A (e.g. irrigation treatment), MS for factor B (e.g. mulch treatment), MS for the

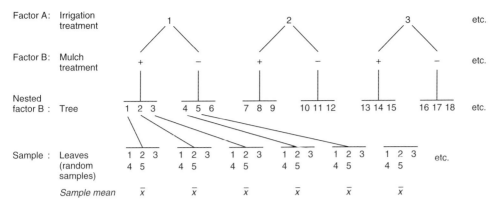

Fig. 9.6. Nested multiple-factor experimental design. Experiment to investigate the effect of irrigation treatments in combination with mulching treatments on leaf water status in trees.

factor A × B interaction (e.g. irrigation × mulching), MS for the nested factor (e.g. trees within main factors) and the residual MS (e.g. for leaves within samples).

The *F*-ratios for the two main treatment factors, the interaction and the nested factor are obtained by comparing the MS for each factor with that of the factor immediately below it in the nested hierarchy. Thus:

$$F\text{-ratio (factor A)} = \frac{\text{MS factor A}}{\text{MS nested factor}}$$

$$F\text{-ratio (factor B)} = \frac{\text{MS factor B}}{\text{MS nested factor}}$$

$$F\text{-ratio (factor A×B)} = \frac{\text{MS interaction A} \times \text{B}}{\text{MS nested factor}}$$

$$F\text{-ratio (nested factor)} = \frac{\text{MS nested factor}}{\text{MS residual}}$$

An explanation of the computation of ANOVA for this type of design is not contained in this book but is a logical extension of the techniques previously described for factorial and simple nested designs. The amount of calculation involved would, in any case, usually necessitate use of a computer statistics package.

9.7 Repeated Measures Designs

A repeated measures experimental design is one in which multiple sets of measurements are made on the same set of subjects. This situation can arise in two main ways. The subjects may be measured initially and then, following imposition of a range of treatments, may be remeasured to ascertain the effect of the treatments. This type of experiment is often referred to as a **within-subjects** design (this term being often used by computer statistical packages). Alternatively, a set of pretreated subjects may be measured repeatedly at different time intervals, either to simply provide more replicate measurements or to examine specifically whether the treatment effect significantly changes with time. In this case the repeated measures design is better referred to as a **serial measurements design**.

9.7.1 Within-subjects design

This type of design is commonly employed in situations where a limited number of subjects is available and a fully randomized replicated factorial experiment is not possible. Often, however, a within-subjects design represents a more appropriate approach than using a fully replicated randomized design irrespective of resource limitations. For example, it may be required to determine the effect of a number of different diets on the milk yield of cows. Within a dairy herd there will obviously be considerable variability in yield between individuals and a randomized approach may lead to uneven distribution of high and low yielders among the treatment groups. If a large number of subjects was available, the problem would be over-come by blocking; however, this is problematic where the number of individuals available is small. Instead, each member of the herd is fed one of the different diets and the yield of each cow measured. The diet of each cow is then changed and the milk yield of each remeasured after an appropriate interval of time. This procedure is repeated until the milk yield of each cow has been determined in response to each different diet. Consequently at the end of the trial a set of replicated milk yield data has been obtained for each diet but each data set is based on the meas-urement of the same individual cows. The design of a within-subjects repeated measures experiment of this type is illustrated in Fig. 9.7.

The within-subjects design has the clear advantage over normal factorial designs of making more economic use of the available experimental subjects. The major dis-advantage, however, is the possibility that the results are influenced by carry-over effects where the effect of one treatment influences the response to a subsequent treat-ment. The possibility of this occurring is, of course, increased as the time interval between treatments is decreased. The only way of counteracting this is to apply the treatments to each subject in a different and random sequence. Thus, in the example in Fig. 9.7, rather than applying the different diets to each cow in the same sequence, as shown in Fig. 9.7(a), it is more appropriate to use a different and randomly selected sequence for each cow, as depicted in Fig. 9.7(b).

9.7.2 Analysis of repeated-measures designs by ANOVA

The data samples collected from a repeated-measures (within-samples) designed experi-ment are all obtained from the same group of subjects and are not, therefore, independ-ent. A normal ANOVA based on independent samples cannot, therefore, be employed (much in the same way that matched sample pairs do not allow use of the normal Student's *t*-test; see section 6.7). The ANOVA is undertaken by first determining the total sum of squares (SS_{total}) in the usual manner but this is then partitioned between a sum of squares component that describes between-subjects variation and a sum of squares component that describes within-subject variation. Since all the treatment meas-urements are made on the same set of subjects, the within-subject variation must include the treatment effect and the residual (random) effect. The $SS_{within-subjects}$ is, therefore,

(a)

Cow	Supplied diet				
	Month 1	Month 2	Month3	Month4	Month 5
1	Diet A	Diet B	Diet C	Diet D	Diet E
2	Diet A	Diet B	Diet C	Diet D	Diet E
3	Diet B	Diet C	Diet D	Diet E	Diet A
4	Diet B	Diet C	Diet D	Diet E	Diet A
5	Diet C	Diet D	Diet E	Diet A	Diet B
6	Diet C	Diet D	Diet E	Diet A	Diet B
7	Diet D	Diet E	Diet A	Diet B	Diet C
8	Diet D	Diet E	Diet A	Diet B	Diet C
9	Diet E	Diet A	Diet B	Diet C	Diet D
10	Diet E	Diet A	Diet B	Diet C	Diet D

(b)

Cow	Supplied diet				
	Month 1	Month 2	Month3	Month 4	Month 5
1	Diet C	Diet E	Diet A	Diet B	Diet D
2	Diet E	Diet B	Diet D	Diet C	Diet A
3	Diet B	Diet A	Diet D	Diet E	Diet C
4	Diet D	Diet C	Diet B	Diet A	Diet E
5	Diet A	Diet D	Diet C	Diet E	Diet B
6	Diet A	Diet E	Diet B	Diet D	Diet C
7	Diet E	Diet B	Diet A	Diet C	Diet D
8	Diet C	Diet D	Diet E	Diet B	Diet A
9	Diet B	Diet A	Diet C	Diet D	Diet E
10	Diet D	Diet C	Diet E	Diet A	Diet B

Fig. 9.7. Example of repeated measures (within-subjects) experimental design. The effect of five treatment diets (A–E) on the milk yield of a small sample of dairy cows ($n = 10$) in which (a) treatment sequence is constant; (b) treatment sequence is randomized. Where it is likely that a 'carry-over' effect occurs (the effect of a given treatment influencing the response to a following treatment), then a randomized treatment sequence should be employed.

further partitioned between the treatment and residual components. The mean squares (MS) are subsequently derived in the normal way, i.e. SS/DF. The F-ratio for the main treatment is then the ratio of the treatment MS to the residual MS. If required, an F-ratio for testing the difference between subjects can also be obtained. Thus:

$$F\text{-ratio (main treatment)} = \frac{MS \text{ treatment}}{MS \text{ residual}}$$

$$F\text{-ratio (subjects)} = \frac{MS \text{ subjects}}{MS \text{ within-subjects}}$$

The full calculation procedure is illustrated in Example 9.6.

Assumptions of ANOVA for repeated measures

As has been previously discussed, it is a requirement of ANOVA that all samples of replicate measurements should possess homogeneous variances. In a

Example 9.6. Design and analysis by ANOVA of a within-subjects repeated-measures experiment.

The effect of five different diet supplements on milk yield in dairy cows.

The design: In a repeated-measures designed trial each of ten cows received five different diet supplements (diets A, B, C, D, E) given over five sequential monthly periods. As there may be natural yield differences between the cows and a 'carry-over effect' may occur from one diet to the next, the order of the treatments was randomized. The design follows that illustrated in Fig. 9.7b.

The data:

											Mean daily milk yield (kg)	
	1st treatment		2nd treatment		3rd treatment		4th treatment		5th treatment		Cow total	Cow
Cow (subject)	Diet	Yield	Diet	Yield	Diet	Yield	Diet	Yield	Diet	Yield	($T_{subjects}$)	mean
Annnie	C	28.4	E	26.4	A	27.3	B	22.7	D	25.0	129.8	25.96
Betty	E	35.0	B	33.1	D	32.1	C	21.5	A	22.8	144.5	28.90
Deardrie	B	30.5	A	34.0	D	38.5	E	29.6	C	22.6	155.2	31.04
Flossie	D	35.7	C	29.4	B	27.3	A	24.6	E	22.9	139.9	27.98
Greta	A	41.1	D	43.4	C	35.4	E	30.9	B	24.6	175.4	35.08
Jemina	A	25.1	E	23.6	B	20.5	D	28.7	C	23.2	121.1	24.22
Mabel	E	25.3	B	24.4	A	24.9	C	27.4	D	23.0	125.0	25.00
Sally	C	30.4	D	37.8	E	28.0	B	26.0	A	24.6	146.8	29.36
Toffee	B	27.2	A	22.5	C	25.4	D	29.0	E	23.4	127.5	25.50
Winifred	D	44.4	C	38.5	E	37.3	A	33.0	B	27.2	180.4	36.08

GT= 1445.6

	Diet treatment totals and means				
	A	B	C	D	E
Total ($T_{treatment}$)	279.9	263.5	282.2	337.6	282.4
Mean	27.99	26.35	28.22	33.76	28.24

No. of observations (N) = 50;

No. of diet treatments ($N_{treatment}$) = 5;   No. of measurements per diet ($n_{treatment}$) = 10

No. of subjects (cows) ($N_{subjects}$) = 10; No. of measurements per subject ($n_{subjects}$) = 5

Null hypothesis: there is no difference in the daily milk yield in cows fed with the different diet supplements.

The analysis:

$\Sigma x^2 = 43,553.32$; $\Sigma T_{treatment}^2 = 421136.6$; $\Sigma T_{subjects}^2 = 212793.4$;

Correction term (CT) = $(\Sigma x)^2/N$ = 41795.19

continued

Example 9.6. Continued.

Sum of squares:

Total	$= \Sigma x^2 - CT$	$= 43553.32 - 41795.19$	$= 1758.13$

$$\text{with DF} = N - 1 = 49$$

Subjects	$= \Sigma T_{subjects}^2 / n_{subjects} - CT$	$= 212793.4/5 - 1795.19 \quad = 763.49$

$$\text{with } DF = N_{subjects} - 1 = 9$$

Within-subjects	$= SS_{total} - SS_{subjects}$	$= 1758.13 - 763.49 \qquad = 994.64$

$$\text{with DF} = DF_{total} - DF_{subjects} \qquad = 40$$

Between
treatments (diets) $= \Sigma T_{treatment}^2 / n_{treatment} - CT \qquad = 421136.6/10 - 41795.19$

$$= 318.47$$

$$\text{with DF} = N_{treatment} - 1 = 4$$

Residual	$= SS_{within\text{-}subjects} - SS_{diets}$	$= 994.64 - 318.47 \qquad = 676.17$

$$\text{with DF} = DF_{total} - (DF_{subjects} + DF_{treatment}) = 36$$

ANOVA table.

Source of variation	SS	DF	MS	F-ratio	F_{crit} $\alpha = 0.05$	$\alpha = 0.01$	Sig.
Subjects (cows)	763.49	9	84.83	$\dfrac{84.83}{24.87} = 3.41$	2.12	2.89	$P < 0.01$
Within-subjects	994.64	40	24.87				
Treatment (diets)	318.47	4	79.62	$\dfrac{79.62}{18.78} = 4.24$	2.63	3.89	$P < 0.01$
Residual	676.17	36	18.78				
Total	1758.13	49					

Conclusion: the F-ratio for the diet treatments $> F_{critical}$ at $\alpha = 0.01$; it is concluded that there is a significant difference in milk yield between diet treatments ($P < 0.01$). The F-ratio for subjects (cows) $> F_{critical}$ at $\alpha = 0.01$; there is a significant variation in milk yield between individual cows ($P < 0.01$).

repeated-measures design in which the sets of measurements are separated by intervals of time, this may become a problematic assumption. For example, milk yields of cows may naturally decrease during the lactation cycle so that variances between measurements become smaller over time. Much the same may happen when determining rates of growth of animals or plants. Furthermore, repeated-measures designs assume that the ranking order of subjects remains basically constant throughout all the measurement samples, i.e. the same individual consistently produces the highest value, another individual consistently produces the second highest value, etc. If this assumption is violated, then there is an increased risk of Type I errors occurring. (Conversely, however, if this assumption holds, then the repeated-measures analysis is actually more powerful than the equivalent one-way ANOVA.)

9.7.3 Analysis of serial measurements

In an experiment where subjects are presented with different treatments initially and then a response measured repeatedly over time, an ANOVA is not usually valid. As already discussed, such an experiment cannot be considered as a factorial experiment because the repeat measurements are made on the same sample of subjects and, furthermore, it is unlikely that samples at each time interval have homogeneous variances. For these reasons, time, which cannot of course be randomized in the way that an applied treatment can, should not be treated as a factor within an ANOVA. A simple solution is, instead, to inspect the difference between the treated subjects at each time interval by performing a series of independent one-way ANOVA tests. This approach is fine unless it is specifically required to determine whether there are significant differences in the data between the time intervals themselves. Under these circumstances a different type of approach is required. One technique available for analysis of repeated measures that does not require any of the assumptions required by ANOVA is that of **multivariate analysis of variance** (MANOVA). This is a complex technique, however, usually requiring use of specialized computer packages, and further consideration of it is beyond the scope of this book; the interested reader is referred to alternative texts, e.g. Zar (2007), Quinn and Keough (2002). More simply, the data could be plotted graphically against time and statistical descriptors of a plotted best-fit line or curve, such as the slope, estimated and compared. The techniques for achieving this are considered in Chapter 12.

10 Non-parametric Sample Comparison Tests

Parameter: quantity constant in case considered, but varying in different cases.

(Concise Oxford English Dictionary)

- Introduction to non-parametric sample comparison testing.
- The Mann–Whitney *U* test of statistical significance for analysing the difference between two independent samples.
- Non-parametric tests for analysing the difference between two non-independent paired samples: the sign test and Wilcoxon's signed-rank test.
- A non-parametric technique for analysing multiple-sample experiments: the Kruskal–Wallis test.
- A non-parametric technique for analysing randomized block designed experiments: Friedman's test.

10.1 Introduction to Non-parametric Significance Testing

All the sample comparison significance tests discussed so far in this book have been parametric tests in that they analyse experiments where sample means and variances are employed as estimates of the population means and variances. The mathematical procedure of such tests relies on the assumption that the samples are extracted from normally distributed populations with equal variances. In situations where these are invalid assumptions, for example, when the data are skewed, when data are in the form of counts or percentages and/or when the variances of the samples to be compared are not statistically homogeneous, an alternative approach is required. What is needed under these circumstances is a test that produces a test statistic which is independent of the data distribution of the population. Such tests are referred to as being 'distribution-free' and, since they do not necessitate employing sample statistics as estimates of population parameters, they are termed **non-parametric tests**.

The general technique used for statistically describing a sample of data that does not rely on the data distribution is to employ the rank values of the data. The ranks attributed to a particular set of data are not affected by the magnitude of the data values or the range and distribution of the values. For example, in a sample of 20 values, the largest value will have rank 1 and the smallest will have rank 20, irrespective of the size of the actual numerical differences between the values. Non-parametric tests that operate on the ranks of the data rather than the actual data do not, therefore, require any assumptions to be made about the type of data or the data distribution involved. In effect, such tests examine the difference between the sample medians rather than the population means, and the null hypothesis may be cited in relation to sample medians.

Since non-parametric tests based on rank values can be used to analyse samples of any type of numerical data, it seems reasonable to ask why one should not use non-parametrical tests all the time and never bother with parametric analyses at all. The reason is fairly obvious. By ranking data, a certain amount of information about the nature of the data is lost; as pointed out above, ranking takes no regard of the magnitude or the frequency distribution of the data. Consequently non-parametric tests are much less sensitive to small differences between data sets; to use correct statistical terminology, *non-parametric tests are less powerful than parametric tests*. Therefore, where data can be shown to adhere to the required assumptions for performing a parametric test, such a test is to be preferred as it will always be more likely to be able to detect smaller differences as being significant than its non-parametric equivalent.

10.2 A Non-parametric Test for Analysing the Difference between Two Independent Samples: The Mann–Whitney *U* Test

10.2.1 The principles of the Mann–Whitney *U* test

The most frequently used non-parametric test for inspecting the difference between two independent samples, is the **Mann–Whitney *U* test**. The basic premise of this test is that if two samples are homogeneous then when the two samples of data are pooled and ranked there will be considerable overlap of the ranks attributed to the data values of each sample. The greater the difference between the samples, however, the less overlap there will be between the ranks of the values of the two samples. This is illustrated in Fig. 10.1. What is required, therefore, is a test statistic which measures the extent of the overlap of the rank values from each of the two samples and which has a known probability distribution under the null hypothesis so that the probability that an observed set of data complies with the null hypothesis can be determined. The Mann–Whitney test statistic, denoted *U*, satisfies this requirement exactly.

The Mann–Whitney test statistic *U* is obtained by examining each value in turn in one of the samples and counting the number of values in the second sample which have a lower value. These counts are summed to give the Mann–Whitney test statistic which is denoted by the symbol *U*. For example, suppose we had two samples, each of five values, as follows:

Sample A: 3, 7, 11, 15, 19

Sample B: 5, 9, 10, 14, 17

If we take the first value in sample A, i.e. 3, then there are no values in sample B that have a lower numerical value than 3. If we take the second value in sample A, i.e. 7, then one value in sample B, i.e. 5, has a lower value than 7. Similarly, for the remaining three values in sample A, i.e. 11, 15, 19, the number of values in sample B that are lower are 3, 4 and 5 respectively. The test statistic *U* is then the sum of 0+1+3+4+5 = 13. Note that in this case the values have effectively been ranked such that the lowest value has rank 1 and the highest value has rank 5; however, it is equally valid to rank from highest to lowest as long as both samples are ranked in the same direction.

If the above procedure is repeated, but now based on the second sample, a second alternative value for U will be obtained. (In the above example, if we count the total number of values in sample A that have a lower rank than each value in sample B we obtain a value of U of 1+2+2+3+4 = 12.) In order to distinguish these two values of U, they are called U_A and U_B. The relation between U_A and U_B is inverse, so that the larger the value of U_A the smaller will be the value of U_B and vice versa. It then follows that the smaller the overlap between the two samples the greater will be the difference between the values of U_A and U_B and therefore the greater the probability of a significant difference existing between the samples. The basis of the calculation of U_A and U_B is illustrated in Fig. 10.1.

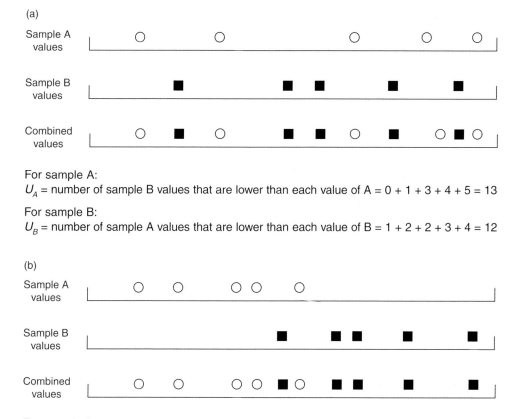

(a)

Sample A values

Sample B values

Combined values

For sample A:
U_A = number of sample B values that are lower than each value of A = 0 + 1 + 3 + 4 + 5 = 13

For sample B:
U_B = number of sample A values that are lower than each value of B = 1 + 2 + 2 + 3 + 4 = 12

(b)

Sample A values

Sample B values

Combined values

For sample A:
U_A = number of sample B values that are lower than each value of A = 0 + 0 + 0 + 0 + 0 = 1

For sample B:
U_B = number of sample A values that are lower than each value of B = 4 + 5 + 5 + 5 + 5 = 24

Fig. 10.1. Determination of the test statistic U for a Mann–Whitney test. (a) Two samples with a small difference between them show a large overlap when values are combined and ranked on the same scale. (b) Two samples with a large difference between them show a small overlap when values are combined and ranked on the same scale. The test statistic U takes the value of U_A or U_B, whichever is the larger.

For all but the smallest sized samples, the calculation of U_A and U_B is a cumbersome process if performed as described above; fortunately there is a mathematical short cut. First, the data of the two samples are combined into one data set and ranked, taking care that any tied values are attributed the same shared rank value. The ranking value of each data item within each sample is summed to give two sum of rank values, termed ΣR_A and ΣR_B respectively. For example, if this procedure is undertaken for the samples A and B mentioned above, we obtain:

Sample A		Sample B	
Value	Rank	Value	Rank
3	1	5	2
7	3	9	4
11	6	10	5
15	8	14	7
19	10	17	9
	$\Sigma R_A = 28$		$\Sigma R_B = 27$

The two alternative test statistics U_A and U_B are then calculated by the derived formulae:

$$U_A = \Sigma R_A - \frac{n_A(n_A + 1)}{2} \quad \text{and} \quad U_B = \Sigma R_B - \frac{n_B(n_B + 1)}{2}$$

where n_A and n_B are the number of values in samples A and B respectively.

A convenient mathematical check of the correctness of the calculation can be applied at this stage because it follows that $\Sigma R_A + \Sigma R_B = N(N+1)/2$, where $N = n_A + n_B$, and also that $U_A = (n_A \times n_B) - U_B$. Therefore for the given samples A and B:

$$U_A = 28 - \frac{5(5 + 1)}{2} = 13 \quad \text{and} \quad U_B = 27 - \frac{5(5 + 1)}{2} = 12$$

As there are two potential test statistics, U_A and U_B, it is necessary to know which to use to perform the Mann–Whitney test. For a two-tailed test, where the direction of the difference is not specified, the larger of the two values can be employed as the test statistic and can be referred to as $U_{observed}$ (but see below). For a one-tailed test, the choice of either U_A or U_B will depend on the direction of the difference that has been predicted to occur. If it is required to test whether sample A is greater than sample B, then U_A will be selected as the test statistic but, if the hypothesis is that sample B is the larger of the two, then U_B becomes the test statistic.

(It should be noted that some texts and some statistical software take an alternative approach in which the smaller of U_A or U_B is employed as the test statistic. The null hypothesis is then rejected when the test statistic is smaller than the critical value. In this case, the critical U value must be read from a different version of the U probability table. While the two approaches are equally valid, the employment of the higher value as the test statistic is adopted here since the procedure is then in common with all other statistical tests described in this book.)

Ranking procedure when two or more values are tied

In cases where two or more data values are tied, it is very important that the correct procedure is used for attributing rank values. All tied values must be attributed the same rank value. For example, if there are two equal values that represent the fifth largest value in a data set, then these are both attributed the rank value of 5.5 on the basis that they use up the ranks 5 and 6 between them. The next largest value would then have rank value 7. Where more than two values are tied, then these must similarly all share the same rank value which is equivalent to the mid-point between the lowest and highest ranks that the values occupy. Thus, if the fifth largest value is shared by four data items, then between them they use up ranks 5, 6, 7 and 8 and each is therefore attributed the rank value 6.5 while the next highest data value has rank value 9. (It is important to note here that the rank function in Microsoft Excel® spreadsheet software does not operate correctly in this respect and, if this software is being employed to rank values in advance of performing a non-parametric data analysis, then a manual correction will need to be applied in the case of tied values.)

10.2.2   The Mann–Whitney U probability distribution

Just as with other test statistics, such as Student's t, the statistic U has a theoretical probability distribution which allows statisticians to relate values of U to the probability of their occurrence given that the null hypothesis is true and thus to formulate a table of critical values of U. To determine the probability distribution of U for any given sample size, it is necessary to know all the possible values that U can take and the probability of obtaining each of these possible values. For example, consider two samples A and B that have been randomly selected from the same population and each contain three values. Since both samples are derived from the same population, we expect no significant difference to exist between them and the six values therefore have an equal chance of occurring in each rank position. Therefore all possible values for U_A and U_B and their theoretical frequency can be determined as in Table 10.1.

Table 10.1. All possible values of Mann–Whitney test statistics U_A and U_B based on two samples of three values.

Sample from which value is extracted — rank →	U_A	U_B	Sample from which value is extracted — rank →	U_A	U_B
A, A, A, B, B, B	0	9	B, A, A, A, B, B	3	6
A, A, B, A, B, B	1	8	B, A, A, B, A, B	4	5
A, A, B, B, A, B	2	7	B, A, A, B, B, A	5	4
A, A, B, B, B, A	3	6	B, A, B, A, A, B	5	4
A, B, A, A, B, B	2	7	B, A, B, A, B, A	6	3
A, B, A, B, A, B	3	6	B, A, B, B, A, A	7	2
A, B, A, B, B, A	4	5	B, B, A, A, A, B	6	3
A, B, B, A, B, A	5	4	B, B, A, B, A, A	8	1
A, B, B, B, A, A	6	3	B, B, B, A, A, A	9	0

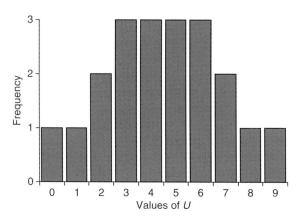

Fig. 10.2. Frequency distribution of the Mann–Whitney U test statistic based on a sample size of 3 for both samples.

A frequency distribution plot of U (i.e. either U_A or U_B), based on all possible rankings of sample A and sample B values, can therefore be constructed as shown in Fig. 10.2.

In the case of small samples sizes, the frequency distribution of U is poorly defined in that a number of different possible values of U have the same probability of occurrence (as shown in Fig. 10.1). Consequently the Mann–Whitney test will be unreliable when comparing small samples. As sample size increases, the definition of the frequency distribution rapidly improves and it is generally considered that once both samples contain five values the Mann–Whitney test should then operate with an acceptable reliability.

Based on the theoretical frequency distribution of U, critical values of U can be determined that cut off the tails representing defined levels of probability of occurrence. For example, where both samples contain five values (i.e. $n_A = n_B = 5$), then the 5% critical value for U is 23, which means that there is a 0.05 probability or less of obtaining a value of $U_{observed}$ that is equal to or larger than 23. Thus if 0.05 had been set as the critical probability for rejecting the null hypothesis, then a value of $U_{observed}$ of ≥ 23 would be considered evidence of a significant difference ($P \leq 0.05$). The two-tailed critical values for the Mann–Whitney U distribution based on different sample sizes are given in Appendix Table A3; note that the table has to be consulted at the sample size of both samples, i.e. n_A and n_B.

In order to perform a one-tailed Mann–Whitney U test, the test statistic $U_{observed}$ needs to be compared with the one-tailed critical values; these can be obtained from the two-tailed U table by consulting the table at twice the required probability. It then follows that, if the observed value of U is equal to or greater than the critical value of U at the selected critical level of probability, then H_0 can be rejected and a significant difference between the samples may be claimed.

10.2.3 Limitations to the use of the Mann–Whitney U test

As already noted, while the sizes of the two samples do not need to be equal, both samples should be fairly large (at least >5) if we are to expect a Mann–Whitney test to reliably detect the presence of significant differences. Furthermore, although the sample data do not need to adhere to a given data distribution, it is important that the two samples to be compared do possess similar shaped distributions and have similar variances. In particular, the ranking procedure involved will not work if the samples are skewed in opposite directions. It is therefore important to identify that a Mann–Whitney test is to be used during the early stages of the experimental design process so that both a sufficient number of replicates and an appropriate sample selection procedure can be planned. If the samples do not possess similar distributions, then the researcher may need to resort to transforming the data prior to analysis.

Example 10.1. Example of a Mann–Whitney U test for testing for a significant difference between two independent non-normally distributed samples.

A test to determine whether there is a significant difference in the percentage leaf predation by leaf moth larvae in plants of two different varieties of gooseberry.

% Leaves infested per plant			
Variety A		Variety B	
Value (%)	Rank	Value (%)	Rank
25	14	15	7
38	21	32	19
31	18	20	11
29	17	8	2
40	22	12	6
26	15	24	12.5[a]
24	12.5[a]	18	10
16	8	11	5
10	4	5	1
35	20	28	16
		9	3
		17	9
$n_A = 10$	$\Sigma R_A = 151.5$	$n_B = 12$	$\Sigma R_B = 101.5$	(N = 22)

[a] Note that there are two tied values of 24. Since these should occupy ranks 12 and 13, they are both attributed the rank value 12.5. The next value therefore has rank 14.

(Mathematical check: $\Sigma R_A + \Sigma R_B = 253 = N(N+1)/2$)

Null hypothesis: there is no difference in the median of the percentage of leaves predated between the two varieties.

$$U_A = \Sigma R_A - \frac{n_A(n_A+1)}{2} \quad = 151.5 - 55 = 96.5$$

$$U_B = \Sigma R_B - \frac{n_B(n_B+1)}{2} \quad = 101.5 - 78 = 23.5$$

continued

Example 10.1. Continued.

Mathematical check: $U_A = 96.5 = (n_A \times n_B) - U_B$

Since $U_A > U_B$, then U_A is employed as the test statistic in a Mann–Whitney two-tailed test of significance.

From tables the critical two-tail U value at $\alpha = 0.05$ and sample sizes of $n_A = 10$ and $n_B = 12$ is 91.0. Since U_A is smaller than 91.0, the null hypothesis can be rejected.

Conclusion: there is a significant difference in leaf predation between the two varieties of gooseberry by leaf moth larvae ($P < 0.05$).

10.2.4 Calculation of Mann–Whitney U test for large sample sizes ($n > 20$) using a normal approximation

The majority of published tables of Mann–Whitney critical U values do not give values for sample sizes of greater than 20. However, if the sample sizes are large ($n > 20$), the distribution of U becomes very close to the normal distribution and the normal distribution can be employed for testing the significance of U.

A mean for the distribution of U (μ_U) is first determined by:

$$\mu_U = \frac{n_A n_B}{2}$$

A value for the normally distributed test statistic z is determined by the standard method of dividing the difference between the observed value and the theoretical mean by the standard error. In this case this is obtained by:

$$z_{observed} = \frac{U_{obs} - \mu_U}{\sqrt{\dfrac{n_1 n_2 (N + 1)}{12}}}$$

where: $N = n_1 + n_2$;
U_{obs} = the larger of U_A or U_B.

Unfortunately, if there are tied ranks present in the data, a rather complicated correction is required to obtain a good approximation of the normal distribution. In this case, a quantity termed Σt is first determined by:

$$\Sigma t = \Sigma \left(t_i^3 - t_i \right)$$

where: t_i = number of ties in a group of tied values

The test statistic $z_{observed}$ is now determined by:

$$z_{observed} = \frac{U - \mu_U}{\sqrt{\dfrac{n_1 n_2}{N^2 - N} \times \dfrac{N^3 - N - \Sigma t}{12}}}$$

The value of $z_{observed}$ is then compared with the critical value of z (see Appendix, Table A1) at the required probability, and the null hypothesis may be either rejected or accepted accordingly.

It may be noted that, while most modern computer statistics packages will perform the Mann–Whitney U test readily, it is not always clear from the output whether the calculated value of U has been referred to the theoretical sampling distribution of U or whether it has been converted to a z value and compared with the standardized z values of the normal distribution.

10.3 Non-parametric Tests for Analysing the Difference between Paired Non-independent Samples

A number of non-parametric tests is available for analysing the difference between two non-normally distributed paired samples. Two of the most commonly used are the **sign test** and **Wilcoxon's signed-rank test**. These are relatively simple tests that inspect whether a particular variable has significantly increased or decreased in magnitude between the paired samples.

10.3.1 The sign test

The sign test is an extremely simple test for inspecting for significant differences between paired samples. It is based only on the direction of the difference between pairs of items, i.e. whether differences are positive or negative, and does not require the input of any quantitative values. It is, therefore, particularly useful when the variable concerned is difficult to quantify. For example, a sample of persons may have been asked to record whether a design modification produces an improvement or reduction in the visual appeal of a particular garden. The data can, therefore, be recorded as a positive effect, a negative effect or no effect at all. If, however, the data are numerical, then it is simply the sign of differences between pairs of items that is employed in the test. For example, a sample of persons may be asked to record how often they visit a certain public garden over a defined period of time and then, following design modifications, record the frequency of visits made to the park during a subsequent similar period. It is then only necessary to record for each person whether the number of visits has increased, decreased or remained the same.

The null hypothesis is that there is no difference between the two populations of measurements from which the samples are taken. If the null hypothesis is true it would be expected that the number of positive differences and the number of negative differences between pairs of items would be approximately the same. If, however, there was a significant difference present, then a difference between the number of positive and negative differences would be expected. The test statistic for the sign test is denoted S, or $S_{observed}$ if you prefer, and is simply the number of positive differences or the number of negative differences, whichever is the greater. Since the outcome from comparing each pair of items can only be '−' or '+' and under the null hypothesis there would be an equal chance of obtaining a '−' or '+' score, then essentially the data are binary and would be expected to follow a binomial distribution with $p = 0.5$. Therefore the test statistic can be compared with a set of theoretical values based on the binomial distribution in order to determine the probability of significance (Appendix: Table A4). The test is illustrated in Example 10.2.

Example 10.2. Comparison of two matched samples by the sign test.

Opinions of a sample of people on the effect on the visual appearance of a public garden due to an alteration in design.

Person	Visual appearance of garden improved	Visual appearance of garden deteriorated	Sign of difference
A	YES	NO	+
B	NO	YES	−
C	NO	YES	−
D	YES	NO	+
E	YES	NO	+
F	YES	NO	+
G	YES	NO	+
H	NO	YES	−
I	YES	NO	+
J	YES	NO	+
K	YES	NO	+
L	YES	NO	+
M	NO CHANGE	NO CHANGE	+/−
N	YES	NO	+
O	NO	YES	−

Number of +ve differences =	10.5
Number of −ve differences =	4.5

(Note that where no difference is recorded the value is split evenly between the +ve and −ve differences.)

Null hypothesis: there is no difference in people's opinion on the visual appearance of the garden before and after changes were made to the garden design.

The test statistic $S_{observed}$	= 10.5
Number of pairs, n	= 15
Critical value of S ($\alpha = 0.05$)	= 12 (read from tables)

Since the test statistic $S_{observed} < S_{critical}$, the null hypothesis is accepted.

Conclusion: there is no significant difference between people's opinion on the visual appearance of the garden before and after changes were made to the garden design ($P < 0.05$).

Two last points should be noted about the sign test which are the subject of some disagreement among statisticians and computer statistics packages. First, it has been assumed here that the test statistic $S_{observed}$ will take the larger value of the number of positive differences or the number of negative differences and the null hypothesis subsequently rejected when $S_{observed} \geq S_{critical}$. It is, however, equally valid for $S_{observed}$ to take the smaller of the two values and then reject the null hypothesis when $S_{observed} < S_{critical}$. While there are some theoretical arguments for the latter approach, it really makes no practical difference to the outcome of the test and, since taking the larger value is in line with all other tests described in this book, this is the preferred approach of this author. Second, sometimes one or more items in the sample may show no change between the two measurements and neither a '–' nor a '+' score can be awarded. A simple and quite valid solution to this is simply to remove such items from the analysis. Alternatively, such items can be split evenly between the categories by awarding half a positive score and half a negative score. It might be argued that the latter approach has the virtue of employing all the data available in the analysis rather than discarding data and this is the approach taken by the present author in Example 10.2. Again, however, which approach is taken makes no practical difference to the outcome of the test.

10.3.2 Wilcoxon's signed-rank test

If quantitative values for the paired samples are available, then, rather than employing the sign test, it is preferable to use a test that makes use of the quantitative information. The Wilcoxon's signed-rank test achieves this by converting the magnitude of the differences between the pairs of items to a rank value.

The test statistic for the Wilcoxon's signed-rank test is denoted by the symbol T. To determine the value of T, first the direction of the numerical difference between each pair of items is recorded (i.e. whether the value has increased or decreased) and then the absolute difference between the paired values determined. These differences are then ranked in ascending order, the smallest difference allotted rank value 1, the next largest difference allotted rank value 2, etc. If there is no change in value between a data pair, then the difference is zero and no rank value is allotted; such data pairs do not therefore contribute to the analysis. If there are two or more difference values that are tied, these must be allotted the same midpoint rank value, using the procedure described for the Mann–Whitney test (see section 10.2). The rank values associated with positive differences and the rank values associated with negative differences are then summed separately to yield two sum of rank values. If there was no significant difference between the two samples then it would be expected that these two values will be approximately equal, while the larger the discrepancy between them the greater is the likelihood that a significant difference is present. The larger of the two sum of rank values may then be selected as the test statistic, $T_{observed}$, and this value is then referred to the known probability distribution of T in order to establish the probability that the data support the null hypothesis.

The probability distribution of T

The probability distribution of T can be ascertained much in the same manner as that employed for the sampling distribution of the Mann–Whitney U statistic. If it is assumed that the two samples of data come from the same population, i.e. the null hypothesis is correct, then one would expect there to be an equal number of positive differences between data pairs as there are negative differences. Similarly, when the magnitude of the differences between data pairs is ranked, each rank value would have an equal chance of representing a positive difference or a negative difference. It is therefore possible to determine, for a given number of matched values, all the possible ranking sequences and thereby all the values that T can take and the number of times that each value of T occurs. (In fact, the number of possible sequences of positive and negative differences is 2^n where n is the number of paired data values.) For example, suppose there were just three data pairs (this is actually far too few pairs for the test to be valid), then the possible sequences of positive or negative differences and the resulting T values are as follows:

Rank position	
1 2 3	T based on +ve differences
+ + +	$T = 1+2+3 = 6$
+ + −	$T = 1+2 \quad = 3$
+ − +	$T = 1+3 \quad = 4$
− + +	$T = 2+3 \quad = 5$
+ − −	$T = 1 \quad = 1$
− − +	$T = 3 \quad = 3$
− + −	$T = 2 \quad = 2$
− − −	$T \quad = 0$

With just three data pairs ($n = 3$), clearly the values of T do not allow a sensible frequency distribution to be plotted. Once $n = 5$, however, the number of repeats of different T values produces a sufficiently well-defined frequency distribution to allow critical values of T to be determined which cut off the tails of the distribution at defined levels of probability and thus allow the performance of a valid significance test. The critical values of T are presented for different sample sizes in a T probability table (Appendix: Table A4).

To conduct the test, the larger of the two sum of rank values is employed as the test statistic $T_{observed}$ and this value is then compared with the critical values of T read at the selected level of probability. If $T_{observed}$ is greater than $T_{critical}$, then the null hypothesis may be rejected and a significant difference between samples can be claimed. The test is illustrated in Example 10.3.

As with all the other two-sample comparison tests described, both the sign test and the Wilcoxon's signed-rank test can be, if required, performed as one-tailed tests, if required, by adjusting the null hypothesis appropriately and referring the test statistic to the one-tailed critical values (which are the equivalent of the two-tailed values read at twice the required critical probability).

10.4 Non-parametric Multiple-sample Analysis

When an experiment produces three or more samples that require comparison but which cannot be assumed to belong to normally distributed distributions, including where the data are skewed or in the form of counts or percentages, it is invalid to employ the parametric ANOVA test. Under these circumstances, a multiple-sample distribution-free analysis is required, of which the most widely employed is the **Kruskal–Wallis** test.

Example 10.3. Comparison of two matched samples by Wilcoxon's signed-rank test.

Effect on the number of visits made to a public garden by a sample of persons over a period of a year due to alteration in garden design.

Person	No. visits made in year prior to change in garden design	No. visits made in year after change in garden design	Sign of difference	Absolute difference	Rank of +ve difference values	Rank of −ve differences values
A	0	8	+	8	8	
B	6	0	−	6		5.5
C	8	1	−	7		7
D	4	23	+	19	11	
E	5	10	+	5	4	
F	1	2	+	1	1.5	
G	23	47	+	24	12	
H	11	7	−	4		3
I	8	20	+	12	10	
J	6	15	+	9	9	
K	0	27	+	27	14	
L	2	28	+	26	13	
M	5	5		0		
N	5	11	+	6	5.5	
O	13	12	−	1		1.5
			Sum of ranks of +ve differences =		88	
			Sum of ranks of −ve differences =			17

(Note that, when zero differences occur, no rank values are attached to these values and n is adjusted to represent the number of pairs with non-zero differences only.)

Null hypothesis: there is no difference in the number of visits made by people to the garden before and after changes were made to the garden design.

The test statistic $T_{observed}$ = 88.0
Number of pairs with non-zero difference values (n) = 14
Critical value of t ($\alpha = 0.05$) = 84 (read from tables)

The test statistic $T_{observed} > T_{critical}$; therefore the null hypothesis is rejected.

Conclusion: there is a significant difference in the number of visits made by people to the garden before and after changes were made to the garden design ($P < 0.05$).

10.4.1 The Kruskal–Wallis test

The test is essentially a multiple-sample analysis based on the **ranked** data and there-fore examines whether samples are taken from populations with the same median rather than the same mean. In effect, it is an extension of the non-parametric Mann–Whitney test that is used to inspect the difference between the medians of two sam-ples. In fact, for two samples, the Kruskall–Wallis and the Mann–Whitney tests are identical. In theory, the Kruskal–Wallis test does require that samples are of similar shaped distribution and range; however, it has been shown to be very robust to all but very large deviations from this assumption.

The initial step in the Kruskal–Wallis test is to combine and then numerically rank every data item in the experiment. (If two or more values are tied, then they are each allotted the same rank value which is the midpoint between the ranking posi-tions they occupy, e.g. if two values tie for the tenth rank, then each is allotted the rank value of 10.5 and the next value is allotted the rank value of 12.) The sum of the ranks (ΣR) is then calculated for each sample. The basic principle of the test is that, if the medians of the samples are not significantly different (i.e. H_0 is correct), then the ranking values will be evenly distributed among the samples and the sum of the ranks (ΣR) will be similar for each sample. If the sample medians are significantly different (i.e. H_0 is not correct), then one or more samples will contain a preponder-ance of low ranks and one or more will contain a preponderance of high ranks. The sum of the ranks will not then be similar for each sample. What is required, therefore, is a statistic which tests the variation in ΣR on the basis that it will be low where the null hypothesis is correct and high where the null hypothesis is incorrect.

Determination of the Kruskal–Wallis test statistic H

In order to account for the possibility of samples having unequal sample size, the test statistic is based on the mean ranks of each sample ($\bar{R}_1$, $\bar{R}_2$, etc.), rather than the sum of ranks (ΣR). If there are no significant differences between samples, then the sam-ple mean ranks should all be similar and should also be similar to the total mean rank, $\bar{R}_T$. A test statistic can therefore be based on the sum of the deviations of each sample mean rank from the total mean rank. Before summing, the deviations are squared to remove negative signs and weighted by multiplying the squared devia-tions by the number of values n on which the sample mean rank is based. This proc-ess is analogous to determining the treatment sum of squares in a parametric ANOVA test but, since we are dealing with rank values, we will call this quantity $SS_{treatment (R)}$. Thus:

$$SS_{treatment(R)} = \Sigma \left[n_i \times \left(\bar{R}_i - \bar{R}_T \right)^2 \right]$$

where: n_i and $\bar{R}_i$ = the size and mean rank of each sample i respectively;
$\bar{R}_T$ = the overall mean rank.

A probability distribution for $SS_{treatment (R)}$ can be determined under the null hypothesis where each data value has an equal chance of having any particular rank. For a particular set of data, if the null hypothesis was correct, we would expect

$SS_{treatment (R)}$ to be close to the mean of this distribution. It can be shown that the mean of this distribution is given by:

$$\frac{N(N+1)}{12} \times (k-1)$$

We now need to derive a test statistic that will enable the determination of the probability of obtaining any particular observed value of $SS_{treatment (R)}$ when the null hypothesis is true. Such a statistic is obtained by dividing the observed value of $SS_{treatment (R)}$ by $N(N + 1)/12$. (This is actually the equivalent of dividing $SS_{treatment (R)}$ by the variance.) This test statistic is denoted H and is thus given by the formula:

$$H = \frac{\Sigma\left[n_i\left(\bar{R}_i - \bar{R}_T\right)^2\right]}{\left[\dfrac{N(N+1)}{12}\right]}$$

Fortunately, this rather complicated appearing formula can be mathematically rearranged to give a more user-friendly form:

$$H = \left[\frac{12}{N(N+1)} \times \Sigma\frac{(\Sigma R)^2}{n}\right] - 3(N+1)$$

where: N = total number of data items in all samples;
        ΣR = sum of ranks within a sample;
        n = number of data values in a sample.

In practice the term $(\Sigma R)^2/n$ is first determined for each sample; all $(\Sigma R)^2/n$ values are then summed to give $\Sigma (\Sigma R)^2/n$ and the calculation can then be completed according to the formula above.

Correction to H when tied ranks are present

Strictly speaking, when tied rank values occur among the data, the calculation of H described here leaves H a little lower than it should be. The error is, however, only appreciable when there are a large number of tied values compared with the total number of values, N. Furthermore, the error is in the right direction in that it makes the test less likely rather than more likely to find a significant difference. Therefore, in the majority of cases, the error makes no difference to the outcome and can usually be ignored. However, a rather complex correction factor can be applied to H if necessary; see Zar (2009) for a further theoretical explanation of this.

To perform the correction, the term Σt is first determined by:

$$\Sigma t = \Sigma\left(t_i^3 - t_i\right)$$

where: t_i = number of ties in a group of tied values.

The corrected value of H is then given by:

$$H_{corrected} = \frac{H}{1 - \left[\dfrac{\Sigma t}{N^3 - N}\right]}$$

Comparison of H with critical values for the rejection of the null hypothesis

Having determined H for a data set, this needs to be compared with critical values from the probability distribution so that a decision can be made to either accept or reject the null hypothesis. The probability distribution of H for a particular set of sample sizes is determined in much the same way as the Mann–Whitney U distribution is obtained, namely by calculating the value for H for every possible different order in which the data values may be ranked. The frequency distribution of H can thereby be constructed and from which the critical values of H that cut off the tails of the distribution that represent the defined probabilities of occurrence can be obtained. This is, of course, an immense calculation (with just three samples of five values each, the number of possible orders in which the data can be ranked is over one million!), but the mathematicians (or should I say computers?) have undertaken this for us and the critical H values can simply be read from a table (see Table A9 in the Appendix). Consulting the table allows values of H that have a probability of occurrence that is equal to or lower than the chosen critical probability for rejecting the null hypothesis to be identified. The table is interrogated at the required critical probability (α), the number of samples in the experiment (k) and the size of the samples (n). If the observed value of the test statistic H is equal to or greater than the critical value, then the null hypothesis can be rejected at the chosen level of probability.

It turns out, however, that the probability distribution of H is very similar to that of another test statistic called chi-squared (χ^2). (Significance tests involving the chi-squared test statistic and the nature of the chi-squared probability distribution will be explained in Chapter 14.) In fact, once the size of all samples reaches five and the total number of samples (k) to be compared reaches five, then the distribution of H is so close to the distribution of the chi-squared, χ^2, that χ^2 values can be reliably used instead of the critical values of H. For this reason, H tables that go much beyond $n = 5$ and $k = 5$ are rarely published. The appropriate critical χ^2 values for performing the Kruskal–Wallis test are given in Table A14 in the Appendix; the table is consulted at the required probability and at the appropriate degrees of freedom, which are given by $k - 1$. It may be noted that the values for χ^2 are always a little larger than the equivalent true values of H so that using the χ^2 values will facilitate a slightly less powerful test but with slightly less chance of making a Type 1 error.

A fully calculated Kruskal–Wallis test is shown in Example 10.4.

10.4.2 Non-parametric multiple comparison testing

If a Kruskal–Wallis analysis leads to the conclusion that significant differences are present, the exact location of the differences may be ascertained by a non-parametric multiple comparison test. The simplest procedure is to use a non-parametric Tukey test, employing the rank sums to establish the critical minimum difference that must

Example 10.4. A Kruskal–Wallis test.

In order to investigate the distribution of a pest caterpillar in the canopy of an orchard, replicate traps were situated in either the lower, middle or upper zones of the canopy and the number of caterpillars collected over a 24 h period was recorded. A Kruskal–Wallis analysis was performed to inspect for significant differences in the number of caterpillars collected in the three canopy zones.

Low canopy		Mid canopy		Upper canopy	
No. caterpillars	(Ranked value)	No. caterpillars	(Ranked value)	No. caterpillars	(Ranked value)
14	(15)	11	(13)	7	(8.5)
12	(14)	5	(4.5)	7	(8.5)
9	(11)	6	(6.5)	2	(1)
8	(10)	6	(6.5)	5	(4.5)
10	(12)	3	(2)	4	(3)
$\Sigma R =$	62		32.5		25.5
$n =$	5		5		5
$\bar{R} =$	12.4		6.5		5.1
$(\Sigma R)^2/n =$	768.8		211.25		130.05

$\Sigma(\Sigma R)$ = 120   Check: $N(N + 1)/2 = 120$

N = 15

$\Sigma((\Sigma R)^2/n)$ = 1110.1

Null hypothesis: there is no difference between the medians of the number of caterpillars collected from each zone of the canopy.

$$H = \left(\frac{12}{N(N+1)}\right) \times \left(\Sigma \frac{(\Sigma R)^2}{n}\right) - 3(N+1) = \left(\frac{12}{15(15+1)} \times 1110.1\right) - 3(15+1) = 7.505$$

H_{crit} read from table (at $\alpha = 0.05$, $n = 5$, $k = 3$) $= 5.78$

Since $H_{observed} > H_{critical}$ the null hypothesis is rejected.

Conclusion: there is a significant difference between the number of caterpillars collected from different zones of the tree canopy ($P < 0.05$).

Application of correction for tied ranks

Since there are two pairs of tied values, a correction to H can be applied.

$$\Sigma t = \Sigma\left(t_i^3 - t_i\right) = \left(2^3 - 2\right) + \left(2^3 - 2\right) = 12$$

$$H_{corrected} = \frac{H}{1 - \left[\dfrac{\Sigma t}{N^3 - N}\right]} = \frac{5.78}{1 - \left[\dfrac{12}{15^3 - 15}\right]} = 5.80$$

Conclusion: there is a significant difference between the number of caterpillars collected from different zones of the tree canopy ($P < 0.05$).

exist between sample medians to allow a significant difference to be claimed. The standard error of the difference between mean ranks is given by:

$$\text{SED} = \sqrt{\frac{N(N+1)}{12}\left(\frac{1}{n_a}+\frac{1}{n_b}\right)}$$

The critical minimum difference is then obtained by: $Q_{crit} \times \text{SED}$, where Q_{crit} is read from a table of critical Q values (Appendix: Table A11) at the required probability and at k number of samples in the experiment. For any pair of samples, if the difference between their mean ranks is greater than the calculated critical minimum difference, then those two samples can be claimed to be significantly different at the stated level of probability. Example 10.5 illustrates this calculation and an appropriate presentation of the results, using the same data from the Kruskal–Wallis test shown in Example 10.4.

10.5 Non-parametric Analysis of Randomized Complete Block Designed Experiments

In the great majority of cases, the sample data arising from randomized complete block designed experiments are assumed to be normally distributed with homogeneous sample variances and are analysed by a parametric ANOVA procedure. Where either of these assumptions breaks down, the most common procedure is to

---

Example 10.5. A non-parametric Tukey-type multiple comparison test.

A test to determine whether there was a significant difference in the number of pest caterpillars collected from different zones of a tree canopy. Data initially analysed by a Kruskal–Wallis test (see Example 10.4).

$$\text{SED} = \sqrt{\frac{N(N+1)}{12}\left(\frac{1}{n_a}+\frac{1}{n_b}\right)} = \sqrt{\frac{15(15+1)}{12}\left(\frac{1}{5}+\frac{1}{5}\right)} \qquad = 2.82$$

Q_{crit} (from tables at $\alpha = 0.05$, $k = 3$) $\qquad\qquad = 2.39$

Critical minimum difference between mean ranks = $2.82 \times 2.394 = 6.75$

Conclusion: any pair of samples having a numerical difference between mean ranks of ≥ 6.75 are significantly different ($P < 0.05$).

Comparison of samples:

Samples	$\overline{R}_a - \overline{R}_b$	Significance
Lower canopy – Middle canopy	12.4 – 6.5 = 5.9	n.s.
Lower canopy – Upper canopy	12.4 – 5.1 = 7.3	$P < 0.05$
Middle canopy – Upper canopy	6.5 – 5.1 = 1.4	n.s.

---

mathematically transform the data so that they then approximates to a normal distribution and/or sample variances become acceptably homogeneous, and ANOVA is then performed on the transformed data (see section 4.2.3 for a description of common transformations for approximating a normal distribution). In the case, however, of a simple single-factor randomized block experiment, there is a non-parametric test available called **Friedman's test** that requires no assumptions concerning the data distribution.

As with most other non-parametric analyses, Friedman's test is based on the ranks of the data values and is, therefore, essentially a test between sample medians. Assuming that every treatment is replicated an equal number of times (n) in each block, then the data are ranked within blocks. Following ranking, the sum of the ranks (ΣR) for each treatment across the blocks is determined and the test statistic, denoted χ_r^2, then calculated by the formula:

$$\chi_r^2 = \frac{12}{abn^2(na+1)}\Sigma(\Sigma R)^2 - 3b(na+1)$$

where: a = the number of treatments;
     b = the number of randomized blocks;
     n = number of replicates of each treatment per block;
     $\Sigma(\Sigma R)^2$ = the sum of the squared rank sums.

The observed value for χ_r^2 is compared with the appropriate critical value obtained from the known probability distribution of χ_r^2 at the required probability and for the number of treatments (a) and the number of blocks (b) (see Appendix Table A10). Where $\chi_{r\ observed}^2$ is $\geq \chi_{r\ critical}^2$, the null hypothesis of no differences between treatments may be rejected. Where values of a and b become greater than 5 and 10 respectively, then the distribution of χ_r^2 becomes very similar to that of the test statistic chi-squared (χ^2) and the chi-squared probability distribution (Appendix Table A14), with degrees of freedom given by $a - 1$, can be used to provide the critical values of χ_r^2. For further discussion of this test, the reader is referred to alternative texts, e.g. Zar (2009), while the worked example provided here (Example 10.6) further illustrates the mathematical procedure.

Example 10.6. Calculation of Friedman's non-parametric randomized block analysis.

The effect of diet supplements on rooting (i.e. food foraging) behaviour in pigs.

In a study of the effect of four diet supplements on the behavioural welfare of pigs, each diet was administered to one of four pigs, housed in each of six replicate pens (blocks). The frequency that each pig engaged in separate pieces of rooting behaviour for more than 1 minute was recorded over a 12-hour period.

continued

Example 10.6. Continued.

Pens ($b = 6$)	Frequency of rooting behaviour per 12 hours Supplementary diets ($a = 4$)				Ranks values per pen Supplementary diets			
	A	B	C	D	A	B	C	D
1	7	11	12	6	2	3	4	1
2	8	10	13	9	1	3	4	2
3	4	6	18	13	1	2	4	3
4	8	13	10	6	2	4	3	1
5	6	7	4	5	3	4	1	2
6	2	8	14	7	1	3	4	2
				$\Sigma R =$	10	19	20	11
				$(\Sigma R)^2 =$	100	361	400	121

$$\Sigma\left(\Sigma R^2\right) = \qquad 982$$

Null hypothesis: there is no difference in frequency of rooting behaviour between pigs fed with the four different supplementary diets.

$$\chi_r^2 = \frac{12}{ba(a+1)}\Sigma\left(\Sigma R\right)^2 - 3b(a+1)$$

$$= \left(\frac{12}{6\times4\times5}\right)\times 982 - (3\times6\times5) \qquad = 8.2$$

$\chi_r^2{}_{\text{critical }(\alpha = 0.05)}$ (from tables at $a = 4$, $b = 6$) $\qquad = 7.6$

Conclusion: $\chi_r^2 > \chi_r^2{}_{\text{critical}}$, therefore H_0 is rejected; pigs fed with the different supplementary diets show a significant difference in frequency of rooting behaviour ($P < 0.05$).

11 Correlation Analysis

Correlate: *have a mutual relationship; analogous; corresponding to each other and regularly used together ... hence* **correlation**.

(Concise Oxford English Dictionary)

- Introduction to the concept of correlation between variables.
- Parametric linear correlation between two normally distributed variables: Pearson's correlation coefficient.
- Significance of correlation and extent of dependency of one variable upon another.
- Non-parametric correlation between non-normally distributed variables: Spearman's correlation coefficient.

11.1 Introduction to Different Types of Correlation

In the preceding chapters we have been concerned with experiments in which a selected variable is measured under different treatment conditions and the subsequent analysis has focused on establishing whether a significant difference occurs between the differently treated populations. However, situations often arise in research where, rather than inspecting for differences, it is required to know whether two or more variables are related to each other. In statistics the term **correlation** specifically applies to the situation where the quantitative value of one variable is related to the value of another variable. For example, in a crop trial it might be required to establish whether the tuber yield of potatoes is correlated to the leaf area of those same potatoes, while in animal production science it may be required to determine the extent to which weight gain of an animal is correlated to the protein content of its food intake. In this chapter the analysis of correlation between two variables will be considered in some detail. While it is possible to undertake a correlation analysis between more than two variables, this becomes a much more complex procedure that is generally beyond the scope of this book.

Two quantitative variables may correlate in a number of different ways. Most commonly, the two variables may display a **positive correlation**, where an increase in one variable is matched by an increase of similar magnitude in the other variable. Often, however, an increase in one variable is matched by a decrease in the other, and such variables therefore display a **negative correlation**. For example, the relationship between lignin content of animal food and its digestibility might be expected to be a negative correlation. Many correlations may be strictly linear, that is, if one variable changes by a given proportion, the other variable changes by the same proportion. If such variables were plotted against each other on a graph, they would produce a straight line and are said to be **linearly correlated**. Frequently, however, two strongly correlated variables when plotted graphically may display a curved relationship rather than a straight line, for example, mammalian fetal length plotted against the period of gestation. Other types of correlation commonly encountered in life sciences include **log-linear relationships**, such as that between concentration of hormone and the growth response of a target tissue, and **exponential relationships** that commonly describe population growth, for example, the number of bacteria in a colony plotted against time.

When two variables are plotted graphically against each other the resulting graph is termed a **scatter plot** and the plotted data values may be referred to as **bivariates**. Examples of a series of different scatter plots are shown in Fig. 11.1. While it may often seem obvious from visualization of a scatter plot that two variables are correlated, frequently it is not certain whether a true correlation exists between two

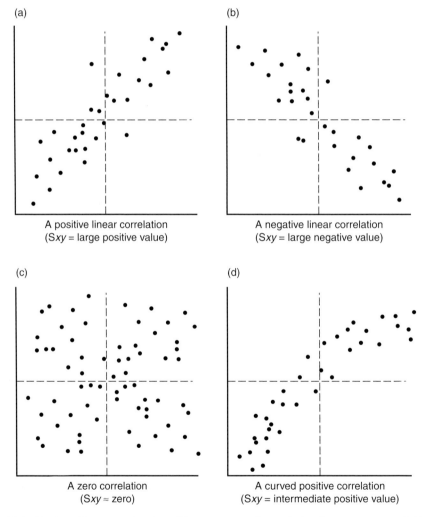

(a)

A positive linear correlation
(Sxy = large positive value)

(b)

A negative linear correlation
(Sxy = large negative value)

(c)

A zero correlation
($Sxy \approx$ zero)

(d)

A curved positive correlation
(Sxy = intermediate positive value)

Fig. 11.1. Scatter diagrams showing different types of correlation between two variables. Each ploted data point arises from the measurement of two variables and is termed a bivariate value. Note that the sum of products (Sxy) is given by $\Sigma(x-\bar{x})(y-\bar{y})$.

variables and furthermore it may be required to assess the strength of an apparent correlation. An objective statistical approach is therefore required.

11.2 Linear Correlation Analysis for Two Normally Distributed Variables

11.2.1 Pearson's product-moment correlation analysis

Where two variables are plotted on a continuous scale on a scatter plot the extent to which they display a linear correlation can be inspected employing a powerful technique called **product-moment correlation analysis**. As we shall see, the analysis produces a statistic termed the **correlation coefficient**, which is given the symbol r. The value of r allows the strength and the significance of the linear correlation between the two variables to be measured. The r coefficient is a sample statistic and provides an estimate of the true population correlation coefficient, which is itself denoted by ρ (Greek letter 'rho' in lower case). Since, therefore, this analysis involves the employment of sample statistics (i.e. sample means and standard deviations) as estimates of the population parameters, this is a parametric test and must only be used when it can be safely assumed that the two sample variables arise from populations that adhere to a normal distribution.

Sum of products (Sxy)

An initial key measure involved in quantitatively assessing linear correlations is a statistic called the **sum of products**. To help understand the sum of products, consider a scatter plot between sample values of two continuous normally distributed variables, such as potato leaf area and tuber yield. When the two variables are plotted against each other on a scatter graph, each x, y point can be described by its deviation from both the mean x value and the mean y value. For example, an observed value for potato leaf area can be subtracted from the mean leaf area of the sample, and the tuber weight of that same plant can be subtracted from the mean tuber weight of the sample. When the deviation of a point from the mean x value is multiplied by its deviation from the mean y value, the outcome is called the 'product of deviations'. Summing the product of deviations for every point on a scatter plot produces the statistical value termed the sum of products, which is represented by the symbol Sxy. The formula for the sum of products is thus:

$$Sxy = \sum(x-\bar{x})(y-\bar{y})$$

The relationship between the sum of products and correlation is illustrated in Fig. 11.2.

So why is the sum of products useful? To answer this question consider a scatter plot between a pair of variables that are positively correlated, as in Fig. 11.1(a). In this case the majority of the data points fall in the bottom left-hand quadrant and in the top right-hand quadrant of a scatter diagram. The data points in the top right-hand quadrant will show positive deviations from both $\bar{x}$ and $\bar{y}$ and will give positive values when the deviations are multiplied together. Data points in the bottom left-hand quadrant will show negative deviations from both $\bar{x}$ and $\bar{y}$, which when

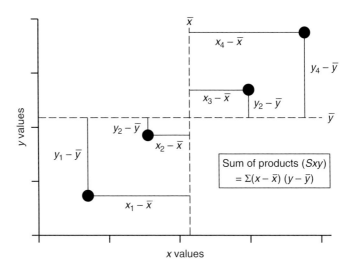

Fig. 11.2. Scatter diagram between two variables ($n = 4$) showing the calculation of the sum of products (Sxy).

multiplied together will also give a set of positive values. Therefore, when the product of deviations is summed, the sum of products will yield a large positive number. Conversely, when two variables are negatively correlated, as in Fig. 11.1(b), the majority of the data points will fall in the top left-hand quadrant and the bottom right-hand quadrant of the scatter plot. In this case, the data points in the top left-hand quadrant will show negative deviations from $\bar{x}$ but positive deviations from $\bar{y}$, which when multiplied together will give a set of negative values. The data points in the bottom right-hand quadrant will show positive deviations from $\bar{x}$ but negative deviations from $\bar{y}$, and will therefore also give negative values when multiplied together. Under these circumstances, the sum of products must yield a large negative value. When two variables are very poorly correlated, the data points will be much more spread out across all quadrants of the scatter diagram, as in Fig. 11.1(c). The sum of positive deviations will, in this case, tend to cancel out the sum of negative deviations and the sum of products will be a very low value that may be either negative or positive. In fact, if the data are totally uncorrelated and the data points are spread perfectly symmetrically across the scatter diagram the sum of products will be zero. In this way the sum of products provides an indication of both the strength and direction of a correlation. The calculation of the sum of products is further illustrated in Fig. 11.2.

Covariance

While the sum of products can be used as a quantitative estimate of the extent of the correlation between two variables, the magnitude of the value will be dependent upon the number of data pairs observed. In order to standardize the sum of products so that it may be compared across samples of different size, it is divided by the sample degrees of freedom. This leads to an important measure of correlation called the **covariance**. Thus:

$$\text{covariance} \quad = \frac{Sxy}{n-1} = \frac{\Sigma(x-\bar{x})(y-\bar{y})}{n-1}$$

where: n = the number of bivariate data items in the sample (= no. of x, y data points on scatter graph).

Pearson's correlation coefficient (r)

The disadvantage of covariance as a measure of correlation is that its value depends upon the magnitude of the measured values of x and y and it is, therefore, difficult to use when comparisons of correlation are required between variables that are measured in different units. This problem is simply overcome by further dividing the covariance by the product of the standard deviations of x and y respectively. This produces a statistic called **Pearson's correlation coefficient** and is denoted by the symbol r. The equation for r thus becomes:

$$r = \frac{\Sigma(x-\bar{x})(y-\bar{y})}{\sqrt{\Sigma(x-\bar{x})^2 \times \Sigma(y-\bar{y})^2}}$$

A slightly easier notation may be used where the sum of products is represented by Sxy and the terms $\Sigma(x-\bar{x})^2$ and $\Sigma(y-\bar{y})^2$, which are the **sum of squares** of the x and y values respectively, are denoted SSx and SSy. The correlation coefficient, r, may then be expressed as:

$$r = \frac{Sxy}{\sqrt{SSx \times SSy}}$$

Following algebraic manipulation this formula can also be expressed in a form that while appearing more cumbersome, is in fact more easily computed, thus:

$$r = \frac{\Sigma XY - \dfrac{(\Sigma X \cdot \Sigma Y)}{n}}{\sqrt{\left(\Sigma X^2 - (\Sigma X)^2/n\right) \times \left(\Sigma Y^2 - (\Sigma Y)^2/n\right)}}$$

where: ΣXY = the sum of the product of x, y data pairs;

n = the number of x, y data pairs.

The greatly convenient mathematical consequence of this formula is to derive a value that will always fall between -1 and $+1$. The greater the absolute value of r, i.e. the nearer to $+1$ or to -1, the stronger the correlation between the two variables. A value of exactly $+1$ indicates a perfect positive correlation while a value of -1 indicates a perfect negative correlation. Examples of scatter diagrams of data sets with different values of r are shown in Fig. 11.3 and a full calculation of r is shown in Example 11.1.

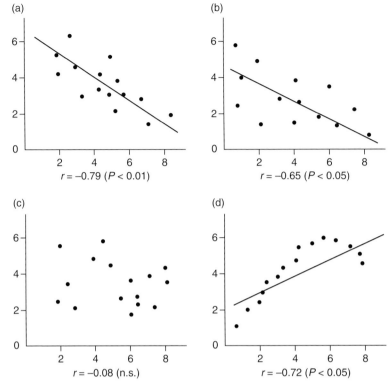

Fig. 11.3. Scatter diagrams between pairs of variables that display different Pearson's correlation coefficient (r) values.

Example 11.1. Determination of Pearson's product-moment correlation coefficient, r.

Relationship between lifetime milk yield of dairy cows and age at which cows first calved.

Age at first calving (months)	Lifetime milk yield (kg)
24	18747
25	18673
26	18456
27	18730
28	17995
29	18300
30	17964
31	17820
32	17842
33	17991
34	17758
35	17650

continued

Chapter 11

Example 11.1. Continued.

A plot of the data on a scatter diagram clearly suggests a negative correlation:

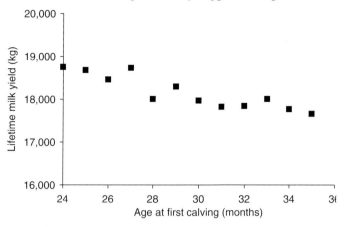

Determination of the correlation coefficient *r*:

Σx = 354 Σy = 217,926 $n = 12$
Σx^2 = 10,586 Σy^2 = 3,959,417,604 Σxy = 6,414,388

$$S xy \quad = \quad \Sigma\, xy - \frac{(\Sigma x . \Sigma y)}{n} \quad = -14,429$$

$$SSx \quad = \quad \Sigma x^2 - (\Sigma x)^2/n \quad = 143; \qquad SSy = \Sigma y^2 - (\Sigma y)^2/n = 1,772,481$$

$$r \quad = \quad \frac{Sxy}{\sqrt{SSx.SSy}} \quad = -0.906 \text{ (with DF = 10)}$$

Conclusion: the high negative value indicates a relatively strong negative correlation between age of first calving and lifetime milk yield.

11.2.2   Determining the reliability of the sample correlation coefficient *r*

Since the correlation coefficient *r* is an estimate of the true population correlation coefficient ρ, we may legitimately ask how reliable *r* is as an estimate. To address this, it is possible to determine both a standard error and confidence limits for the sample correlation coefficient. (The specific mathematical derivations of the appropriate formula need not worry us here but are described elsewhere if the reader is interested, e.g. Zar, 2009.)

Standard error of r

The standard error of the sample correlation coefficient (SE_r) is given by:

$$SE_r \quad = \quad \sqrt{\frac{1 - r^2}{n - 2}}$$

where: *r* = sample correlation coefficient;
 n = no. of *x*, *y* data pairs.

Confidence limits for r

There are several alternative techniques for determining the confidence limits for the correlation coefficient; possibly the most straightforward is that provided by Zar (2009) and uses two-tailed critical values of the test statistic F. These may be obtained from Table A5.3 in the appendix.

Assuming that 95% confidence limits are to be determined, then the lower confidence limit CL_{low} is given by:

$$CL_{low} = \frac{(1 + F_{0.05})r + (1 - F_{0.05})}{(1 + F_{0.05}) + (1 - F_{0.05})r}$$

and the upper confidence limit CL_{high} is given by:

$$CL_{high} = \frac{(1 + F_{0.05})r - (1 - F_{0.05})}{(1 + F_{0.05}) - (1 - F_{0.05})r}$$

where: r = sample correlation coefficient:
  $F_{0.05}$ = two-tailed F value at $P = 0.05$ and at degrees of freedom $V_1 = V_2 = n - 2$.

The calculation of the standard error and confidence limits for a sample correlation coefficient is shown in full in Example 11.2 based on the data of Example 11.1.

Example 11.2. Determination of standard error, confidence limits, significance of linear correlation and coefficient of determination.

Relationship between lifetime milk yield of dairy cows and age at which cows first calved (data from Example 11.1).

Age at first calving (months)	Lifetime milk yield (kg)
24	18747
25	18673
26	18456
27	18730
28	17995
29	18300
30	17964
31	17820
32	17842
33	17991
34	17758
35	17650

$r = -0.906$ $(n = 12)$; $r_{criti. (0.05, DF = 10)} = 0.576$ (from table)

Conclusion: the absolute value of r is greater than the critical value. The null hypothesis is rejected and it is concluded there is a significant negative correlation between age of first calving and lifetime milk yield $(P < 0.05)$.

continued

Example 11.2. Continued.

The standard error (SE$_r$), and lower and upper 95% confidence limits (95%CL) for the sample correlation coefficient

$$SE_r = \pm\sqrt{\frac{1-r^2}{n-2}} = \pm\sqrt{\frac{1-0.906^2}{12-2}} = \pm 0.13$$

$$95\%CL_{low} = \frac{(1+F_{0.05})r+(1-F_{0.05})}{(1+F_{0.05})+(1-F_{0.05})r} = \frac{(1+3.72)0.906+(1-3.72)}{(1+3.72)+(1-3.72)0.906} = 0.690$$

$$95\%CL_{high} = \frac{(1+F_{0.05})r-(1-F_{0.05})}{(1+F_{0.05})-(1-F_{0.05})r} = \frac{(1+3.72)0.906-(1-3.72)}{(1+3.72)-(1-3.72)0.906} = 0.974$$

Student's t-test of sample correlation:

$$t_{observed} = \frac{r}{\sqrt{\frac{1-r^2}{n-2}}} = \frac{-0.906}{\sqrt{\frac{1-0.906^2}{12-2}}} = -6.769$$

$t_{critical\ (0.05,\ DF\ =\ 10)} = 2.228$ (from table)

Conclusion: the absolute value of $t_{observed}$ is greater than the critical value. The null hypothesis is rejected and it is concluded there is a significant negative correlation between age of first calving and lifetime milk yield ($P < 0.05$).

Coefficient of determination for the relationship between lifetime milk yield of dairy cows and age when cows first calved.

The coefficient of determination = $r^2 = -0.906^2$ = 0.821

Conclusion: 82.1% of the variation in lifetime milk yield in the trial is accounted for by its correlation to the age at which dairy cows first calved; the remainder is due to other unidentified factors.

11.2.3 The frequency distribution of *r* and testing the statistical significance of a correlation

Having assessed the relative strength of a correlation through the magnitude of the correlation coefficient *r*, we may now ask whether the correlation is statistically significant. The procedure adopted is the familiar one of determining whether there is sufficient statistical evidence to allow the null hypothesis, which states that the two variables are not linearly related, to be rejected.

The sample correlation coefficient *r* is a random variable and has a known probability distribution that depends on the population parameter ρ and the sample size *n*. As with the probability distributions of other test statistics, the distribution of *r* can be derived by a mathematical approach; however, this is highly complex and description of the derivation is well outside the scope of the present text. The distribution can be

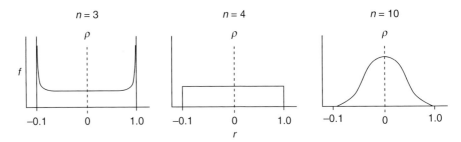

Fig. 11.4. The frequency distribution of the sample Pearson's correlation coefficient r, when the population correlation coefficient (ρ) is 0.

envisaged, however, as arising from a process in which r is calculated for every possible sample of x, y data pairs of a given sample size n taken from a population and the values displayed as a frequency plot. What is interesting about the distribution is that with small sample sizes both high and low r values are as likely to occur as middle values; in fact, where $n = 4$ and the population correlation coefficient ρ is zero, all values between −1 and +1 are equally likely. This is illustrated in Fig. 11.4. Consequently, in cases of small sample size ($n \leq 5$), one cannot conclude that a high r value necessarily represents a strong correlation between variables. Consequently, it is important that r is tested for significance in a formal statistical manner. There are two approaches that are commonly used for this, one in which the sample correlation coefficient r is employed as the test statistic or alternatively a Student's t-test can be employed.

Correlation coefficient r test

The probability of significance is assessed by comparing the observed value of r with critical values that cut off a fixed proportion of the tails of the r probability distribution. The critical values of r are read from a table (Table A12 in the Appendix) at the required level of probability and at the appropriate degrees of freedom. The degrees of freedom are given by the number of paired observations (n) less the number of variables being correlated, i.e. $n - 2$. The null hypothesis of no correlation can be rejected where $r \geq r_{critical}$ and it may then be concluded that a significant linear correlation is present between the two variables at the level of probability α at which $r_{critical}$ was obtained. Note that the test is usually conducted as a two-tailed test and, therefore, where the correlation is negative the sign is ignored for the purposes of inspecting for significance. If required, the test can be performed as a one-tailed test, in which case the sign of r must agree with the one-tailed hypothesis to allow the null hypothesis to be rejected and the one-tailed critical r value is equivalent to the two-tailed value determined at twice the critical probability.

Student's t-test

If no correlation exists between two variables, then the population correlation coefficient ρ should be equal to zero. However, even where there is absolutely no correlation between the variables, the sample correlation coefficient r is most unlikely to equal zero because random variation will mitigate against an absolutely even dispersal of values. The presence of a significant correlation can therefore be inspected by using a Student's t-test to determine whether the r value observed is significantly different from the value zero. Following the normal procedure for a Student's t-test, a test statistic is obtained by dividing the difference between the population and sample value (i.e. $\rho - r$) by the standard error, SE_r. Since the hypothesized value for ρ is zero and using the formula for SE_r given in the last section, this becomes:

$$t_{observed} = \frac{r}{\sqrt{\dfrac{1 - r^2}{n - 2}}}$$

The value of $t_{observed}$ can be referred to the Student's t distribution (Table A2 in the Appendix), which is consulted at the required critical probability and at degrees of freedom of $n - 2$. If $t_{observed}$ is $\geq t_{critical}$, the null hypothesis can then be rejected and it may be concluded that sufficient evidence exists to support a valid claim of a significant linear correlation.

It should be noted that, while employing Student's t-test works well for testing whether r differs from zero, it cannot be applied to test whether r differs from some other given value of ρ. This is because r only adheres to a normal distribution when ρ is equal to zero (see Fig. 11.4); thus, when ρ is equal to a non-zero value, the use of the Student's t statistic becomes invalid. In such cases, the recommended method is to perform a transformation, termed Fisher's transformation, that converts r to a statistic that estimates a normally distributed variable (denoted ζ) and then allows a z-test to be performed. This technique will not be considered further here but is again described more fully by Zar (2009).

The determination of the significance of correlation through both the correlation coefficient r and Student's t-test is shown in Example 11.2.

11.2.4 The coefficient of determination r^2

Having ascertained that a significant linear correlation exists between two variables, it is often useful to further determine the proportion of the change in one variable that is directly accounted for by its correlation to the other variable. Clearly it will be very rare that the change in one variable is due entirely to the change in the other and that no other factors influence the relationship. In Example 11.1 it would be expected that the lifetime milk yield of cows would be affected by many additional factors other than age of first calving, for example, nutrition, dairy management and random variability. In order to assess the extent to which the change in one variable is attributable to the change in a second variable the **coefficient of determination, r^2**, is determined.

The theory behind the calculation of r^2 will not be described here (this statistic will be considered in more detail in Chapter 12 on linear regression analysis); however, as the symbol indicates, r^2 is the square of the product-moment correlation coefficient r. The value will fall between 0 and +1 but can be expressed as a percentage if desired.

An example of the calculation of the coefficient of determination r^2 and its interpretation is shown in Example 11.2.

11.2.5 The problem of analysing non-linear correlations

The product-moment correlation analysis only reliably detects associations that are linear. When two variables are plotted against each other on a scatter diagram often a strong relationship is apparent but rather than being linear it is curved. As previously mentioned, log-linear and exponential relationships are particularly common in biological sciences and will show up as curved responses on a scatter diagram. Other relationships may show a linear response for part of the data set but linearity may break down as the variables increase in magnitude. For example, the relationship between crop leaf area and yield is usually linear up to a certain point and then will gradually curve off as leaf area reaches an optimum and may actually decrease with further increase in leaf area. Such a relationship is referred to as **curvilinear**. A product-moment correlation analysis of such data may reveal only weak or insignificant correlations when clearly the variables are actually very strongly related.

In such cases it may be appropriate to analyse only that portion of the data that falls around a straight line on a scatter diagram. Alternatively, curved lines can sometimes be mathematically converted to straight lines by the process of **data transformation**, in which the data for either one or both of the variables are mathematically converted to alternative values by application of a constant mathematical function. For example, many curved data are readily converted to linear data by converting them to their $\log_{10}$ values. The subsequent correlation analysis is then based on the transformed values rather than the original values. This process will be discussed in more detail in Chapter 12, which describes fitting trend lines to data on a scatter plot by linear regression analysis. If data transformation cannot resolve the problem or, for one reason or another, is not desirable, another possible approach is to employ a non-parametric method of data analysis, as described in the next section.

11.3 Non-parametric Correlation Analysis

11.3.1 Spearman's rank correlation analysis

Spearman's rank correlation analysis represents an alternative non-parametric approach for correlation analysis. As with most non-parametric statistical analyses, it is based on the ranks of the data values. Consequently it does not rely on any assumption concerning the distribution pattern of the data values, nor does it assume that any correlation present is necessarily linear. It can be employed, therefore, in a wide range of different situations, including the analysis of both count and measured data as well as ranked (i.e. ordinal) data. While it will operate quite safely with normally distributed data, if the variables are normally distributed and a linear relationship is expected, then Spearman's rank correlation is less powerful than Pearson's product-moment correlation analysis since it has a lower probability of finding a significant correlation when one is truly present.

The aim of the analysis is to produce a test statistic based on the data ranks that can be used to describe and determine the significance of the correlation between two variables. First, the data for each variable are ranked separately and the difference (d) in rank between each x, y data pair is determined. The premise of the test is then relatively simple: if two variables are strongly correlated, the x value and the y value of each data pair should have similar ranks; the poorer the correlation the larger will be the difference in the rank values of each x, y pair. If the differences (d) between the rank values for each x, y pair are squared to remove negative values and then summed, this will produce a value (Σd^2) which is related to the degree of correlation. For a perfect correlation, each x and its corresponding y value will have exactly the same ranks and Σd^2 will therefore equal zero, while the largest value of Σd^2 will be produced by a perfect negative correlation. This is illustrated in Fig. 11.5.

Σd^2 has, however, limited use as a measure of correlation because its magnitude will naturally tend to increase with sample size and it does not, therefore, allow easy comparison or interpretation. Just as the sum of products (Sxy) was standardized to produce Pearson's correlation coefficient (r), which always falls between –1 and +1, similarly Σd^2 can be 'standardized' to produce the same effect. The manipulation of Σd^2 to this end is a lovely piece of mathematical 'trickery'. It can be shown that the maximum value for Σd^2, which occurs under a perfect negative correlation, is given by $(n^3 - n)/3$ where n is the sample size. Therefore Σd^2 is standardized by dividing by $(n^3 - n)/3$, which forces the value to lie between 0 for a perfect positive correlation, and 1 for a perfect negative correlation. If this value is then doubled, to force it to lie between 0 and 2, and subtracted from 1, it produces the desired outcome of a value that occurs between –1 for a perfect negative correlation and +1 for a perfect positive correlation, while zero indicates absolutely no correlation. This statistic is termed Spearman's correlation coefficient and is denoted by the symbol r_s. Thus:

$$r_s = 1 - \frac{2\Sigma d^2}{\dfrac{(n^3 - n)}{3}} \qquad = 1 - \frac{6\Sigma d^2}{n^3 - n}$$

where: d = the difference in rank between each x, y data pair;
n = number of x, y data pairs.

11.3.2 The frequency distribution of r_s and testing the statistical significance of a rank correlation

As with Pearson's coefficient, we can understand the frequency distribution of Spearman's correlation coefficient r_s by imagining that every possible x, y data sample of a given size n is extracted from a population that satisfies the null hypothesis, i.e. the population Spearman's correlation coefficient (ρ_s) = 0. For each sample, r_s can be determined and, because many of the values will occur a number of times, the r_s values can be plotted in the form of a frequency distribution. Where the sample size is ≤4, the distribution is very poorly resolved with only a very limited number of values of r_s possible and there is, therefore, no possibility of rejecting the null hypthoses at a significance level of less than 5%. Only with a sample size of five are there enough possible values of r_s for the probability of obtaining the most extreme values of –1 and +1 to fall below 0.05 and thus permit rejection of the null hypothesis. As sample size increases above 5, the resolution of the frequency distribution of r_s increases greatly, the tails become increasingly more defined and the probability of obtaining extreme values is discernibly reduced to lower and lower values. It then becomes possible to sensibly determine critical values of r_s that cut off given proportions of the tails of the distribution, which can be tablulated and against which observed values of r_s can be compared.

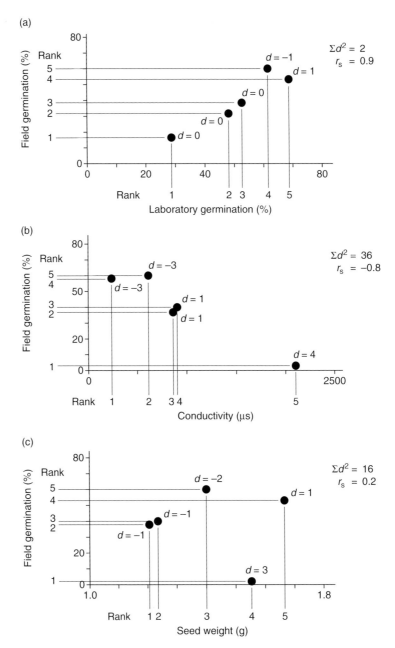

Fig. 11.5. The determination of Spearman's rank correlation coefficient r_s for five bivariate data values ($n = 5$). (a) Correlation between laboratory germination rate and field germination rate in bean seeds. (b) Correlation between conductivity of seed immersing solution and field germination rate of bean seeds. (c) Correlation between mean seed weight and field germination rate in bean seeds. For each plotted point, d is the difference between the rank of the x value less the rank of the y value. Note that Σd^2 takes a low value when there is a strong positive correlation, is high under a strong negative correlation and has a medium value when there is no correlation. A standardization procedure converts Σd^2 into Spearman's correlation coefficient r_s, which lies between −1 (perfect negative correlation) and +1 (perfect positive correlation).

The critical values of r_s are given in Table A13 in the Appendix. The values are consulted at the required critical probability and the sample size n. In the normal way for a two-tailed test, if the absolute value of $r_s \geq$ critical value, then the null hypothesis of no correlation may be rejected at the level of probability at which the critical value was obtained. If required, the test can be performed as a one-tailed test where the direction of the expected correlation is specified and the critical value of r_s is then the equivalent of the two-tailed value read at twice the critical probability. The full calculation of Spearman's rank correlation coefficient is shown in Example 11.3.

Example 11.3. Determination of the Spearman's correlation coefficient, r_s.

The correlation between electrical conductivity of immersing aqueous solution and bean seed germination rate.

In ageing seed, degradation of the testa and internal membranes leads to leakage of ions out of the seed that should increase the electrical conductivity of an immersing solution. This gives a potential laboratory test for determining seed viability. In order to examine the validity of the technique, a number of seed batches of different age were immersed in water and the electrical conductivity of the solutions measured. The germination rate of these same seed batches was subsequently determined. The data were as follows:

Seed batch	Conductivity (μs)	Rank	Germination (%)	Rank	d	d^2
1	2100	19.5	6	5	14.5	210.25
2	900	11	38	12	−1	1
3	550	3	60	21.5	−18.5	342.25
4	910	12	40	13	−1	1
5	225	1	58	19.5	−18.5	342.25
6	1480	15	18	8	7	49
7	750	8	49	14	−6	36
8	885	10	56	17	−7	49
9	2800	23	0	1	22	484
10	565	5	64	23	−18	324
11	1625	17	21	9	8	64
12	1510	16	24	10	6	36
13	670	7	52	16	−9	81
14	640	6	57	18	−12	144
15	2950	24	2	2.5	21.5	462.25
16	500	2	60	21.5	−19.5	380.25
17	2100	19.5	9	6	13.5	182.25
18	1750	18	12	7	11	121
19	930	14	32	11	3	9
20	825	9	58	19.5	−10.5	110.25
21	2350	21	5	4	17	289
22	2395	22	2	2.5	19.5	380.25
23	555	4	71	24	−20	400
24	915	13	50	15	−2	4
$n = 24$						$\Sigma d^2 = 4502$

continued

Example 11.3. Continued.

Null hypothesis: there is no correlation between conductivity of immersing solution and the field germination rate of bean seed.

Calculation of Spearman's rank correlation coefficient, r_s:

$$r_s = 1 - \frac{6\sum d^2}{n^3 - n} \quad = \quad 1 - \frac{6 \times 4502}{24^3 - 24} \quad = -0.96$$

$r_{s\ critical}$ from tables (at $\alpha = 0.05$, $n = 24$) = 0.409

Conclusion: there is a significant negative correlation between germination rate and conductivity of immersing solution ($P < 0.05$).

12 Inspecting Relationships by Simple Linear Regression Analysis

Linear: 1. *of or in line* 2. *involving one dimension.*

Relation: *what one person or thing has to do with another.*

Relationship: *state of being related.*

(Concise Oxford English Dictionary)

- Introduction to exploring linear relationships and fitting trend lines by simple linear regression analysis.
- The simple linear regression model.
- Simple linear regression by the method of 'least squares'.
- Testing the significance and reliability of a best-fit regression line.
- Comparing two regression lines.

12.1 Introduction to Linear Regression Analysis

Many experiments consist of the investigator controlling the level of one variable and measuring a responding variable in order to ascertain the relationship between the two and thereby allow future responses to be predicted. For example, the rooting of stem cuttings may be measured in response to a controlled variation in the concentration of an applied plant growth hormone or the digestibility of an animal food may be measured in response to controlled changes in lignin content. The relationship between an independent variable and a responding variable may also arise from observational studies such as that between the age at which cows first calved and lifetime milk yield. Following such an investigation, it will usually be required to plot the response against the controlled or independent variable on a scatter graph in order to establish whether a relationship exists and, if so, it will then be appropriate to plot the 'best-fit' trend line through the data points to demonstrate the form of the relationship.

If the relationship between two variables is linear then a technique called **simple (or ordinary) linear regression analysis** can be used to model the relationship and determine the equation of the 'best-fit' line that can be plotted through the bivariate data points on a scatter graph. The best-fit straight line between a set of linearly related data points obtained by this technique is then called the **line of regression**. In this chapter the method of simple linear regression analysis will be discussed in depth, including methods for assessing the reliability and statistical significance of the regression line and for the comparison of two regression lines.

In more complex cases, a measured variable, such as crop yield, may be determined in response to a number of different controlling variables simultaneously, such as the level of a number of different soil nutrients. Assuming the underlying relationship remains linear, **multiple regression analysis** can be employed to produce a model that describes the relationship between a measured variable and the combination of a number of controlling variables. On the other hand, two variables in biology may often be highly correlated but not in a strictly linear manner; for example, a scatter diagram of growth of an organism plotted against the level of an exogenous growth factor may often reveal an obvious curved relationship. Regression techniques for dealing with multiple variables and curved relationships will be discussed in Chapter 13.

12.2 The Simple Linear Regression Model

Initially it is important to understand what is meant by the term 'linear' in the context of linear regression analysis. We have in fact already come across the concept of linear mathematical models when describing the analysis of multiple-sample experimental designs (see section 8.5) and learnt that in statistical analysis the term linear applies to the form of the mathematical equation underlying the analysis rather than to the overt relationship between the measured variables. While biologists will generally consider a linear relationship to infer a situation where the quantitative response to a factor is linear and produces a straight line on a graph, in mathematics a linear equation is one which consists of a series of components that are summed to produce the solution. The familiar equation that describes straight lines on a graph is of this sort but in fact the more complex polynomial equations that describe curved lines on graphs are also linear equations.

All straight lines drawn on a graph can be described by a simple linear equation which takes the form:

$$y = \alpha + \beta x$$

where: α is the intercept of the line with the Y-axis and β is the slope of the line.

Where two variables have a perfect linear relationship, each x, y data point will fall on a straight line; however, in reality random variation will generally cause values to stray either side of a best-fit straight line plotted through the data points, as shown in Fig. 12.1. Therefore, where a linear relationship exists between two variables, the relationship between the measured variable Y and the independent variable X can be expressed as:

$$y_i = \alpha + \beta x_i + \varepsilon_i$$

where: y_i = the value of the Y-variable that occurs in response to a given value x_i of the X-variable;

α = the intercept of the line with the Y-axis (i.e. the predicted value of y when x = zero);

β = the slope of the line;

ε_i = the random error involved in making the measurement y_i.

Now consider the situation where we have a sample of measurements of the X- and Y-variables from which we wish to make an estimate of the population parameters α and β. If we have a number of y values for each value of x, then the sample variance of these y values, s_y^2, will give an estimate of the variance of the random errors, ε_i. The assumption is now made that the error associated with the measurement of each sample of y values for every possible value of x is the same and this then allows the s_y^2 term to be ignored. Therefore the sample regression equation that enables values of y to be predicted for given values of x becomes:

$$\hat{y} = a + bx$$

where: $\hat{y}$ = the predicted value of the dependent Y-variable for a given value x of the
      X-variable;
      a = the estimated intercept of the line with the Y-axis;
      b = the estimated slope of the line.

The aim of simple linear regression analysis is, therefore, to determine the values of a and b from the sample data which will provide the best estimates of the population parameters α and β and thus determine the equation of the line of best fit between the two variables.

12.3 Simple Linear Regression by the Method of Least Squares

The most common technique employed for determining the line of regression is called the **method of least squares**. This technique is based on a particular set of assumptions and should only be used when certain conditions apply. These conditions need to be clearly understood before proceeding with the analysis.

12.3.1 The conditions for use of least squares regression

The main conditions for the use of the least squares method of linear regression are:

1. One of the variables is considered to be **independent** and is plotted along the horizontal X-axis. The second variable, which is measured as a response to changes in the former, is the **dependent** variable and is plotted along the vertical Y-axis.

2. A definite relationship must exist between the Y variable and the X variable (although it is not necessarily implied that the relationship is causal).

It is invalid to plot a best-fit line through the data points on a scatter plot unless the variables are significantly linearly correlated. Therefore, a Pearson's correlation analysis is usually performed prior to determination of a regression line.

Since determination of the regression line for two variables will potentially allow values of the dependent Y variable to be predicted for different values of the independent X variable, the latter is also commonly referred to as the **predictor** variable. In a manipulative experiment, the independent (or predictor) variable will have non-random values that are fixed by the investigator, e.g. temperature of a controlled growth-room, while the dependent variable has random values obtained by measurement, e.g. plant dry weight. In this case, the underlying regression model describing the relationship between the variables is termed a **Model I** regression. In observational studies,

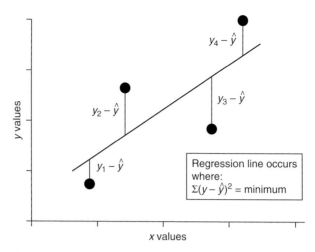

Fig. 12.1. Determination of a line of regression by the method of least squares.

however, both variables will consist of random values, e.g. age of first calving and lifetime milk yield in cows. Where both variables are random, this is termed a **Model II** regression. In some Model II regressions, because the levels of neither of the variables are fixed by the investigator, it may in fact not be clear which variable, if either, should be considered as independent and which one as dependent. When the aim of a Model II regression is simply to facilitate predictions of the Y-variable, then the method of least squares is still generally valid. When, however, it is required to determine estimates of the parameters of the underlying model II regression equation, i.e. α and β, then the method of least squares will not be appropriate.

The method of least squares is a parametric technique in which the slope of the regression line of the population is estimated from sample data; therefore, the usual assumptions required by parametric statistical techniques are required. Thus the third set of conditions for linear regression analysis are:

3. For any single defined value of the independent X-variable, the theoretical population of associated y values must be normally distributed and the variances of different populations of y values that correspond to each different value of x must be statistically homogeneous.

If any of the above assumptions fail, then the method of least squares becomes invalid. While it is usually straightforward to check assumptions **1** and **2** above, it is more problematic to be assured that the assumptions of normality and homogeneous variances (**3** above) are complied with. In many cases there will only be one y value, or just a very few y values, measured for each value of x, so that checking mathematically for a normal distribution and for homogeneous sample variances is not possible. The investigator may need to consider in a theoretical manner, therefore, whether these assumptions can be expected to hold if large samples of replicated y measurements were to be made under the prevailing experimental conditions. Some help in this can be obtained by checking the distribution of the deviations of each plotted x, y value from the value predicted by the regression equation, i.e. the frequency distribution of the **residuals**. We will come to this technique in due course. It may be commented, however, that the method of least squares has been shown to be relatively robust and will often produce a reliable outcome for data that depart a little from the assumptions relating to a normal distribution.

12.3.2 Determination of the line of regression by the method of least squares

The method of least squares is based on the vertical deviations of each plotted data point from the best-fit line. The best-fit line is defined where the sum of these deviations is as small as it can possibly be. By definition, however, half the vertical deviations from a best-fit line will be positive and half will be negative so that their numeric total will be zero. The familiar practice is therefore applied of squaring the deviations to remove negative signs before summing them. For each value of the X-variable, the corresponding value of the Y-variable that is predicted to lie on the regression line is denoted $\hat{y}$ and the term $\Sigma(y - \hat{y})^2$ is referred to as the **residual sum of squares**. The regression line is thus obtained where the residual sum of squares, $\Sigma(y - \hat{y})^2$, is at a minimum. This is illustrated in Fig. 12.1.

The slope b of a best-fit regression line is termed the regression coefficient. We now arrive at one of those points in practical statistics where the mathematics really do become extremely complex (and well beyond the scope of this book to explain); fortunately however, the outcome is relatively straightforward. It can be shown (through application of matrix mathematics) that the regression coefficient b for a sample of x, y values that fulfils the requirement that the sum of residuals is minimal is given by the sum of x, y products (Sxy) divided by the sum of squares of x (SSx). Both Sxy and SSx were employed in the determination of Pearson's correlation coefficient and their derivation should already be familiar, thus:

$$b = \frac{\Sigma(x - \bar{x})(y - \bar{y})}{\Sigma(x - \bar{x})^2} = \frac{Sxy}{Sxy}$$

where: Sxy is the sum of products of x and y;
        SSx is the sum of squares of x.

It is a further logical assumption, and in fact can be mathematically proved, that the data point representing the mean value of the Y-variable plotted against the mean value of the X-variable must fall on the regression line. Therefore the value of the intercept a of the regression line can be obtained by rearranging the regression line equation and substituting the known values of b, $\bar{x}$ and $\bar{y}$, thus:

$$a = \bar{y} - \left(b \times \bar{x}\right)$$

Once a and b are known, any value of x can be substituted into the regression equation and the predicted value $\hat{y}$ that lies on the regression line can be determined. If this procedure is undertaken for two x, $\hat{y}$ pairs at opposite ends of, but lying within, the observed data range, then the straight line joining these two points will represent the best-fit regression line through the data.

A full calculation of a linear regression by the method of least squares is shown in Example 12.1. Once plotted, the regression line can be used to predict values of the dependent variable for given values of the independent variable. For example, the regression line in Example 12.1 could provide a basis for predicting the milk yield of cows that have different first calving dates. It is, however, invalid to make predictions based on extrapolations beyond the range of the observed values since there is no information that the calculated regression line is valid beyond this range.

Example 12.1. Determination of a simple linear regression by the method of least squares.

Relationship between age of first calving and lifetime milk yield of dairy cows (data as in Example 11.1).

Data table

Age at first calving (months)	Lifetime milk yield (kg)
24	18,747
25	18,673
26	18,456
27	18,730
28	17,995
29	18,300
30	17,964
31	17,820
32	17,842
33	17,991
34	17,758
35	17,650

A plot of the data points on a scatter diagram indicates a negative linear relationship; however, it is required to know whether the correlation is significant and where the best-fit straight line should be plotted.

Calculation of regression statistics: the regression coefficient r is determined as described previously (see Example 11.1).

Σx = 354	Σy = 217,926	n = 12
Σx^2 = 10,586	Σy^2 = 3,959,417,604	Σxy = 6,414,388
SSx = 143	SSy = 1,772,481	Sxy = -14,429

$$r = \frac{Sxy}{\sqrt{SSx.SSy}} = -0.906$$

Conclusion: there is a negative correlation between age at first calving and lifetime milk yield. The gradient of the best straight line, b, is calculated:

$$b = \frac{Sxy}{SSx} = \frac{-14,429}{143} = -100.90$$

The intercept with the Y-axis, a, is calculated: $a = \bar{y} - (b \times \bar{x}) = 21,137.11$

The equation for the best-fit straight line is thus: $y = 21,137.11 - 100.90x$
The regression line can now be plotted through the data points on the scatter graph.

continued

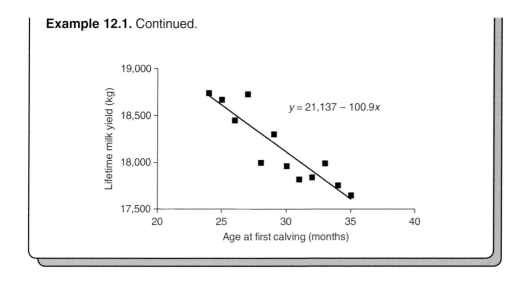

Example 12.1. Continued.

12.3.3 Testing the significance of a regression line

Clearly it is feasible to calculate a regression line for any set of bivariate data plotted on a scatter diagram whether or not the data appear to be strongly linearly related. In many cases, where the data are highly random and scattered, common sense will tell us that to plot a regression line is a meaningless exercise; however, there will be occasions when the validity of fitting a regression model is unclear. It would be useful, therefore, to be able to test the statistical significance of the regression line, in other words, to test whether the Y-variable shows a significant linear dependency on the X-variable. This can be achieved by two approaches, a Student's t-test and an analysis of variance (ANOVA) test.

Student's t-test of significance of regression

The underlying premise of this test is that, if there is absolutely no dependency of the Y-variable on the X-variable, then the regression line through the points must lie parallel with the X axis and consequently must have a true regression coefficient, β, of zero. A Student's t-test can then be used to determine whether the observed regression coefficient, b, significantly differs from a value of zero. Recalling that the Student's t-test statistic is generally determined by the difference between the value of the estimating statistic and the hypothesized value divided by the standard error of the estimate, and the standard error of the regression coefficient is given by $\sqrt{RMS/SSx}$ (see section 12.3.5), then t is obtained by:

$$t = \frac{b - 0}{\sqrt{\dfrac{RMS}{SSx}}}$$

where: RMS = residual mean square;
SSx = sum of squares of x.

The null hypothesis that there is no difference between the observed regression coefficient b and a value of 0 is tested by comparing the observed value of t with the critical t value obtained at the required probability and $n-2$ degrees of freedom. In cases where $t \geq t_{critical}$, it may then be concluded that the Y-variable shows a significant linear dependency on the X-variable. (Note that the Student's t-test can be further used to determine whether the observed regression coefficient, b, departs significantly from any other known or theoretical value of β.)

Analysis of variance of regression

The null hypothesis for an ANOVA test of a regression line is that there is no linear dependency of the Y-variable on the X-variable. (Alternatively this can be considered as a statement that the regression coefficient β equals zero.) The ANOVA test follows the same basic procedure as a one-way ANOVA in which the total variation is partitioned between that due to the linear dependency of the Y-variable upon the X-variable (the equivalent of the treatment effect in a normal one-way ANOVA) and that due to residual effects. The ratio of one to the other produces an F value that can be used to test the significance of the linear regression.

The total variation in the data is given by the sum of squares of the independent Y-variable, and is thus obtained by:

$$SS_{total} = SS_y = \Sigma(y - \bar{y})^2$$

The amount of variability in the y values that results directly from a linear dependency upon the X-variable is measured by the **regression sum of squares** ($SS_{regression}$). Regression SS is the sum of the squared deviations of every point $\hat{y}$ on the regression line from $\bar{y}$, i.e.:

$$SS_{regression} = \Sigma(\hat{y} - \bar{y})^2$$

Solving this formula would involve a long calculation; fortunately, it can be shown that $SS_{regression} = (Sxy)^2/SSx$, where Sxy = the sum of products. Since the regression slope $b = Sxy/SSx$, this becomes:

$$SS_{regression} = b. Sxy$$

The residual SS ($SS_{residual}$) is given by the sum of the squared deviations of every point y from its predicted point on the regression line $\hat{y}$, i.e. $SS_{residual} = \Sigma(y - \hat{y})^2$. Again, solving this formula would involve a long calculation; however, $SS_{residual}$ is obtained far more easily by subtraction of the regression SS from the total SS, thus:

$$SS_{residual} = SS_{total} - SS_{regression}$$

Note that when every single data point falls on the regression line the total SS will equal the regression SS and residual SS will equal zero.

The degrees of freedom (DF) associated with the total variability is $n - 1$. The DF associated with the regression SS is $p - 1$, where p is the number of parameters in the regression model that are being estimated. For a simple linear regression between two

Table 12.1. Format of an ANOVA table for testing the significance of a simple linear regression.

Source of variation	Sum of squares (SS)	Degrees of freedom (DF)	Mean square (MS)	F-ratio
Regression	$SS_{regression}$	1	$\dfrac{SS_{regression}}{DF_{regression}}$	$\dfrac{MS_{regression}}{MS_{residual}}$
Residual	$SS_{residual}$	$n - 2$	$\dfrac{SS_{residual}}{DF_{residual}}$	
Total	SS_y	$n - 1$		

variables, there are two model parameters, α and β, therefore the regression DF is always 1. The residual DF are given by $n - p$ and therefore become $n - 2$ for a simple linear regression. The regression and residual mean squares are then calculated by dividing the respective sum of squares by their associated degrees of freedom (i.e. MS = SS/DF). The F-ratio test statistic is then given by:

$$F\text{-ratio} = \frac{\text{regression MS}}{\text{residual MS}}$$

An ANOVA table for the linear regression may then be constructed and will have the format shown in Table 12.1.

Under the null hypothesis of no linear dependency so that the population regression coefficient β equates to zero, the value of the F-ratio should be close to 1, whereas the greater the F-ratio is above 1 the more improbable it becomes that the null hypothesis is correct. The significance of the regression is tested by comparing the observed F-ratio with the critical value obtained from F tables at the required probability and at the regression (V_1) and residual (V_2) degrees of freedom. If the observed F-ratio is equal to or exceeds the critical F value, the null hypothesis may be rejected on the basis that the probability of obtaining the observed F-ratio under the null hypothesis was sufficiently low, and a significant linear relationship between the variables may thus be claimed.

The determination of both an ANOVA and a Student's t-test of regression are shown in Example 12.2.

Example 12.2. Calculation of the significance of a regression.

Relationship between age of first calving and lifetime milk yield of dairy cows (data from Example 12.1).

$SSx = 143$; $SSy = 1,772,481$; $Sxy = -14,429$; $b = -100.902$; $n = 12$

Student's t-test: H_0: the regression coefficient $\beta = 0$

$$t = \frac{b - 0}{\sqrt{\dfrac{RMS}{SSx}}} = \frac{100.902}{\sqrt{\dfrac{31,656.6}{143}}} = 6.78$$

$t_{critical\ (P = 0.01;\ DF = 10)} = 3.17$

continued

Example 12.2. Continued.

Conclusion: there is a significant difference between the observed regression coefficient b and a value of zero ($P < 0.01$), implying that there is a significant linear dependency of y (milk yield) upon x (age when first calved).

ANOVA test: H_0 = there is no linear dependency of the Y-variable (lifetime milk yield) upon the X-variable (age at first calving).

Sum of squares:

Total SS	=	SSy	= 1,772,481.00
Regression SS	=	$b \times Sxy$	= 1,455,914.96
Residual SS	=	$SS_{total} - SS_{regression}$	= 316,566.04

ANOVA table

Source of variation	SS	DF	MS	F-ratio	$F_{crit \, (P = 0.01)}$
Regression	1,455,916.37	1	1,455,916.37	45.99	10.04
Residual	316,564.63	10	31,656.46		
Total	1,772,481.00	11			

Conclusion: there is a significant linear dependency of lifetime milk yield upon age when cows first calved ($P < 0.01$).

12.3.4 Testing the goodness-of-fit of a regression line: the coefficient of determination r^2

Having determined the best-fit regression line for a set of x, y values and shown that the regression model is significant, the next obvious question is how closely the data points actually fit the regression line. If the linear correlation is perfect, i.e. $r = 1$, then the variance due to the regression must equal the total variance, thus the ratio of $SS_{regression}$ to SSy must equal 1; where there is no linear correlation, this ratio becomes zero. Therefore the goodness-of-fit of data points to the regression line can be measured by the ratio $SS_{regression}/SSy$. It can be mathematically shown that the ratio $SS_{regression}/SSy$ is equal to r^2, i.e. the square of Pearson's correlation coefficient, which is the same term as that introduced in Chapter 11 (section 11.2) as the **coefficient of determination**. The value of r^2, which must lie between 0 and 1 (and is often expressed as a percentage), gives the proportion of the variance in the values of the dependent Y-variable that is explained by the regression. For example, the value of r^2 for the data shown in Example 12.1 is 0.82. This can be interpreted by saying that 82% of the variability in variation in lifetime milk yield of the observed cows is accounted for by the linear dependency of milk yield upon the age at which dairy cows first calved, the remainder of the variation is due to other unidentified factors.

12.3.5 Testing the reliability of a regression line

We have already discussed in detail how, in situations where samples are employed to represent populations, it is normal practice to estimate the reliability of the samples, by which we mean how close the sample statistics are to the true population parameters which they represent. This argument also applies to the determination of linear regression statistics. If a regression model has been constructed using a set of sample data and especially if it is intended to use the regression equation to make predictions, then it is important to be able to assess how close we might expect the sample regression coefficient b and intercept a to be to the true population regression coefficient β and the true intercept α respectively. This is achieved by determination of the standard error and the confidence limits for these statistics.

Standard error and confidence limits for the regression coefficient b

In Example 12.1 a regression coefficient b for the relationship between age of first calving and lifetime milk yield was determined from a sample of ten cows. Imagine that the regression coefficient was determined again but from a different sample of ten cows but which had the same set of values for age of first calving. Due to the variable characteristics of the cows, this second sample would almost certainly produce a different set of values for milk yield from the first sample and therefore produce a new value of b that might be very similar to or could be quite different from the first value. Both values of b are estimates of the overall population regression coefficient, β, but we have no idea which of the two is the best estimate. Now imagine that we repeat the determination of b based on a very large, indeed an infinitely large, number of samples of ten cows to produce a very large number of different values of b. Under normal circumstances we would expect this set of values of b to follow a normal distribution and the mean of these values, $\bar{b}$, to be equal to the true but unknown population value, β. Further, it can be shown that the variance of this infinitely large set of b values will be equal to the population variance divided by the sum of squares of x, i.e. σ^2/SSx. In practice, however, we only have one regression line based on just one sample; therefore, the population variance, σ^2, needs to be estimated from the sample statistics. The random variation between y values, once the effect of the regression is removed, is given by $\Sigma(y - \hat{y})^2/DF$, which is termed the **residual mean square** (RMS), and it is this value that can now be used as an estimate of σ^2. (The value for the RMS can be obtained from the analysis of variance of the regression discussed in section 12.3, 'Testing the significance of a regression line'). Therefore, substituting σ^2 for the RMS and remembering that standard deviation is the square root of variance, then the standard error of the regression coefficient b (SE_b) is given by:

$$SE_b = \sqrt{\frac{RMS}{SSx}}$$

The confidence limits around the sample regression coefficient b give the range within which we would expect the true but unknown value of population coefficient β to occur with a given level of certainty. The confidence limits around b are obtained

---

in the normal manner by multiplying SE by the appropriate Student's t value, which is obtained from Student's t distribution at the required probability and at degrees of freedom $n - 2$. Thus:

$$95\%\,CL = SE_b \times t_{(P = 0.05,\, DF = n-2)}$$

Standard error and confidence limits for the regression intercept a

Without going through the detailed mathematical explanation, it can be shown by a similar argument that the standard error of the regression intercept, SE_a, is given by:

$$SE_a = \pm\sqrt{RMS \times \left(\frac{1}{n} + \frac{\bar{x}^2}{SSx}\right)}$$

and the 95% CL for the intercept are given by:

$$95\%\,CL = SE_a \times t_{(P=0.05,\, DF = n-2)}$$

12.3.6 Testing the reliability of predictions

While it is pertinent to present the standard error and confidence limits as estimates of reliability for the calculated values of the regression statistics b and a, it is perhaps even more important that these estimates be applied to any predictions made of the y values for given values of x based on the regression statistics. There are, however, three different scenarios under which such predictions may be made:

1. First, for a given value of the X-variable, it may be required to predict the mean of the population of all possible values of the Y-variable that could occur. In relation to Example 12.1, this would be equivalent to asking, for example, what the mean lifetime milk yield is of all cows in the population that produce their first calf at 30 months. Without discussing the derivation of the equation here, it can be shown that the standard error for a predicted mean value of the Y-variable ($\hat{y}_\mu$) for a given value of the X-variable (x_i) based on a linear regression analysis is given by the following equation:

$$SE\hat{y}_{(\mu)}\sqrt{RMS \times \left(\frac{1}{n} + \frac{(x_i - \bar{x})^2}{SSx}\right)}$$

where: $SE\hat{y}_{(\mu)}$ = standard error of the predicted mean value of y for all items that have a value of the X-variable of x_i;

RMS = residual mean square = $\Sigma(y - \hat{y})^2/DF$;

SSx = sum of squares of x.

2. Second, it may be required to predict the mean y value for a sample of given size and of a given x value that is taken from the population. In relation to Example 12.1, this would be equivalent to asking what the predicted mean lifetime milk yield is of a selected sample of ten cows in the population that all produced their first calves at 30 months. If the size of the selected sample is denoted m, then the standard error for a predicted mean value of the Y-variable ($\hat{y}_\mu$) for a sample with a given value x_i of the X-variable is given by the following equation:

$$SE\hat{y}_{(m)} = \sqrt{RMS \times \left(\frac{1}{m} + \frac{1}{n} + \frac{(x_i - \bar{x})^2}{SSx} \right)}$$

where: $SE\hat{y}_{(m)}$ = standard error of the predicted mean value of y for a sample of m items all with a value of the X-variable of x_i.

3. Third, it may be required to predict the y value for a single item with a given x value that is taken from the population. In relation to Example 12.1, this would be equivalent to asking what the predicted mean lifetime milk yield is of any single cow in the population that produced their first calf at 30 months. This is the equivalent of scenario 2 but where the sample size m equals 1; therefore the standard error in this case is given by:

$$SE\hat{y} = \sqrt{RMS \times \left(1 + \frac{1}{n} + \frac{(x_i - \bar{x})^2}{SSx} \right)}$$

where: $SE\hat{y}$ = standard error of the predicted value of y for any item with a value of the X-variable of x_i.

In all cases the confidence limits for the predicted mean value $\hat{y}$ can then be obtained by multiplying the standard error by Student's t value obtained at the required probability and at DF = $n - 2$. Many computer statistics packages produce plots in which the upper and lower confidence limits based on a prediction of the mean value of y (scenario 1 above) are plotted above and below the regression line on a scatter graph. This is useful in that it shows graphically the interval around each predicted y value within which there is a 95% probability that the true value of y lies and thus indicates the reliability of the predictions. Unsurprisingly, the confidence interval becomes minimal where x is equal to $\bar{x}$ and becomes increasingly larger as predictions for y are made for values of x that are increasingly further away from the mean of x. (This can be seen in the scatter plot shown in Example 12.3) It may also be noted that that the confidence interval for the mean of y at a given x (scenario 2 above) will be narrower than the prediction interval for a single y at a given x (scenario 3).

The full calculation of the standard error and confidence limits for the regression statistics, for predicted values of y and the construction of a graphical plot of the confidence interval around the regression line are shown in Example 12.3.

Example 12.3. Determination of confidence limits for a linear regression.

Relationship between age of first calving and lifetime milk yield of dairy cows (based on the data from Example 12.1).

SSx = 143; SSy = 1,772,481; Sxy = −14,429; b = −100.902;

RMS (from ANOVA) = 31,656.60

continued

Example 12.3. Continued.

Standard error of regression coefficient $= \pm\sqrt{\dfrac{RMS}{SSx}} = \pm\sqrt{\dfrac{31{,}656.60}{143}} = \pm 14.88$

95% CL for regression coefficient $= SE \times t_{(P=0.05;\ DF=n-2)} = 14.88 \times 2.23 = \pm 33.15$

Confidence limits for predicted values of y:

Age at first calving (months)	Predicted lifetime milk yield (kg) based on regression model: $y = 21{,}137.11 - 100.90x$	$\pm SE$ $SE\hat{y}_{(\mu)} =$ $\sqrt{RMS \times \left(\dfrac{1}{n} + \dfrac{(x_i - \bar{x})^2}{SSx}\right)}$	$\pm 95\%$ CL $= SE \times t$	Upper 95% CL limit	Lower 95% CL limit
24	18715.46	96.62	451.09	18930.72	18500.20
25	18614.56	84.39	438.74	18802.57	18426.55
26	18513.66	73.14	428.60	18676.62	18350.70
27	18412.76	63.42	420.84	18554.05	18271.46
28	18311.85	56.00	415.58	18436.62	18187.08
29	18210.95	51.90	412.93	18326.58	18095.32
30	18110.05	51.90	412.93	18225.68	17994.42
31	18009.15	56.00	415.58	18133.92	17884.38
32	17908.24	63.42	420.84	18049.54	17766.95
33	17807.34	73.14	428.60	17970.30	17644.38
34	17706.44	84.39	438.74	17894.45	17518.43
35	17605.54	96.62	451.09	17820.80	17390.28

Upper and lower 95% confidence limits around predicted values of y on the regression line are now plotted graphically:

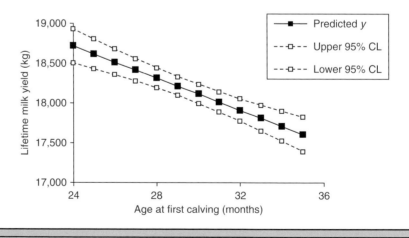

12.3.7 Simple linear regression with replicated values

In many experiments, replicate measurements, of the responding variable may be made for each fixed value of the controlled variable. For example, in an experiment to determine the effect of the plant hormone indole butyric acid (IBA) on the production of adventitious roots in stem cuttings, each IBA concentration was tested on a number of replicate cuttings and the root mass produced subsequently recorded separately for each cutting. The obvious procedure is to plot the mean value of y against each value of x and then to perform linear regression analysis based on the $x, \bar{y}$ data pairs (e.g. mean root mass in response to hormone concentration). While this simple approach is usually safe when there is equal replication, it does ignore the variance of the measured y values around each value of $\bar{y}$ and these variances cannot be expected to be equal when the number of values of y for each value of x is not the same.

The correct approach is to treat each x, y data pair as a separate piece of information so that each make an equal contribution to the determination of the regression statistics. For example, if a root-promoting hormone solution of concentration 1.0 mg/l was applied to three stem cuttings that subsequently produced a root mass of 3.0, 3.2 and 3.5 g respectively, these would be treated as three separate x, y values (i.e. $x = 1.0$, $y = 1.99$; $x = 1.0$, $y = 2.31$; $x = 1.0$, $y = 1.84$) that would be plotted as separate points on a graph and each would be involved in the determination of the best-fit regression line. This is illustrated in Example 12.4. Once the regression statistics have been determined, the significance of linearity may be tested by analysis of variance, exactly as described in the previous section.

Example 12.4. Simple linear regression with replicated values.

Relationship between concentration of applied plant root-promoting hormone IBA and mass of roots produced on stem cuttings.

Data table:

Hormone concentration (mg/l) (*X*-variable)	Root weight (g) (*Y*-variable)					
	Replicate plant					Mean
	1	2	3	4	5	
0.1	0.08	0.06	0.25	0.13		0.13
0.2	0.15	0.36	0.55	0.32	0.25	0.33
0.5	0.84	1.08	1.32	0.94	0.85	1.01
1.0	1.99	2.31	1.84			2.05

To test the linearity of the relationship between hormone concentration and root mass a simple least squares regression analysis is performed, treating each dry weight measurement as an independent bivariate value.

continued

Example 12.4. Continued.

Calculation of regression statistics:

Σx	= 6.90	Σy	= 13.32	n	= 17	
Σx^2	= 4.49	Σy^2	= 18.61	Σxy	= 9.03	
SSx	= 1.6894	SSy	= 8.1746	Sxy	= 3.6267	

The gradient of the best straight line, b, is calculated:

$$b = \frac{Sxy}{SSx} = \frac{3.6267}{1.6894} = 2.147$$

The intercept with the Y axis, a, is calculated: $a = \bar{y} - (b \times \bar{x}) = -0.088$

The regression line can now be plotted on a scatter graph:

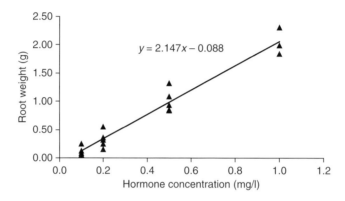

Analysis of variance of the regression: H_0 = there is no linear dependence of root mass on the hormone concentration.

ANOVA table

Source of variation	SS	DF	MS	F-ratio	$F_{crit(P = 0.01)}$
Regression	7.7853	1	7.7853	299.978	8.68
Residual	0.3893	15	0.0260		
Total	8.1746	16			

Conclusion: there is a significant linear dependency of Y (root mass) upon X (concentration of applied rooting hormone) ($P < 0.01$).

12.3.8 The problem of extreme values

The presence of just a few extreme values in a data set will have, as we saw in Chapter 2, a large effect on biasing the descriptive statistics. Similarly, a major problem for regression analysis is that extreme values of the independent variable have a disproportionately large influence on the outcome of the analysis, that is, the further away an observed x value is from the mean of the X-variable, the greater is its influence upon the regression coefficient and the predicted y values. The degree of influence that extreme values of the X-variable have on the regression line is referred to as

(a) **Data values plotted:**

x	y
5	4
7	8
14	22
18	11
25	7
32	18

(b) **Data values plotted:**

x	y
5	4
7	8
14	22
18	11
25	7
32	18
100	96

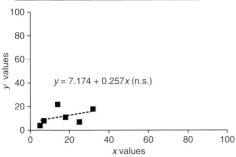
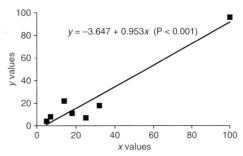

Fig. 12.2. The influence of an extreme value on regression analysis. (a) A non-significant linear regression. (b) The presence of a single extreme X-variable value causes the linear regression to become highly significant. Note that in (b) the 7th value ($x = 100$, $y = 96$) has a leverage value (h_i) of 0.93, suggesting a very strong influence on the regression line.

leverage. This is illustrated in Fig. 12.2, where in scatter graph (a) six plotted data values apparently show little relationship and although a regression line is plotted through the points the regression is insignificant ($P = 0.40$). In scatter graph (b) there is, however, a single extreme value ($x = 100$, $y = 96$) which causes a threefold increase in the regression coefficient, which then becomes highly significant ($P < 0.001$). In this latter case, it might be that a significant linear relationship really is present between the X- and Y-variables; however, it would be very risky to come to this conclusion based on the data observed since it is just one pair of x, y values that provides the main basis for arriving at this conclusion. At the very least, we should check the y measurement for the outlying x value and, better still, we should obtain more observations of the Y-variable for values of the X-variable that are intermediate between the majority of observed values and the extreme value.

Although we will not go into mathematical details here, methods are available for measuring the leverage that each x value has upon the fitted regression line. The most common way of quantitatively expressing the leverage associated with a particular value in a regression is through calculation of a term denoted h_i, which measures the contribution that an observed value makes to its own predicted value. Although the derivation is complex (see, for example, Quinn and Keough, 2002), the formula is fairly straightforward:

$$h_i = \frac{1}{n} + \frac{(x_i - \bar{x})^2}{SSx}$$

where: h_i = the leverage associated with the ith value in a bivariate data set;
n = number of bivariate values on the scatter plot;
SSx = the sum of squares of x.

The leverage value will fall between $1/n$ and 1 and the commonly accepted rule is that for simple linear regression a leverage of greater than $4/n$ indicates a value with sufficiently high influence on the regression to render it of major concern to the investigator.

Most modern statistics software packages will automatically compute and check the leverage of the data values in a regression analysis and produce a warning message for any values for which the leverage is unduly high. It is then the decision of the investigator whether to retain the identified culprit values or remove them from the data set and rerun the regression analysis. What this does go to show is the importance of always viewing the scatter graph of any bivariate data set that is subjected to a regression analysis and not to simply accept the outcome of the statistical analysis without question.

12.4 Testing the Assumptions for Simple Linear Regression by the Analysis of Residuals

The assumptions required by simple linear regression analysis were explained at the beginning of section 12.3. The requirement that values of y for each value of x are from normally distributed populations with homogeneous variances is particularly problematic to verify, especially since we invariably have available to us only a single y value for each value of x. If, however, we have a reasonable number of bivariate data values, then the deviations of the observed y values from the $\hat{y}$ values predicted by the regression model, i.e. the residuals, can be employed to check these assumptions. The residuals measure the random error present in making each measurement and, if there is no bias in the data so that the assumptions of normality and homogeneity of variance are valid, then the residuals should themselves be evenly scattered and show a normal distribution.

The distribution of the residuals can usually be verified by producing a series of graphical plots. In cases where the assumptions hold good, a plot of the residuals $(y - \hat{y})$ against the x values will display an even spread of positive and negative values about a zero line, while the presence of non-homogeneity will show up as an irregular spread of the values. Similarly, if the bivariate values come from a normally distributed population, then the histogram of the residuals will tend towards the typical symmetrical bell-shaped normal curve, while a departure from normality will be revealed by a non-symmetrical skewed curve. A third rather more complex graphical plot of residuals produced by many statistical software packages is the **normal probability plot**, the principles of which were explained in Chapter 3 (section 3.3, 'Testing for a normal distribution'). In a normal probability plot, the observed residuals are plotted against the standardized values (i.e. z values) that would be expected based on a perfect normal distribution. If the residuals adhere to a normal distribution, such a plot will produce a straight line. Examples of these plots are shown in Fig. 12.3.

Since the residuals are measuring the random error present in the data, we cannot expect any of these plots to produce perfect responses. However, regression analysis

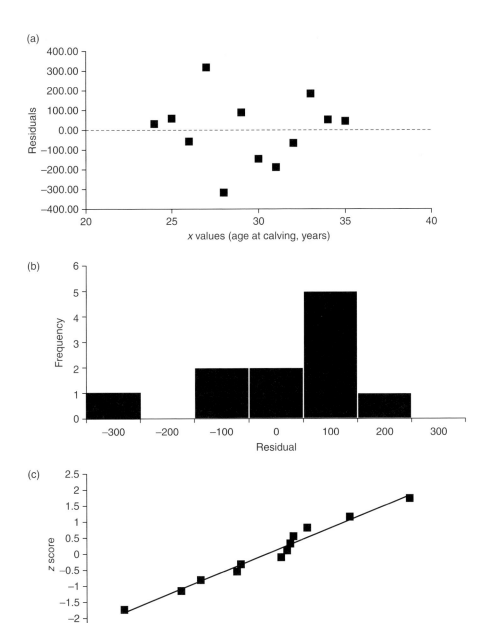

Fig. 12.3. Residual plots to determine adherence of a bivariate data set to a normal distribution. Data from Example 12.1. (a) Plot of residuals against x values. Normal distribution would be indicated by an even spread of data points above and below zero. (b) Histogram of residuals. Normal distribution would be indicated by a symmetrical bell-shaped curve. (c) Normal probability plot of residuals. Normal distribution would be indicated by data points falling on a straight line.

by least squares is generally considered a robust parametric method; therefore only major discrepancies from the expected forms of these plots need concern us and usually no further analysis beyond visual inspection of the plots is required.

12.5 Calculation of Linear Regression When Both Variables are Random: Reduced Major Axis Regression

It is often required to fit a line of best fit through points on a scatter diagram that represent two random variables where neither is controlled by the investigator. As previously mentioned, this situation is referred to as a **Model II regression**. The bivariate data in Example 12.5 concerning the viability of seed and the conductivity of solutions immersing the seed are of this type since the values of neither of these variables could be controlled by the investigator. If a least squares regression analysis is conducted on this type of data, the outcome will be biased and will tend to produce a low value for the regression coefficient b. The reason for this is that the method of least squares is

---

Example 12.5. Calculation of regression analysis of random bivariate data by reduced major axis method.

Correlation between electrical conductivity of immersing aqueous solution and seed germination rate.

Data table:

Seed batch	Conductivity (µs) (X-variable)	Germination (%) (Y-variable)	Seed batch	Conductivity (µs) (X-variable)	Germination (%) (Y-variable)
1	2100	6	13	670	52
2	900	53	14	640	57
3	550	62	15	1220	33
4	910	48	16	500	60
5	225	69	17	2100	9
6	1480	44	18	1750	12
7	750	49	19	930	32
8	885	56	20	825	58
9	2356	0	21	2350	5
10	565	64	22	2395	2
11	1625	21	23	555	71
12	1510	24	24	915	50

Calculation of regression statistics:

$\bar{x}$ = 1196.08   ;   standard deviation of x values (s_x) = 672.83

$\bar{y}$ = 39.04   ;   standard deviation of y values (s_y) = 23.39

Slope, $b' = \dfrac{s_y}{s_x}$ = − 0.035 *   Intercept, $a' = \bar{y} - b' \times \bar{x}$ = 80.62

continued

Example 12.5. Continued.

Equation of regression line: $y = 80.62 - 0.035x$

The regression line can now be plotted through the data points on the scatter plot:

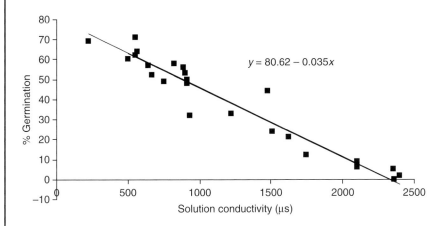

$y = 80.62 - 0.035x$

*Note that a negative sign is applied to the regression coefficient as the scatter plot of these data clearly indicates a negative slope.

based on the vertical distances from each plotted data value to the best-fit line (see Fig. 12.1) and therefore assumes the error associated with the x values is zero. If the x values are fixed by the investigator, then this assumption is valid; however, if the x values represent a set of random observations or measurements that are being used to represent a population, then each x value will have an error associated with its measurement, which, in principle, should be accounted for. In practice, however, where the error associated with each x value is random and the regression is only to be used to make predictions of y for given values of x, then the method of least squares will remain valid. Where, however, the error in measuring x would be expected to vary in proportion to the value of x and/or it is required to identify the underlying regression model, then an alternative to the method of least squares should be employed.

Out of a number of techniques that are available the simplest and possibly the most widely applied is the method known as **reduced major axis (RMA) regression**. The principle of this technique, rather than minimizing just the sum of the vertical differences between each data point and the best-fit line, is to minimize the sum of the area of the triangles formed by both the vertical and horizontal lines between each data point and the best-fit line. This is depicted in Fig. 12.4.

The slope of the regression line determined by RMA regression is denoted b' (in order to distinguish it from b, the slope of a least squares regression line) and is simply given by the ratio of the standard deviation of the y values to the standard deviation of the x values:

$$b' = \frac{s_y}{s_x}$$

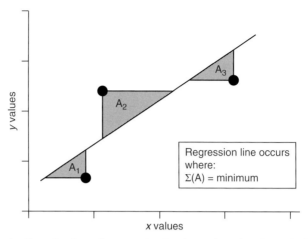

Fig. 12.4. Determination of a line of regression by the reduced major axis (RMA) technique. The regression line occurs where the sum of the triangular areas $A_1, A_2 \ldots A_n$ reaches a minimum.

However, it is very important to note that this calculation *does not give the sign of the slope* but will always give a positive value because standard deviations are always positive. The direction of the slope must be identified by inspection of the scatter diagram. In any case, where the direction of the slope is not obvious, the undertaking of a regression analysis will be of doubtful validity. Following determination of b, the intercept of the regression line with the Y axis, a', is determined just as in the least squares method:

$$a' = \bar{y} - b' \times \bar{x}$$

The final equation for the regression line produced by RMA regression is, therefore,

$$\hat{y} = a' + (b' \times x)$$

The regression line is then fitted through the data points on the scatter diagram by selecting two values of x, calculating their respective $\hat{y}$ values from the equation and joining these points with a straight line on the graph. The calculation of a regression by the reduced major axis technique is shown in Example 12.5. (For further discussion of this rather controversial subject, see Quinn and Keough, 2002.)

12.6 Comparing Linear Regression Lines

It may occur that an experiment produces more than one set of bivariate data and each set has been analysed to produce a linear regression equation. We may then ask whether the linear relationship is significantly different between the data sets. Continuing with the example of lifetime milk yield in response to age of first calving in cows, this may have been examined in a number of different breeds of dairy cattle.

The question that then arises is whether the linear relationship is the same in each breed or whether each breed has a different quantitative response between milk yield and calving age. This question can be examined by determining whether there is significant difference between the regression lines.

There are two steps to comparing regression lines. First, the slope of the regressions may be compared by examining if there is a significant difference between the regression coefficients, b. If it is shown that there is no difference between the regression slopes, then the lines may be concluded to be parallel. In this case, it may then be required to test whether the elevations, i.e. the vertical positions of the lines on the scatter plot, are significantly different. If there are just two data sets to be compared, then both the slopes and the elevations can be compared by employing Student's t-test. If there are multiple data sets to be compared, the situation becomes much more complex and requires a complex technique called **analysis of covariance**. In this section, we will examine the comparison of two data sets; readers interested in multiple comparisons between three or more samples are referred to alternative texts, e.g. Zar, 2009.

12.6.1 Comparing two regression slopes

Two regression slopes can be compared by a Student's t-test where the null hypothesis is that there is no difference between the regression coefficients. As usual for a t-test, the test statistic is given by dividing the numerical difference between the two statistics to be compared by the standard error of the difference, thus:

$$t = \frac{b_1 - b_2}{\text{SED}_{b_1 - b_2}}$$

where: b_1 and b_2 = regression coefficients of the two respective slopes;
$\text{SED}_{b1 - b2}$ = standard error of the difference between the regression coefficients.

The standard error of the difference between regression coefficients is given by:

$$\text{SED}_{b1-b2} = \sqrt{\frac{\text{RMS}_c}{\text{SS}x_1} + \frac{\text{RMS}_c}{\text{SS}x_2}}$$

where: RMS_c = combined residual mean square for the two samples;
$\text{SS}x_1, \text{SS}x_2$ = sum of squares of x for the two samples respectively.

The combined RMS is, in turn, given by:

$$\text{RMS}_c = \frac{\text{SS}_{\text{residual1}} + \text{SS}_{\text{residual2}}}{\text{DF}_{\text{residual1}} + \text{DF}_{\text{residual2}}}$$

The degrees of freedom for the test are the sum of the residual degrees of freedom for the two samples; therefore:

$$\text{DF} = (n_1 - 2) + (n_2 - 2)$$

The observed value of t is then compared with the critical value and the null hypothesis accepted or rejected accordingly.

If it is shown that there is no significant difference between two regression slopes, then the two lines may be assumed to adhere to a common regression line that has a slope b_c, which is calculated as:

$$b_c = \frac{Sxy_1 + Sxy_2}{SSx_1 + SSx_2}$$

where: Sxy_1; $SSxy_2$ = sum of x, y products for the two samples respectively.

12.6.2 Comparing two regression elevations

If two regression lines are shown to have the same slope, it may then be required to test whether their elevations are significantly different. It may be considered that a test of the elevations of two regression lines is equivalent to testing whether the intercepts of the regression lines differ. While the values of two intercepts, a_1 and a_2, can be compared by a t-test, the danger is that the intercepts may lie far from the range of the observed x values on which the regression is based and therefore this involves an extrapolation beyond the observed data. It is therefore recommended to employ the following method, which compares the vertical position of the lines over the range of the observed values of the X-variable.

In order to determine Student's t statistic for this test, the combined sample values for the sum of squares of the x values (SSx_c), for the sum of squares of the y values (SSy_c), for the sum of x, y products (Sxy_c) and for the residual sum of squares ($SS_{residual_c}$) all need to be determined initially. These are obtained as follows:

$$SSx_c = SSx_1 + SSx_2$$
$$SSy_c = SSy_1 + SSy_2$$
$$SSxy_c = SSxy_1 + SSxy_2$$
$$SS_{residual_c} = SSy_c - (Sxy_c^2 / SSx_c)$$

The combined residual degrees of freedom, $DF_{residual_c}$, is given by: $n_1 + n_2 - 3$. The combined residual mean square (RMS_c) is then obtained by:

$$RMS_c = SS_{residual_c} / DF_{residual_c}$$

The test statistic is finally calculated as:

$$t = \frac{(\bar{y}_1 - \bar{y}_2) - b_c(\bar{x}_1 - \bar{x}_2)}{\sqrt{RMS_c \times \left(\frac{1}{n_1} + \frac{1}{n_2} + \frac{(\bar{x}_1 - \bar{x}_2)^2}{SSx_c} \right)}}$$

where: b_c = the common regression coefficient (see above).

The null hypothesis is that there is no significant difference in the elevation of the regression lines. The critical value of t is found at the appropriate probability and at the combined residual DF. The observed value of t is then compared with the critical value and the null hypothesis of no difference between the elevations of the regression lines can be rejected where $t_{observed} \geq t_{critical}$.

If no significant difference is found between either the slopes or the elevations of the two regression lines, then it may be concluded that both regressions estimate the same population. In this case, if judged appropriate, the two data sets may be combined on a single plot and described by a single common regression line.

Example 12.6 illustrates the procedure for comparing regression lines based on the relationship between milk yield and age of first calving in two breeds of dairy cow. In this example there was no significant difference in the regression lines but a significant difference in the elevations ($P < 0.05$). This indicates that while there is no interaction between age of first calving and lifetime milk yield, one breed of cow does consistently produce a higher lifetime milk yield than the other breed.

Example 12.6. Comparison of two regression lines.

Relationship between age of first calving and lifetime milk yield in two breeds of dairy cows.

Data table:

Age at first calving (months)	Lifetime milk yield (kg)	
	Breed A	Breed B
24	18747	19401
25	18673	18804
26	18456	18457
27	18730	18366
28	17995	18472
29	18300	18600
30	17964	17982
31	17820	18104
32	17842	18321
33	17991	18040
34	17758	18150
35	17650	17946
Regression statistics		
$\bar{x}$	29.5	29.5
$\bar{y}$	18160.5	18386.92
SSx	143	143
SSy	1772481	1879549
Sxy	−14429	−13378.5
β	−100.902	−93.5559
α	21137.11	21146.82
Residual SS	316564.6	627910.7
Residual DF	10	10

continued

Example 12.6. Continued.

Regression lines are plotted on a scatter graph:

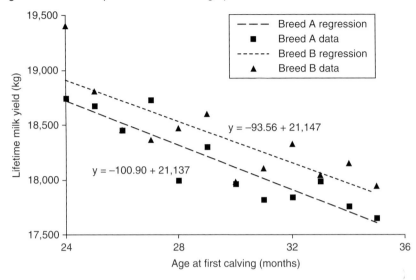

Comparison of slope of the regression lines by Student's *t*-test:

Null hypothesis: there is no difference between the regression slopes for the two breeds of dairy cow.

$$\text{Combined RMS} \quad = \text{RMS}_c = \frac{SS_{\text{residual A}} + SS_{\text{residual B}}}{DF_{\text{residual A}} + DF_{\text{residual B}}} \quad = \quad 47{,}223.7674$$

$$\text{Combined SED} \quad = \text{SED}_{b_A - b_B} = \sqrt{\frac{\text{RMS}_c}{SSx_A} + \frac{\text{RMS}_c}{SSx_B}} \quad = \quad 25.6997$$

$$t_{\text{observed}} = \frac{b_A - b_B}{\text{SED}_{b_A - b_B}} \quad = \quad -0.286$$

$$t_{\text{critical}} \ (P = 0.05; DF = 20) \quad = 2.086$$

Conclusion: the absolute value of $t_{\text{observed}} < t_{\text{critical}}$ at $P = 0.05$; therefore null hypothesis, is accepted. There is no difference in the regression slopes for the two breeds of dairy cow.

Comparison of elevation of the regression lines by Student's *t*-test:

Null hypothesis: there is no difference between the elevations of the regression lines for the two breeds of dairy cow.

$$\text{Common regression slope} \quad = \quad b_c = \frac{Sxy_A + Sxy_B}{SSx_A + SSx_B} \quad = \quad -97.2290$$

continued

Example 12.6. Continued.

SSx_c	$= SSx_A + SSx_B$	$= 286$
SSy_c	$= SSy_A + SSy_B$	$= 3,652,029.92$
Sxy_c	$= Sxy_A + Sxy_B$	$= -27,807.5$
$SS_{residual_c}$	$= SSy_c - (Sxy_c^2 / SSx_c)$	$= 948,333.9158$
$DF_{residual_c}$	$= n_1 + n_2 - 3$	$= 21$
RMS_c	$= SS_{residual_c} / DF_{residual_c}$	$= 45,158.7579$

$$t_{observed} = \frac{(\bar{y}_A - \bar{y}_B) - b_c(\bar{x}_A - \bar{x}_B)}{\sqrt{RMS_c \times \left(\dfrac{1}{n_A} + \dfrac{1}{n_B} + \dfrac{(\bar{x}_A - \bar{x}_B)^2}{SSx_c} \right)}}$$

since $\bar{x}_A = \bar{x}_B$ then $t_{observed} = \dfrac{(18,160.5 - 18,386.92)}{\sqrt{47,223.7674 \times \left(\dfrac{1}{12} + \dfrac{1}{12} \right)}}$ $= -2.552$

$t_{critical}$ $(P = 0.05; DF = 21)$ $= 2.081$

Conclusion: the absolute value of $t_{observed} > t_{critical}$ at $P = 0.05$; therefore null hypothesis is rejected. There is a significant difference in the regression elevations for the two breeds of dairy cow $(P < 0.05)$.

13 Multiple Regression and Non-linear Regression Analysis

Regression: return of curve, relapse, reversion.

Regression curve: giving best fit to inexact data.

(Concise Oxford English Dictionary)

- An introduction to multiple and non-linear regression.
- The multiple linear regression model.
- Testing the significance, goodness-of-fit and assumptions of multiple linear regression.
- Regression analysis of non-linearly related data.
- Applying data transformations to 'straighten' the data.
- Curvilinear regression analysis.
- Truly non-linear regression analysis: growth models.

13.1 Introduction

In the previous chapter, the method of simple linear regression analysis for describing the relationship between two variables was considered in some detail. However, it really is only in very controlled experiments that a measured variable responds to just a single factor; in many cases the measured variable is responding to a number of factors operating simultaneously. For example, a field trial may be conducted to identify the main determinants of tuber yield in potatoes. While it might be considered that plant density would have the major role in determining tuber yield, in the field we would expect yield to be influenced by a range of additional factors such as nitrogen supply, water supply, mean daily temperature and so on and it is the combination of the levels of these factors that determines the final response. If such factors are delivered as experimental treatments at fixed levels, then their individual effects on the responding variable would be analysed by multiple-factor ANOVA. If, however, the factors are in the form of a number of continuous predictor variables and we are attempting to describe the relationship between a responding measured variable and the combination of these predictor variables, then a method of **multiple regression analysis** is required.

Although multiple regression analysis may be used to describe a fairly complex set of relationships, it is still assumed that the relationship between each of the independent X-variables and the responding Y-variable is linear. However, as discussed at the start of the previous chapter (section 12.1), two or more strongly correlated variables may display a curve rather than a straight line when plotted

---

 © C. Ireland 2010. *Experimental Statistics for Agriculture and Horticulture* (C. Ireland)

on a scatter graph. Clearly it would be invalid to determine the best-fit regression line through such data and instead a method of **non-linear regression** is then required. In this chapter we will introduce some of these more complex aspects of regression analysis.

13.2 Multiple Regression Analysis

When attempting to unravel the relationship between a response variable and a number of possible predictor variables by multiple regression analysis, there are two questions that we are generally asking: first, whether there is a significant and predictive relationship between the measured variable (Y) and the combination of multiple predictor variables (X_A, X_B, X_C, ... X_K); and, secondly, if there is a relationship, which of the predictor variables has the most influence in determining the observed value of the Y-variable. Given that the data adhere to certain assumptions, then multiple regression analysis functions to resolve these two key questions.

13.2.1 The multiple linear regression model

If the assumptions that are required to enable a bivariate data set to be analysed by simple linear regression can also be applied to a multivariate data set, then multiple linear regression may proceed by a logical extension of the simple linear regression model. The main assumptions are that, for any given combination of values of the X predictor variables, the theoretical population of dependent y values is normally distributed and the variances of different populations of y values are statistically homogeneous. (Some further important assumptions beyond this will be discussed in due course.) Recalling that the linear regression model for a population of bivariate values takes the form $y = \alpha + \beta x + \varepsilon$, this equation can be extended to describe the additive effect of further predictor variables. Thus, where there are K predictor variables present, then:

$$y_i = \alpha + \beta_1 x_{A,i} + \beta_2 x_{B,i} + \ldots + \beta_k x_{K,i} + \varepsilon_i$$

where: y_i = the value of the dependent Y-variable for a given value x_i of each of the predictor variables X_A, X_B ... X_K;
α = the 'intercept' with the Y-axis;
β_i = the regression coefficient associated with each respective X predictor variable;
ε_i = a random error term associated with the measurement of y for a given set of values x_i of each of the predictor variables.

The regression coefficients (β_1, β_2, ..., β_k) are referred to as the **partial regression coefficients** and together determine the 'slope' of the overall regression line. However, because we are no longer considering the relationship between just two variables but rather the relationship between three or more variables, the concept of a best-fit line becomes problematic. We need, therefore, to think in three (or more) dimensions and statisticians refer not to a regression line but instead to the best-fit 'regression plane'. More straightforwardly, each partial regression coefficient can be considered as a measure of the change in the Y-variable that would occur for each unit change in the associated X-variable were all the remaining X-variables to be held constant. Similarly, while the notion of a physical intercept on a multi-variable scatter plot is rather difficult to conceptualize, mathematically α is simply the mean value of the Y-variable when the value of all the X predictor variables is zero. The variance of the random error ε_i terms is given by the variance of the population of y values for each given set of x values and is estimated by the sample variance s_y^2. If the sample variances are assumed to be homogeneous, then s_y^2 can be ignored and the sample multiple regression equation becomes:

$$\hat{y}_i = a + b_1 x_{A,i} + b_2 x_{B,i} + \ldots + b_k x_{K.i}$$

where: $\hat{y}_i$ = the predicted value of the Y-variable for a given value x_i of each of the predictor variables X_A, X_B ... X_K;

a = the 'intercept' with the Y-axis;

b_i = the regression coefficient associated with each respective X predictor variable.

It should be noted that this model treats the predictor variables (X_A, X_B, X_C, etc.) as independent factors that have an additive effect on the responding Y-variable. If, however, these factors are themselves correlated with each other and/or they interact with each other to affect the response, then the multiple linear regression model becomes very much more complex.

The values of the equation parameters can be determined by a complex computation based on the method of least squares and involving matrix algebra. If it is necessary for the reader to delve further into this mathematics, then more advanced texts should be consulted (e.g. Fox, 1997; Draper and Smith, 1998); however, most users will be content to let the computer undertake the hard work. Most modern statistical software packages will present the partial regression coefficients (b_1, b_2 ... b_k) and the intercept constant a together with their associated standard errors (remember the values of b and a are estimates of the population parameters β and α). They will usually also present a matrix displaying the individual correlation coefficients r and/or r^2 between each of the individual independent variables involved in the multiple regression analysis. Just as with simple linear regression analysis, once the values of b_1, b_2 ... b_k and a are known, then a new value of the Y-variable can be predicted for any given combination of values of the X predictor variables by substituting the new x values into the equation. As with predictions using simple linear regression models, extrapolative predictions involving x values that lie outside the range of the observed data should be firmly avoided since there is no evidence that the regression model is valid beyond the observed data.

A relatively simple example of multiple linear regression is given here (Example 13.1), in which the yield of potatoes is related to three variables, the levels of soil nitrogen (N), soil phosphate (P) and soil potassium (K). The analysis was undertaken using the GenStat statistics package. The output also includes an analysis of variance test to determine the significance of the regression model.

Chapter 13

Example 13.1. A multiple linear regression analysis.

The response of potato tuber yield to soil nitrogen, phosphate and potassium content.

In an attempt to establish the nature of the relationship between potato yield and soil nitrogen, phosphate and potassium (N:P:K) content, tuber fresh weight was determined for crops grown in plots with differing relative N:P:K levels. The results obtained were as follows:

Plot	Total soil nitrogen (g/kg)	Total soil phosphate (mg/kg)	Total soil potassium (mg/kg)	Tuber yield (t/ha)
1	0.95	6.62	0.13	17.4
2	1.55	6.51	0.33	28.3
3	1.20	5.23	0.15	20.8
4	0.85	6.90	0.14	17.6
5	1.30	5.88	0.27	25.6
6	1.30	6.28	0.28	29.5
7	1.35	6.95	0.29	27.9
8	1.45	5.77	0.26	28.5
9	0.90	5.68	0.22	22.3
10	1.00	5.22	0.19	21.7
11	0.75	6.34	0.16	18.8
12	1.25	5.47	0.22	20.0
13	1.30	5.88	0.21	20.2
14	0.90	5.26	0.19	19.4
15	1.45	5.40	0.32	27.8
16	1.05	6.00	0.25	24.3
17	1.10	5.48	0.23	22.0
18	1.00	5.33	0.21	20.7
19	0.90	6.11	0.22	19.5
20	1.40	5.55	0.29	26.0

Scatter plots and simple linear regressions

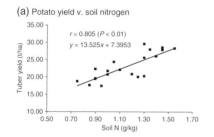

(a) Potato yield v. soil nitrogen

$r = 0.805$ ($P < 0.01$)

$y = 13.525x + 7.3953$

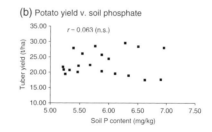

(b) Potato yield v. soil phosphate

$r = 0.063$ (n.s.)

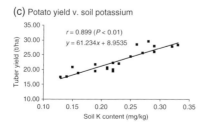

(c) Potato yield v. soil potassium

$r = 0.899$ ($P < 0.01$)

$y = 61.234x + 8.9535$

continued

Example 13.1. Continued.

Multiple linear regression (derived by computer analysis using GenStat program)

Table of partial correlations coefficients (r) between predictor variables:

	Nitrogen	Phosphate	Potassium
Nitrogen	1.000		
Phosphate	0.078	1.000	
Potassium	−0.789	−0.065	1.000

Estimated constant (a) = 4.141 (SE = ± 4.771)

Table of partial regression coefficients (b) and associated statistics

	b	± SE	Student's t	Significance	Tolerance[a]
Nitrogen	4.442	2.754	1.613	0.126	0.377
Phosphate	0.502	0.723	0.694	0.498	0.994
Potassium	47.020	11.158	4.214	0.001	0.378

[a]This statistic is used as a test for collinearity; see text for explanation.

Multiple linear regression model:

$y = 4.4141 + 4.442 \; x_{nitrogen} + 0.502 \; x_{phosphate} + 47.020 \; x_{potassium}$

Analysis of variance of the multiple regression:

Null hypothesis: potato tuber yield is not linearly related to the three predictor variables ($\beta_1 = \beta_2 = \beta_3 = 0$).

ANOVA Table.

Source of variation	SS	DF	MS	F-ratio	Significance
Regression	250.033	3	83.344	27.533	<0.01
Residual	48.433	16	3.027		
Total	298.465	19			

Adjusted r^2 = 0.807

Conclusion: the potato tuber yield is significantly linearly related ($P < 0.01$) to the three predictor variables (soil nitrogen, phosphate and potassium) through the stated multiple regression model.

13.2.2 Testing the significance of a multiple linear regression model

Before employing a calculated multiple regression model to predict values of a dependent variable for a given combination of values of two or more predictor variables, we will wish to know how well the derived model fits the data. If the observed data fail to fit the model appreciably, then employment of the model to describe the relationship between the variables is clearly invalid. When discussing simple linear regression analysis, it was shown that employment of both Student's t-test and ANOVA provided the means for this (Chapter 12, section 12.3, 'Testing the reliability of a regression line') and the same approach can be used for a multiple linear regression model. If there is no relationship between a dependent variable and a given predictor variable, then the partial regression coefficient for this predictor variable will in theory be zero. Therefore Student's t-test can be used to examine whether each partial regression coefficient significantly differs from zero. The Student's t test statistic is provided in this case by:

$$t_{observed} = \frac{b_i}{SE_{b_i}}$$

where: b_i = the partial regression coefficient for the predictor variable i;

SE_{b_i} = the standard error of partial regression coefficient b_i.

The degrees of freedom are given by $n - (p + 1)$ where p is the total number of independent variables.

Rather than testing the significance of each partial regression coefficient separately, ANOVA allows all the coefficients to be tested at once, the null hypothesis being that the estimated slopes $\beta_1, \beta_2, \beta_3 \dots \beta_k$ all equal zero. The ANOVA proceeds by the same principles used in simple linear regression, that is, the total variation is partitioned between that explained by the linear regression and that due to random residual effects. If the null hypothesis is true, then, by the now familiar arguments, the regression variance ($MS_{regression}$) and residual variance ($MS_{residual}$) will be similar and the F-ratio ($MS_{regression}/MS_{residual}$) will approximate to 1. If the null hypothesis is false, then at least one of the partial regression slopes will not be equal to zero so that the regression variance now becomes larger than the residual variance and the F-ratio becomes >1. The formulae for determining the regression MS and the residual MS and the format of the ANOVA table are shown in Table 13.1. Whether the observed F-ratio is sufficiently large to allow the null hypothesis to be rejected at a given level of probability is tested by comparing with the appropriate critical F value obtained from tables in the normal manner.

Table 13.1. Format of an ANOVA table for testing the significance of a multiple linear regression model.

Source of variation	Sum of squares (SS)	Degrees of freedom (DF)	Mean Square (MS)	F-ratio
Regression	$SS_{regression} = \Sigma(\hat{y} - \bar{y})^2$	p	$\dfrac{SS_{regression}}{DF_{regression}}$	$\dfrac{MS_{regression}}{MS_{residual}}$
Residual	$SS_{residual} = \Sigma(y - \hat{y})^2$	$n - p - 1$	$\dfrac{SS_{residual}}{DF_{residual}}$	
Total	$SS_{total} = SSy$	$n - 1$		

In Example 13.1 Student's *t*-test and ANOVA were employed to assess the significance of the multiple regression model involving the relationship between potato tuber yield and the soil levels of N, P and K. It will be noted that only the partial regression coefficient for potassium appears to be significant ($P < 0.01$), suggesting that potassium is the major predictor variable controlling yield; however, the overall multiple regression model is highly significant ($P < 0.01$), suggesting this to be a reliable model for the prediction of yield for given combined levels of N, P and K.

13.2.3 Testing the goodness-of-fit of a multiple linear regression model

The goodness-of-fit of a set of bivariate data to a regression model is assessed by the coefficient of determination r^2, which may be more specifically interpreted as a measure of the proportion of the variance in the Y-variable that is explained by the regression. The value of r^2 can be obtained by $SS_{regression}/SS_{total}$ (see Chapter 12, section 12.3.4). However, r^2 does not alter when more predictor variables are added to the model although the addition of extra predictor variables may either have improved or reduced the goodness-of-fit. For this reason, it cannot be used to compare multiple regression models with different numbers of predictor variables and is of little use, therefore, when trying to establish which of a number of possible models provides the best fit. An acceptable measure of goodness-of-fit for a multiple regression is obtained, however, by subtracting the ratio of the residual to the total mean squares from 1, which then produces a statistic termed **adjusted r^2**. Adjusted r^2 is thus given by:

$$r^2_{adjusted} = 1 - \frac{MS_{residual}}{MS_{total}} = 1 - \frac{SS_{residual}/n-p-1}{SS_{total}/n-1}$$

The value of adjusted r^2 will always lie between zero and one and may be expressed as % value if desired. Thus, in Example 13.1, the adjusted r^2 is 0.807, which may be interpreted by stating that 80.7% of the variance in potato yield is due to its multiple linear dependency on the levels of nitrogen, phosphate and potassium.

Unlike r^2, the adjusted r^2 value will increase if a further predictor variable added to a regression model improves the fit and will decrease if an additional predictor variable reduces the fit. Therefore, where data have been collected for a number of possible different predictor variables that may affect a particular response variable, the adjusted r^2 may be used to ascertain which combination of predictor variables in the multiple regression model gives the best fit to the data. The predictor variables can in turn be added and removed from the model until the maximum value of $r^2_{adjusted}$ is found. This procedure, termed **stepwise multiple regression,** is well discussed, together with some other measures of goodness-of-fit, by Quinn and Keough (2002).

13.2.4 Testing the assumptions of a multiple linear regression model

Diagnostic checks of the validity of the underlying assumptions required by multiple regression analysis can be performed in much the same way as for simple linear regression analysis (see section 12.4). The requirement that values of the Y-variable for each value of each X-variable are from normally distributed populations with homogeneous variances can be assessed through the pattern of the residuals, that is the deviation of each predicted y value from that predicted by the model equation, i.e. $(y - \hat{y})$. When the histogram of residuals is plotted, a departure from a normal distribution within the data will be revealed by a non-symmetrical skewed curve. Further, the existence of any extreme x values that have an unduly large degree of influence on the model can be identified through measurement of the **leverage** associated with each observation (see 12.3.7). The calculation in this case is based on the means of all the predictor variables and is rather complex; however, it is usually standard output from specialized statistics software packages. It is generally considered that a leverage value $> 2(p/n)$ would identify a value of a predictor variable that has undue influence on the multiple regression model.

13.2.5 The problem of collinearity between predictor variables in multiple regression analysis

It is a requirement of multiple linear regression analysis that the predictor variables should be independent of each other and thus have an additive effect on the responding variable. In crop and animal research, however, very often this will not be the case. Consider Example 13.1, in which potato tuber yield was related to the three predictor variables of soil nitrogen, phosphate and potassium. While in very carefully controlled pot or hydroponic experiments it may be possible to control each of these independently, in the field, where soils may have supported crop rotations over a number of years and been subjected to blanket fertilizer application, it is most unlikely that the levels of N, P and K are totally independent of each other. In other words, a soil lacking in nitrogen is also likely to be lacking in phosphate and potassium. In multiple regression analysis, the existence of a mutual correlation between two or more of the predictor variables is termed **collinearity**. For complex mathematical reasons, the existence of collinearity causes inaccuracies in the determination of the multiple regression model and, in particular, causes the estimates of the partial regression slopes to become unreliable. One of the more overt effects of this is that small changes made to the data or deleting/adding one of the predictor variables can change the estimated regression coefficients considerably and may even change their sign. This problem is discussed in further detail by Quinn and Keough (2002), who point out that, while the existence of collinearity does not necessarily prevent fitting a significant multiple regression model to a data set and using it to make effective predictions, it does mean that if a second sample is taken from the same population this might produce quite different estimates of the multiple regression parameters.

There are two fairly straightforward ways for detecting the existence of collinearity among the predictor variables. First, the matrix of partial correlation coefficients between the predictor variables is examined (remember that such matrices are commonly produced by modern statistical software packages). Collinearity will be indicated by a large correlation value between a pair of predictor variables. A second method is to determine the **tolerance** value associated with each predictor variable. Tolerance is determined by $1 - r^2$, where r is the correlation coefficient obtained by a simple linear regression analysis between the predictor variable of interest and the other predictor variables in the model. Again, modern statistical software packages will usually give this as standard output. Tolerance values of less than 0.1 (corresponding to $r > 0.95$) are generally considered to be indicative of significant collinearity. It may be noted in Example 13.1 that, while the observed partial correlation coefficient between nitrogen and potassium of 0.789 is fairly high, the tolerance values are all well above 0.1 and therefore do not suggest the presence of any appreciable collinearity that is likely to compromise the derived multiple regression model.

When collinearity is shown to be present, dealing with it is highly problematic. A number of techniques have been suggested by a range of authors and are summarized well by Quinn and Kough (2002); however, none are totally satisfactory in that they all result in biased estimates of the regression coefficients. It is beyond the scope of the current text to discuss these further, especially as none are considered definitive by statisticians. In some ways, the easiest solution is simply to remove from the analysis any predictor variables that show a high degree of correlation with any of the other predictor variables. While this might give the impression that valuable information is being discarded, in fact a correlated predictor variable cannot contribute any further information useful to the analysis and its removal is actually quite appropriate.

13.2.6 Interaction between predictor variables in multiple regression analysis

The possibility that interaction between treatment factors may determine the magnitude of a responding variable was first encountered when discussing multiple-factor analysis of variance in Chapter 8. It was shown that two treatment factors, which may or may not produce a response on their own, when in combination may be able to produce an extra interactive effect. Similarly, in multiple regression analysis two predictor variables which each have an independent effect on the responding variable may interact to produce an extra level of response. Taking a simplistic approach, the interaction effect may then be treated as a further independent predictor variable in the multiple regression model. Thus, if we have two interacting predictor variables X_A and X_B, then their interaction can be denoted by $X_A \times X_B$ and incorporated into the multiple regression model as a further additive component:

$$y_i = \alpha + \beta_1 x_{A,i} + \beta_2 x_{B,i} + \beta_3 \left(x_{A,i} \times x_{B,i} \right) + \varepsilon_i$$

The statistical significance of the partial regression coefficient for the interaction term, i.e. β_3, can be determined by a Student's t-test where the null hypothesis is that β_3 is equal to zero. Rejection of the null hypothesis indicates that the interaction between the variables X_A and X_B is significant. This implies that the partial regression slope of the responding variable Y on the first predictor variable X_A depends on the value of the second predictor variable X_B. Similarly, the partial regression slope of the responding variable Y on the second predictor variable X_B depends on the value of the first predictor variable X_A. Analysis of variance can confirm, or otherwise, the significance of the overall fit of the model equation to the observed data.

Example 13.2. A multiple linear regression analysis with interaction between predictor variables.

In a Euonymus (Spindle Tree) stem cutting rooting trial, the weight of root produced 2 months after application of a combination of the root-promoting hormones indole butyric acid (IBA) and naphthaleneacetic acid (NAA) was determined with the following results:

IBA (mg/l)	NAA (mg/l)	Root weight (g)	IBA (mg/l)	NAA (mg/l)	Root weight (g)
0.1	0.1	2.71	1.5	0.1	8.83
0.1	0.5	2.50	1.5	0.5	10.11
0.1	1.0	3.12	1.5	1.0	11.50
0.1	1.5	2.95	1.5	1.5	13.10
0.1	2.0	2.83	1.5	2.0	15.33
0.5	0.1	4.03	2.0	0.1	11.23
0.5	0.5	3.87	2.0	0.5	13.33
0.5	1.0	4.21	2.0	1.0	15.26
0.5	1.5	5.50	2.0	1.5	17.42
0.5	2.0	4.49	2.0	2.0	19.69
1.0	0.1	5.35			
1.0	0.5	5.92			
1.0	1.0	6.54			
1.0	1.5	7.35			
1.0	2.0	7.78			

Scatter plots and simple linear regressions:

(a) IBA v. root weight

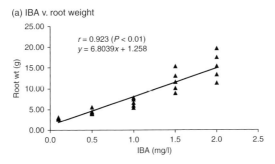

$r = 0.923$ ($P < 0.01$)
$y = 6.8039x + 1.258$

(b) NAA v. root weight

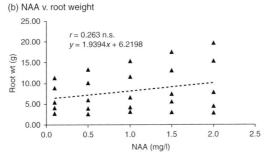

$r = 0.263$ n.s.
$y = 1.9394x + 6.2198$

continued

Example 13.2. Continued.

Multiple linear regression (derived by computer analysis using Genstat program)

Estimated constant (a) = 1.736 (SE = ±4.771)

Table of partial regression coefficients (b) and associated statistics

	b	±SE	Student's t	Sig.
IBA	4.396	0.515	8.53	<0.01
NAA	−0.468	0.515	0.91	n.s.
IBA × NAA (interaction)	2.360	0.420	5.62	<0.01

Multiple linear regression model:

$$y = 1.736 + 4.396x_{IBA} - 0.468x_{NAA} + 2.360(x_{IBA} \times x_{NAA})$$

Analysis of variance of the multiple regression

Null hypothesis: root weight is not linearly related to the predictor variables ($\beta_1 = \beta_2 = \beta_3 = 0$).

ANOVA table.

Source of variation	SS	DF	MS	F	Sig.
Regression	607.31	3	202.44	215.06	<0.001
Residual	19.77	21	0.94		
Total	627.08	24			

Conclusion: the root weight is significantly linearly related ($P < 0.01$) to the predictor variables (IBA, NAA, IBA × NAA) through the stated multiple regression model.

As before, the mathematics are highly complex but modern statistical software can readily run the analysis. This is illustrated in Example 13.2, involving a study of the rooting response of stem cuttings of *Euonymus europaeus* (spindle tree) to application of two root-promoting hormones, indole butyric acid (IBA) and naphthalene-acetic acid (NAA). Using a computer statistics package, a general linear model is entered in the form: $y = X_{IBA} + X_{NAA} + (X_{IBA} \times X_{NAA})$, where X_{IBA} and X_{NAA} represent the predictor variables (IBA and NAA) and $X_{IBA} \times X_{NAA}$ represents their interaction. The output shows that only IBA produces an independent significant effect; however, there is a very significant interaction, i.e. the rooting response to the first rooting promoter hormone (IBA) depends on the level of the second rooting hormone (NAA). The output gives the parameters for the multiple regression model and ANOVA shows that the multiple regression is significant.

13.3 Regression Analysis of Non-linearly Related Data

Where graphically plotted bivariate data show a curved relationship, it is clearly inappropriate to plot a best-fit regression line and an alternative approach is required. In terms of their statistical treatment, two distinct types of non-linear relationship can be distinguished. First, it is quite common for the relationship between two variables to be described using a linear regression model although the bivariate data points plot a curve rather than a straight line on a scatter plot. This may be achieved either by applying a mathematical transformation to the values, which has the effect of straightening the data, or by applying a polynomial mathematical model which describes a curvilinear response but remains a linear equation, i.e. each of the equation parameters are additive. We might consider such cases as being 'pseudo non-linear'. On the other hand, where the relationship between two variables cannot be described by a linear equation and neither transforming the data nor employing a polynomial model is effective (or indeed valid), this is referred to as being a truly or **intrinsically non-linear** relationship.

13.3.1 Data transformation for linear regression analysis

It is often found in biological relationships that, where two variables display a curved relationship, if the values associated with the dependent Y-variable are transformed by applying a constant mathematical function a linear relationship may then be revealed. As long as the transformation does not cause the sample variances to become non-homogeneous, a simple regression analysis can then be performed using the transformed values. Common mathematical transformations that may achieve this under different circumstances include $\log_{10}$, which is used for a range of general curved responses, $\log_e$ (i.e. natural log) for data that show an exponential relationship, and square root for data expressed in the form of proportions or ratios.

Example 13.3 demonstrates the use of a simple $\log_{10}$ transformation. It is known that there is a very strong correlation between the numbers of photoperiodic short days

---

Example 13.3. A $\log_{10}$ transformation of data prior to performing linear regression analysis.

The relationship between flowering and the number of short days received by *Kalanchoe* plants.

No. of short days	No. of florets produced per plant	Log_{10} floret number
0	1	0.0
1	3	0.477
2	5	0.699
3	7	0.845
4	11	1.041
6	22	1.342
8	63	1.800
10	94	1.973
12	182	2.260
15	401	2.603
	$r = 0.88$	$r = 0.99$

continued

Example 13.3. Continued.

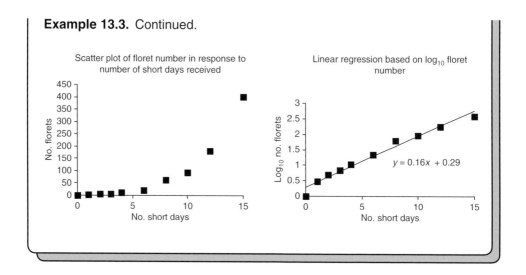

Scatter plot of floret number in response to number of short days received

Linear regression based on $\log_{10}$ floret number

$y = 0.16x + 0.29$

received and the number of florets produced by the pot plant *Kalanchoe*. A graphical plot of the number of florets (Y-variable) against the number of short days received (X-variable) indicates a strongly correlated but curvilinear relationship. However, when the $\log_{10}$ of the flower number is plotted against the number of short days received, a clear linear relationship is revealed. The data in this example therefore display a log-linear relationship, and a full correlation analysis and least squares regression analysis can subsequently be performed using the $\log_{10}$-transformed y values.

13.3.2 Polynomial regression analysis

Where data clearly show a curvilinear relationship, it is preferable where possible to describe this by a best-fit curve rather than by transforming the data and fitting a regression line. While the mathematics involved in determining a best-fit curve through a set of bivariate data is complex, the ready availability of computers and reasonably friendly statistical software packages do today allow best-fit curves to be determined fairly easily.

The most straightforward method for fitting a curve through a set of plotted data is to use a technique called **polynomial regression**. In order to understand the technique let us return to the equation of a straight line: $y = \alpha + (\beta \times x)$ where the constants α and β are the intercept with the Y-axis and the slope of the line respectively. This type of equation is called a **first-order polynomial**. If, however, a further term is added to this equation so that it takes the form:

$$Y = \alpha + \left(\beta_1 \times X\right) + \left(\beta_2 \times X^2\right)$$

this now plots a curve rather than a straight line and is referred to as a second-order polynomial. (Readers with a mathematical knowledge will realize that this is a quadratic equation and, therefore, this model is often referred to as a quadratic regression.

Because, however, the parameters α, β_1 and β_2 are additive, in mathematical terms it remains a linear model.)

Calculating estimates of the constants α, β_1 and β_2 from the observed data values is a complex process but can be easily obtained using appropriate computer software. Once estimates of α, β_1 and β_2 have been obtained, then predicted values for y that lie on the curve can be obtained for a range of x values and the curve plotted through the data points on the scatter graph. It is very important, however, that the curve is not extrapolated as there will be no evidence that the relationship holds beyond the observed values of x. Again, many computer software packages will plot the best-fit curve based on the derived polynomial equation.

Occasionally the apparent relationship between the x, y data is an 'S-shaped' or sigmoid curve. In this case, the equation can be further extended to a third-order (or 'cubic') polynomial model in order to determine the best-fit sigmoid curve:

$$Y = \alpha + \left(\beta_1 \times X\right) + \left(\beta_2 \times X^2\right) + \left(\beta_3 \times X^3\right)$$

The types of curves produced by these polynomial models are depicted in Fig. 13.1. Note that, as with the truly linear model, the sign of the β values determines the direction of the curve. While in theory it would be possible to further extend the model to a fourth-order polynomial, this is rarely necessary in practice.

A fully worked example of the use of a third-order polynomial to describe a curvilinear relationship is provided in Example 13.4.

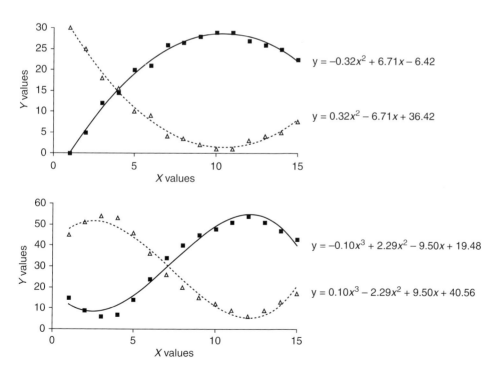

Fig. 13.1. Examples of best-fit curves and regression equations for curvilinear data produced by (a) a second-order polynomial model; (b) a third-order polynomial model.

Testing the significance of a polynomial regression

As with linear regression, the overall significance of a polynomial regression model can be determined by an ANOVA procedure. The null hypothesis tested is that the relationship between the variables in the population does not adhere to the stated polynomial model.

The total variation in the data is given by the sum of squares of the independent Y-variable:

$$SS_{total} = SSy = \Sigma(y - \bar{y})^2$$

For each observed x value, the difference between the observed y value and the y value predicted by the polynomial equation, i.e. $(y - \hat{y})$, is determined and the residual sum of squares then obtained by:

$$SS_{residual} = \Sigma(y - \hat{y})^2$$

The regression sum of squares is then determined by simple subtraction:

$$SS_{regression} = SS_{total} - SS_{residual}$$

The degrees of freedom (DF) associated with the total variability are $n - 1$. The regression DF are given by the number of parameters (p) in the polynomial equation that have been estimated less 1. For a second-order polynomial α, β_1 and β_2 are estimated; therefore the regression DF are $3 - 1$. For a third-order polynomial the regression DF are $4 - 1$. The residual DF are then the total DF – regression DF. The regression and residual mean squares are calculated by SS/DF and the ratio of the regression MS to the residual MS then gives the F test statistic in the normal way:

$$F = \frac{\text{regression MS}}{\text{residual MS}}$$

The observed F-ratio is then compared with the appropriate critical value from the F distribution at the required probability and at the regression and residual DF. Where the observed F-ratio $\geq F_{critical}$, the null hypothesis may be rejected and it may be concluded that the stated polynomial regression model provides a significant fit to the observed data.

Testing whether a polynomial regression gives a significantly better fit than a linear regression

Occasionally bivariate data will be obtained such that, when plotted on a scatter graph, it is unclear whether a line or a curve provides the best description of the relationship. Under these circumstances, it is possible to test statistically whether a linear or a polynomial model provides the better fit. First, a linear regression is performed and the regression SS obtained as previously described (Section 12.3, 'Testing the significance of a regression line'). Subsequently, the alternative polynomial regression SS and the residual mean square based on a polynomial model are obtained as described above. An F-ratio test statistic is then given by:

$$F = \frac{SS_{\text{regression}}(\text{polynomial}) - SS_{\text{regression}}(\text{linear})}{RMS(\text{polynomial})}$$

To test whether the polynomial regression provides a significantly better fit than the linear model, the observed F-ratio is compared with the appropriate critical value read from F tables at 1 and $n - 1$ degrees of freedom.

The calculation of the ANOVA for testing the significance of a curvilinear model and for comparing a linear and curvilinear model is illustrated in Example 13.4.

Example 13.4. A polynomial regression analysis.

The relationship between canopy attained by a potato crop and the planting density.

A field trial was conducted to determine the relationship between the canopy area achieved by potato crops and the planting density. The results obtained were as follows:

Planting density (× 10^3/ha)	Green leaf area index (LAI) of potato crop
10	1.3
20	1.9
30	2.5
40	3.2
50	4.2
60	4.8
70	5.1
80	5.0

A simple linear regression analysis gives the model: $y = 0.8321 + 0.0593x$.

A scatter plot of the data indicates, however, a possible curvilinear relationship. A third-order polynomial is therefore applied and the computer-generated best-fit model is: $y = -0.0000232x^3 + 0.00267x^2 - 0.0186x + 1.286$.

The polynomial model gives the following plot:

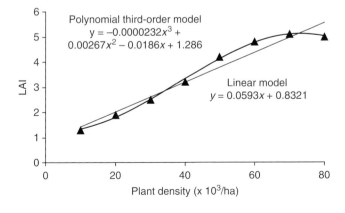

continued

Example 13.4. Continued.

Test of significance of the polynomial regression by ANOVA: H_0 = data do not fit the polynomial regression model.

Table of residuals:

Planting density (x)	LAI (y)	Predicted LAI based on polynomial model ($\hat{y}$)	Residual ($y - \hat{y}$)
10	1.3	1.3437	−0.0437
20	1.9	1.7967	0.1033
30	2.5	2.5053	−0.0053
40	3.2	3.3302	−0.1302
50	4.2	4.1320	0.0681
60	4.8	4.7712	0.0288
70	5.1	5.1086	−0.0086
80	5.0	5.0047	−0.0047

$$\Sigma(y - \hat{y})^2 = 0.0351$$

Sum of squares:

$$SS_{total} = SSy = \Sigma(y - \bar{y})^2 \qquad = \qquad 15.4800$$

$$SS_{residual} = \Sigma(y - \hat{y})^2 \qquad = \qquad 0.0351$$

$$SS_{regression} = SS_{total} - SS_{residual} \qquad = \qquad 15.4449$$

ANOVA table.

Source of variation	SS	DF	MS	F-ratio	Significance
Regression	15.4449	3	5.1483	586.56	<0.001
Residual	0.0351	4	0.0088		
Total	14.4800	7			

Conclusion: H_0 rejected: the data significantly adhere to the stated polynomial regression model ($P < 0.001$).

Test of significance to compare the polynomial regression with a linear model: H_0 = no difference in the adherence of the data to the linear and the third-order polynomial regression models.

The alternative linear regression analysis gave the model: $y = 0.0593x + 0.8321$ with a calculated $SS_{regression}$ of 14.7621.

$$F = \frac{\text{polynomial } SS_{regression} - \text{linear } SS_{regression}}{\text{polynomial RMS}} = \frac{15.4449 - 14.7621}{0.0088} = 77.59$$

$$F_{critical\ (P = 0.05;\ DF = 1,\ 7)} = 5.59$$

Conclusion: H_0 is rejected ($P < 0.01$): the third-order polynomial regression provides a significantly better fit than a linear model.

13.3.3 Truly non-linear regression models

Certain relationships are truly non-linear, that is, they cannot be modelled using a linear equation and neither transforming the data nor fitting a polynomial model will produce a satisfactory description of the relationship. A common example in life sciences is **growth models** where the increase in bulk of an organism is described as a function of time. Frequently, growth displays an exponential relationship with time which is described by a non-linear model taking the general form:

$$y = \beta_0 e^{\beta_1 t} + \varepsilon$$

where: β_0 = initial bulk at time $t = 0$
β_1 = exponential growth rate
e = exponential constant
ε = random error (with an assumed variance of σ^2)

The problem with fitting such a model is that y continues to increase with time, whereas the growth of a living organism generally declines as the organism matures and will eventually cease or may even become negative; for example, the dry matter of a plant will reduce as the plant enters the senescence phase of the growth cycle. Most growth patterns are therefore S-shaped (sigmoid) curves when plotted against time. One common solution to this is to employ the so-called **logistic regression model** that takes the form:

$$y_t = \frac{\beta_2}{1 + e^{\beta_0 + \beta_1 t}} + \varepsilon$$

where: y_t = predicted value of the Y-variable (e.g. plant weight) at time t;
β_0 = expected value of y at time $t = 0$;
β_1 = measure of the growth rate;
β_2 = expected value of y for a very large value of t (i.e. the asymptotic value of y);
e = exponential constant;
ε = random error (with an assumed variance of σ^2).

The parameter β_2 represents the limiting or maximum value for the Y-variable and will need to be estimated from a plot of the observed data. If it is assumed that the variance of the error term is homogeneous across all values of time t and thus can be ignored, an initial value for the parameter β_0 can then be found by solving the logistic equation where $t = 0$ and substituting y_t for the observed initial value y_1, thus:

$$\beta_0 = \log_n \left(\frac{\beta_2}{y_1} - 1 \right)$$

An initial estimate for β_1 can then be determined by substituting the second observed value for y (i.e. y_2) and inputting the newly estimated value for β_0, thus:

$$\beta_1 = \log_n \left(\frac{\beta_2}{y_2} - 1 \right) - \beta_0$$

This procedure gives starting values for the parameters for the logistic equation which can be used in a first attempt to determine the best-fit regression model. Through a reiterative process, the parameters may be adjusted until the sum of residuals is minimized and a truly best-fit regression is obtained. Confidence limits and the significance of the regression can subsequently be determined by methods similar to those used in ordinary least squares regression; however, these will not be discussed further here. (For a more detailed description of non-linear regression procedures see, for examples Draper and Smith 1998; Freund *et al.*, 2006).

Example 13.5 illustrates a very typical employment of a non-linear model in animal production where a logistic regression procedure is used to describe the growth of pigs.

Example 13.5. Intrinsic non-linear regression analysis using a logistic growth model.

The growth of pigs in relation to time.

The live weight of a sample of pigs was determined from birth ($t = 0$) on a weekly basis. The data were as follows:

Time (weeks from birth)	Live weight (kg)	Time (weeks from birth)	Live weight (kg)
1	3.4	9	18.0
2	4.0	10	26.4
3	7.5	14	36.4
4	7.8	16	52.6
5	8.9	18	72.3
6	11.3	20	93.6
7	14.1	24	115.7
8	14.5		

Estimate of asymptotic value $\beta_2 = 200$ (determined from visual inspection of data)

Estimate of $\beta_0 = \log_n \left(\frac{\beta_2}{y_1} - 1 \right) = \log_n \left(\frac{200}{3.4} - 1 \right) = 4.06$

Initial estimate of $\beta_1 = \log_n \left(\frac{\beta_2}{y_2} - 1 \right) - \beta_0 = \log_n \left(\frac{200}{4.0} - 1 \right) - 4.06 = -0.17$

By reiteration β_1 is adjusted to -0.19 to minimize the residuals $\Sigma (y - \hat{y})^2$

This gives a best-fit model: $\hat{y}_t = \dfrac{200}{1 + e^{4.06 - 0.19t}}$

continued

Example 13.5. Continued.

The data and best-fit curve may now be plotted:

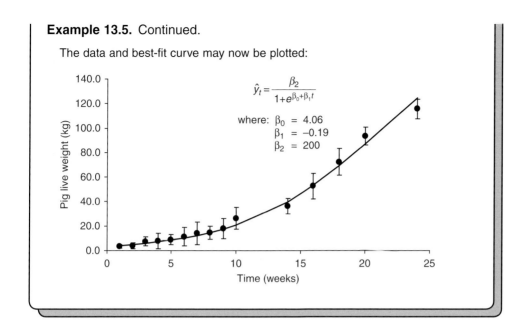

$$\hat{y}_t = \frac{\beta_2}{1+e^{\beta_0+\beta_1 t}}$$

where: $\beta_0 = 4.06$
$\beta_1 = -0.19$
$\beta_2 = 200$

The reader will have realized that both multiple and non-linear regression analysis are complex techniques that have been described here only rather superficially. While it is hoped that some useful information has been conveyed, the reader requiring more in-depth explanations will need to consult the specialized literature on this subject (see previously quoted references).

14 Analysis of Frequency Data

Frequency: commonness of occurrence… rate of recurrence… number of repetitions in given time.

(Concise Oxford English Dictionary)

- Introduction to the analysis of discontinuous frequency data using chi-squared tests.
- Chi-squared 'goodness-of-fit' test.
- Using chi-squared to fit data to distribution models.
- Chi-squared tests of association.
- Testing the homogeneity of frequency data sets.
- The Kolmogorov–Smirnov goodness-of-fit test.

14.1   Introduction to the Analysis of Frequency Data

Data can often take the form of the frequency of items that fall into particular categories, e.g. the number of weed plants of different species that occur in a crop field or the number of times that pigs are observed to perform recognized pieces of behaviour over a given time period. This type of data, which is based on counts of items, is termed **nominal data**. Such data are discontinuous and would be typically presented using a frequency bar chart, or, if the data are in the form of proportions, possibly by a pie chart. When analysing such data, it is often required to compare the frequency that observed items occur within a number of categories with a theoretical or already known set of frequencies. For example, in a breeding trial, we may want to know if the observed frequency of occurrence of some distinct phenotypes adheres to a particular ratio as predicted by the laws of genetics. The possibility that an observed set of frequencies matches a given set of frequencies is examined by a statistical significance test referred to as a **goodness-of-fit test**.

Alternatively, it is commonly required to compare the observed frequencies that a sample of items fall within a number of categories with the frequencies observed in a different sample. For example, in a trial designed to establish whether pig behaviour was associated with the type of pen in which pigs were housed, it was required to compare the frequency that a sample of pigs performed recognized pieces of behaviour when housed in one type of pen with the frequencies observed in a different sample of pigs housed in a different type of pen. Similar questions also commonly arise in the context of questionnaire or survey data. For example, it may be asked whether the frequency of males replying 'yes' or 'no' to

© C. Ireland 2010. *Experimental Statistics for Agriculture and Horticulture* (C. Ireland)

a particular question is the same as the frequency of females replying 'yes' or 'no' to the same question. In such cases the necessary statistical test will examine whether the frequencies in one set of categories is significantly associated with the frequencies observed in another set of categories and the analysis is thus termed a **test of association.**

Both goodness-of-fit tests and tests of association are commonly based on a test statistic called **chi-squared** denoted by the symbol χ^2 (the Greek letter 'chi' in upper case). Chi-squared tests will be described in some detail in this chapter. An alternative type of goodness-of-fit test that is preferred under certain circumstances is the Kolmogorov–Smirnov test and this will also be described in the final section of the chapter.

14.2 Chi-squared Significance Tests

In general terms, chi-squared is a summary statistic that expresses the extent that the frequency with which the observed data fall within a set of categories differs from a theoretical or expected frequency. The equation for chi-squared is fairly intuitive and is based on the numerical difference between the observed frequency (O) and expected frequency (E) summed across all the categories. However, when the difference between the observed and the expected number of items is summed for all categories, the positive differences are cancelled out by the negative differences and the solution is zero, i.e. $\Sigma(O - E) = 0$. Therefore the common expedient is taken of squaring the differences before summing in order to remove negative signs. It is also necessary to apply a weighting factor to the difference between the observed and expected values for a particular category in order to take into account the sample size; otherwise categories having different expected frequencies cannot be fairly compared. This is achieved by simply dividing the squared difference between the observed and expected frequencies by the expected frequency, i.e. $(O - E)^2/E$. The values obtained for each category are then summed to obtain the chi-squared value. The formula for chi-squared is thus:

$$\chi^2 = \Sigma \frac{(O - E)^2}{E}$$

where: O = observed frequency;
E = expected frequency.

It should be apparent that the larger the disagreement between the observed and theoretical frequencies the larger will be the value of the χ^2 statistic. However, increasing the number of categories will also automatically increase the χ^2 value. Therefore, in performing the test, the number of categories (k) must be taken into account. This is achieved through the degrees of freedom for the test, the derivation of which will be addressed later. In general terms, the null hypothesis will state that the observed and expected frequencies across all categories are the same. The observed χ^2 value is then compared with critical χ^2 values obtained from the chi-squared probability distribution at the appropriate DF in order to inspect the probability of the data supporting the null hypothesis.

14.2.1 The chi-squared distribution

While it is not essential to understand the theoretical basis of the chi-squared frequency distribution in order to perform a chi-squared test, some familiarity with the underlying theory does help the user to employ the test appropriately and, in particular, to appreciate the important limitations to the use of the test.

So what is a chi-squared distribution? To understand this we need to return to the normal distribution. When a set of data adheres to a normal distribution, the likelihood of obtaining any particular value x on a random basis is determined by dividing the difference between the population mean μ and x by the standard deviation σ. This process, referred to as standardization, produces a value termed z. Thus:

$$z = \frac{(x - \mu)}{\sigma}$$

The calculated z value can then be compared with a theoretical distribution to determine the probability that the value x belongs to the given normal distribution (see section 3.3.2). Now suppose that, instead of having a single value for x, we had a sample containing a number of random x values. In order to measure the average deviation of all the items in the sample from the population mean, we could sum the z values for each individual x value. However, in a random sample it is more or less certain that some of the deviations will be positive while others will be negative and these will tend to cancel each other out; therefore, the deviations are squared and standardized by dividing by σ^2 before summing. The result of this calculation is a value that is called chi-squared, χ^2. Thus:

$$\chi^2 = \Sigma \frac{(x - \mu)^2}{\sigma^2}$$

As with the z value, chi-squared (χ^2) has a theoretical frequency distribution and can therefore be employed as a test statistic. While we will not worry about the exact mathematical basis of the χ^2 distribution, it is possible to understand how it may be derived. Consider, as we have done before, the possibility of taking every possible sample of a given size n from a population. For each of these samples a value for χ^2 could be calculated and the values then plotted as a frequency distribution. However, since χ^2 will increase as sample size increases, this means that there will be a different frequency distribution for each possible sample size, remembering that the size of a single sample is correctly given by its degrees of freedom (DF), i.e. $n - 1$. If all items in a sample have exactly the same value as the mean, then χ^2 will be zero (this is, of course, a very unlikely occurrence) and as samples become larger χ^2 increases. To illustrate this point, the chi-squared distribution for three samples with respective DF of 2, 5 and 10 are plotted in Fig. 14.1.

From the theoretical distribution of chi-squared, critical values can be derived which cut off particular proportions of the distribution. Since χ^2 cannot be smaller than zero and because we are generally only interested in whether the calculated χ^2 value is sufficiently large to indicate a discrepancy between the sample (observed values) and population (expected values), then we are only interested in the upper right-hand tail of the χ^2 distribution. This is illustrated in Fig. 14.1, which shows the

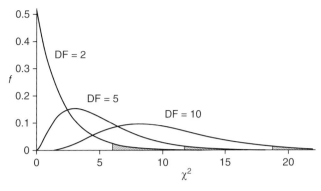

Fig. 14.1. The frequency distribution of chi-squared (χ^2) for samples with 2, 5 and 10 DF. Hatched regions represent the 5% portion of the right-hand tail of the distribution curve that is cut off by the critical χ^2 value at $P = 0.05$.

portion of the right-hand tail of three χ^2 distributions that are cut off by the 5% critical χ^2 value. The critical values of χ^2 that cut off given proportions of the upper tail are given in Appendix Table A14. A calculated value of χ^2 derived from a data set can then be compared with these critical values to determine whether to reject or accept the null hypothesis.

It will be apparent from the above that chi-squared is a continuous variable and is based to a large extent on the theory of the normal distribution. However, in goodness-of-fit tests and tests of association, chi-squared is applied to the analysis of discontinuous nominal data. It can be mathematically shown, however, that the distribution of the $\Sigma(O - E)^2/E$ variable does follow that of the continuous χ^2 distribution very closely, and use of the χ^2 distribution in tests of discrete nominal data leads to very few problems, given that a few limitations to the use of chi-squared as a test statistic need to be recognized (see section 14.6). It is, however, sometimes required to clearly distinguish between the theoretical chi-squared values and the discrete values based on the $\Sigma(O - E)^2/E$ calculation and this can be achieved by employing different symbols to represent each. For example, many statisticians use the symbol χ^2 to represent the theoretical value and X^2 to represent the calculated value; however, in this book we will simply refer to the calculated value of chi-squared as $\chi^2_{observed}$.

14.3   Chi-squared Goodness-of-fit Tests

Goodness-of-fit tests are used to test whether the observed frequency of a particular nominal variable significantly conforms to, or 'fits', a theoretical frequency distribution. For example, following a cross between two parent plants of the ornamental tree species *Acer platenoides*, seedlings showed either green-coloured leaves or red-coloured leaves. If leaf colour in *A. platenoides* is controlled by a single gene that is segregated according to Mendel's laws of inheritance, then the number of green- and red-leaved seedling plants should adhere to a fixed ratio of 3 to 1. To test the theory it will be required to examine whether the frequency of the two leaf colours observed in the progeny statistically conforms to the expected frequencies predicted by Mendel's laws.

The first step in a goodness-of-fit test, as in all statistical significance testing, is to formulate the null hypothesis. The null hypothesis for a goodness-of-fit test is that the observed frequencies (O) and the theoretical frequencies (E) are the same. The chi-squared test statistic ($\chi^2_{observed}$) is then determined by the formula:

$$\chi^2_{observed} = \Sigma \frac{(O - E)^2}{E}$$

The observed chi-squared value is then compared with a critical value which is read from the chi-squared probability distribution table (see Appendix: Table A14) by consulting along the rows to obtain the required critical probability (α) and down the columns to obtain the appropriate degrees of freedom. If the observed chi-squared value is greater than (or equal to) the critical value, then the probability of obtaining the observed data given the null hypothesis is true must be less than the critical probability at which the χ^2 distribution was consulted and the null hypothesis may therefore be rejected. Usually the null hypothesis is inspected at a level of probability of 5% ($P = 0.05$) in the first instance and, if H_0 is rejected, the test may then be repeated at a higher level of certainty if required.

Degrees of freedom for a goodness-of-fit test

Unlike parametric tests, the degrees of freedom (DF) for chi-squared significance tests are not based on the sample size but on the number of categories (k) on which the frequency values are based. The DF are obtained by subtracting from k the number of constants that are used to generate the expected frequencies. In a normal goodness-of-fit test only one constant is required for this, namely the total number of observations made (which is subsequently partitioned according to the predicted ratios to produce the expected values); hence DF is usually given by $k - 1$.

A fully calculated goodness-of-fit test involving the frequency of red-leaved and green-leaved *Acer* in a sample of seedlings ($n = 100$) is shown in Example 14.1. In this case it is hypothesized that there should be a 3:1 ratio of green-leaved to red-leaved seedlings according to Mendel's laws of inheritance, and the goodness-of-fit test is performed to establish whether the observed frequency of 82:18 statistically adheres to this ratio. The chi-squared value of 3.84 is smaller than the critical value for rejecting the null hypothesis at a probability of 0.05; therefore the null hypothesis is accepted and it is concluded that there is no significant difference between the observed and predicted frequencies.

14.3.1 Goodness-of-fit test with more than two categories

Clearly, examining the segregation of items between just two categories is the simplest case of a goodness-of-fit test; however, extending the analysis to any larger number of categories is readily achieved. For example, in addition to green or red leaf colour in *Acer platenoides*, the leaf shape may also have been described in terms of being either smooth-edged or serrated. Mendel's laws of inheritance may then be used to predict the frequency with which *Acer* seedlings now fall into one of four categories

Example 14.1. A chi-squared goodness-of-fit test.

Frequency of leaf colour in *Acer platenoides* plants.

The leaf colour of a sample of 100 seedlings produced by a cross between a green-leaved and a red-leaved *Acer platenoides* tree was counted. It is expected that a 3 to 1 ratio of green- to red-leaved seedlings will be produced based upon Mendel's laws of genetics. The observed and the expected frequencies were as follows:

	Leaf colour (k = 2)		
	Green	Red	*N*
Observed frequency	82	18	100
Expected frequency (based on a 3:1 ratio)	75	25	

Null hypothesis: there is no difference between the observed ratio of green- to red-leaved seedlings and the expected ratio.

Calculation of chi-squared:

$$\chi^2_{observed} = \Sigma \frac{(O-E)^2}{E} = \frac{(82-75)^2}{75} + \frac{(18-25)^2}{25} = 2.61$$

Degrees of freedom:

DF = $k-1$ = $2-1$ = 1

Comparison with critical chi-squared:

The critical value for χ^2 read from tables (at α = 0.05 and DF = 1) = 3.84 $\chi^2_{observed}$ < critical χ^2, therefore H_0 is accepted.

Conclusion: there is no significant difference between the frequency of green- and red-leaved *Acer* seedlings observed in the sample and an expected 3:1 ratio. The observed frequencies of leaf colour do 'fit' the expected frequencies and indicate, therefore, that this character has been inherited in compliance with Mendel's laws of genetics.

Note: in this example where DF = 1, it would have been appropriate to apply Yates's correction for continuity (see section 14.6.3). This would give a reduced χ^2 value of 1.71. In this particular example, this correction would not affect the final conclusion reached.

and the chi-squared goodness-of-fit test then used to examine whether the observed frequencies in each category statistically agree with the expected frequencies. This is illustrated Example 14.2. In this case the chi-squared value is larger than the critical value at $P = 0.05$ and it is concluded that there is a significant difference between the observed and expected frequencies.

Example 14.2. A chi-squared goodness-of-fit test for four data classes.

Frequency of leaf colour and leaf shape in *Acer platenoides* seedlings.

In a study of the inheritance of green and red leaf colour and smooth-edged and serrated leaf shape in *Acer platenoides*, parent plants were crossed and a progeny produced. According to Mendel's laws of genetics, it is predicted that the progeny should show a 9:3:3:1 ratio of the four possible combinations of phenotypes. To test the prediction, a sample of 250 seedlings from the progeny were observed and the following data were obtained:

| | Leaf colour and shape ($k = 4$) | | | | |
| | Green | | Red | | |
	Smooth-edged	Serrated	Smooth-edged	Serrated	N
Observed frequency	152	39	53	6	250
Expected frequency (9:3:3:1)	140.625	46.875	46.875	15.625	

Null hypothesis: there is no difference between the observed ratio of phenotypes and the expected ratio.

Calculation of chi-squared:

$$\chi^2_{observed} = \Sigma \frac{(O - E)^2}{E}$$

$$= \frac{(152 - 140.625)^2}{140.625} + \frac{(39 - 46.875)^2}{46.875} + \frac{(53 - 46.875)^2}{140.625} + \frac{(6 - 15.625)^2}{140.625}$$

$$= 8.97$$

Determination of degrees of freedom:

DF = $k - 1 = 4 - 1 = 3$

Comparison with critical chi-squared: the critical value for χ^2 read from tables (at $\alpha = 0.05$ and DF = 3) = 7.81. $\chi^2_{observed}$ > critical χ^2, therefore H$_0$ is rejected.

Conclusion: there is a significant difference between the frequency of the four possible phenotypes in the sample and the expected 9:3:3:1 ratio ($P < 0.05$). The observed frequency of leaf colour and leaf shape indicates that these characteristics have not been inherited in compliance with Mendel's laws of genetics.

14.3.2 Subdivision of a chi-squared goodness-of-fit test

Once a goodness-of-fit test has led to the conclusion that a significant difference is present between the observed and expected data frequencies, it will often be important to ascertain whether such differences are present for all the data categories present or for just a limited number of the data categories. In Example 14.2, it was concluded that the observed frequency of *Acer* leaf colour and leaf shape differed significantly from the theoretical ratio of 9:3:3:1. To biologists this might seem a somewhat surprising result since Mendel's laws would normally be expected to govern the outcome of genetic crosses of this type. The investigator may, therefore, wish to look into this a little further and ask whether a significant difference lies between the observed and predicted ratio of all four phenotypes or whether the difference is due to simply one phenotype differing from its expected frequency. A subdivision of the chi-squared analysis can be performed in order to answer this question. The technique simply involves selecting one particular category, usually the one with the largest discrepancy between observed and expected frequencies, and removing these data from the analysis. The chi-squared is then recalculated for the remaining categories. This procedure can be performed for each category in turn and the contribution made by each category to the overall discrepancy between the observed and expected frequencies can thereby be identified.

The analysis in Example 14.2 shows that the greatest discrepancy between observed and expected frequencies is for the red-coloured/serrated leaf phenotype and this category must, therefore, contribute to the significant difference between the overall observed and expected frequencies. If this category is removed from the analysis, the remaining categories can then be tested to inspect whether they still show differences between the observed and predicted frequencies. The calculation of this subdivision of chi-squared is shown in Example 14.3. The new chi-squared value is 2.54, which is smaller than the critical value read at $P = 0.05$ and the reduced DF value of 2 (since k now becomes 3). It is concluded, therefore, that there is no significant difference between the observed and expected frequencies of the remaining categories once the red-coloured/serrated leaf category is removed from the analysis.

The analysis shown in Example 14.3 could be further continued by testing whether the frequency ratio of the red-coloured/serrated leaved seedlings compared to all other phenotypes lumped together conforms to the expected 1 to 15 ratio. In this case a chi-squared test reveals a significant difference ($P < 0.05$). Therefore it can be finally concluded that the non-conformity of the overall data frequency to an expected 9:3:3:1 phenotypic ratio is due solely to the red-coloured/serrated leaf phenotype. This particular leaf phenotype does not appear, therefore, to be inherited strictly according to Mendel's laws of inheritance.

14.3.3 Using goodness-of-fit tests to assess the fit of a specified data distribution

Since the chi-squared statistic can be used to test the goodness-of-fit of an observed frequency with a theoretical frequency, it can in principle be used to test whether a set of observed data adheres to a theoretical data distribution such as

---

Example 14.3. Subdivision of a chi-squared goodness-of-fit test.

Frequency of leaf colour and leaf shape in *Acer platenoides* seedlings (data from Example 14.2).

If the data for the red-coloured/serrated leaf category are removed from the analysis in Example 14.2, the remaining categories should theoretically fit a 9:3:3 ratio. The data table is amended by deleting this category and then recalculating the totals and the expected frequencies.

	Leaf colour and shape ($k = 4$)				
	Green		Red	Red	
	Smooth-edged	Serrated	Smooth-edged	Serrated*	N
Observed frequency	152	39	53	(6)	244
Expected frequency (9:3:3)	146.4	48.8	48.8		

*Data for red-coloured/serrated leaf category removed from the chi-squared analysis.

Null hypothesis: there is no difference between the observed ratio of the three phenotypes and the expected ratio.

Calculation of chi-squared:

$$\chi^2_{observed} = \Sigma \frac{(O-E)^2}{E} = \frac{(152-146.4)^2}{146.4} + \frac{(39-48.0)^2}{48.8} + \frac{(53-48.0)^2}{48.8} = 2.54$$

Degrees of freedom:

$$DF \quad = \quad k-1 \quad = \quad 3-1 = 2$$

Comparison with critical chi-squared:

The critical value for χ^2 read from tables (at $\alpha = 0.05$ and DF = 2) = 5.99
$\chi^2_{observed}$ < critical χ^2, therefore H_0 is accepted.

Conclusion: there is no significant difference between the observed and expected frequencies for the three leaf phenotype categories green-coloured/smooth edged, green-coloured/serrated, red-coloured/smooth edged.

the binomial, Poisson or normal distribution. The probability equation for the distribution under inspection is used to predict the frequency of occurrence of data items within defined class intervals and a goodness-of-fit test then used to compare the observed frequencies with the predicted frequencies. The null hypothesis is that there is no difference between the observed and predicted frequencies and, if the chi-squared value allows the null hypothesis to be accepted, then the data can be concluded to significantly fit the probability distribution in question. It should be noted that, while this technique is commonly used to examine the goodness-of-fit of discontinuous distributions such as the binomial and Poisson distributions, it does not perform well in the case of the continuous normal distribution and using it to identify normal distributions is not generally recommended.

While the general procedure for using a chi-squared goodness-of-fit test to fit a data distribution is as described above, the degrees of freedom for such a test require a little more consideration. In this case the degrees of freedom become the number of categories less 1 $(k - 1)$ and less the number of parameters (p) in the distribution model equation that have to be estimated from the sample data in order to determine the expected values. In the case of the binomial distribution, this will depend on whether the probability P of a binomial event occurring is already known or whether it is estimated from the observed data. If the former, the degrees of freedom are simply $k - 1$, while, if it is necessary to estimate P from the data, then the degrees of freedom become $k - 2$. For the Poisson distribution we will normally have to estimate the population mean μ from the data (see section 3.2.2) so the degrees of freedom become $k - 2$.

Example 14.4 illustrates the use of the goodness-of-fit test to inspect whether a set of data concerning the frequency of weed plants within sampling areas in a wheat field significantly adheres to a Poisson distribution.

Example 14.4. Use of chi-squared goodness-of-fit test to inspect if a data set adheres to a Poisson distribution.

The distribution of weed plants in sample quadrats in a wheat field.

The frequency of weed plants in 85 sample quadrats in a wheat field was recorded. To examine whether the data adhere to a Poisson distribution, a goodness-of-fit test was performed. The predicted frequencies based on the Poisson probability distribution (see section 3.2.2) were calculated and are shown in the table and figure below along with the calculation of the chi-squared statistic.

Number of weed plants present in sample quadrats in a wheat field

No. weed plants/ quadrat	Frequency (observed)	Total	Poisson distribution (probability)	Predicted frequency (expected)	$(O - E)^2/E$
0	1	1	0.00	0.42	0.820
1	3	3	0.03	2.21	0.280
2	7	14	0.07	5.89	0.211
3	9	27	0.12	10.44	0.198
4	13	52	0.16	13.88	0.056
5	15	75	0.17	14.77	0.004
6	12	72	0.15	13.10	0.092
7	10	70	0.12	9.95	0.000
8	7	56	0.08	6.62	0.022

continued

Example 14.4. Continued.

No. weed plants/ quadrat	Frequency (observed)	Total	Poisson distribution (probability)	Predicted frequency (expected)	$(O - E)^2/E$
9	3	27	0.05	3.91	0.213
10	2	20	0.02	2.08	0.003
11	1	11	0.01	1.01	0.000
12	1	12	0.01	0.45	0.687
13	1	13	0.00	0.18	3.659
14	0	0	0.00	0.07	0.069

$k = 15$ $\qquad\qquad\qquad\qquad\qquad\qquad\qquad \chi^2_{observed} = 6.314$

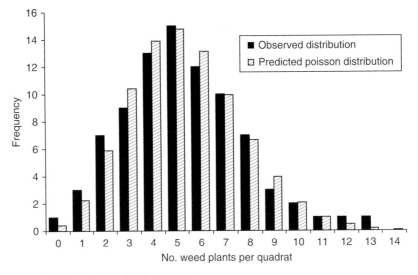

Comparison with critical chi-squared:

the critical value for χ^2 read from tables (at $\alpha = 0.05$ and DF = 13) = 22.362. $\chi^2_{observed} <$ critical χ^2; therefore H_0 is accepted.

Conclusion: there is no significant difference between the observed and expected frequencies; therefore, the data can be assumed to conform to a Poisson distribution.

14.4 Tests of Association (Contingency Table Tests)

In goodness-of-fit tests, the hypothesis that an observed frequency distribution adheres to a theoretical or predicted distribution is tested and, because the hypothesis being examined is based on an external model, it is referred to as an 'extrinsic hypothesis'. In many cases, however, data are collected for two or more sets of categories simultaneously and it is required to inspect whether the observed frequency of occurrence of one set of categories is associated with the observed frequency of occurrence of another set of categories. For example, in response to a questionnaire, members of the public may have stated whether or not they are in favour of the use of genetically modified (GM) crops. In addition, they may also have indicated their type of employment. The analyst may then wish to inspect the hypothesis that the opinion of persons on the use of GM crops is associated with their type of employment, in other words, whether the proportion of persons in favour and against GM crops is statistically different among persons engaged in different types of employment. In this case, the hypothesis being examined is generated from the data itself and is thus termed an 'intrinsic hypothesis'. In order to inspect such a hypothesis, a **chi-squared test of association** is performed.

14.4.1 The procedure for calculating a chi-squared test of association

To perform a chi-squared test of association, a **contingency table** is first constructed to show the frequency of items in each of the categories. In a contingency table the number of rows is denoted r and the number of columns c, and there are therefore $r \times c$ cells in the table. The generalized null hypothesis for a contingency table chi-squared test is that the frequency of observations found in the rows is not associated with the frequency found in the columns. In the chi-squared test of association shown in Example 14.5, there are four categories of employment that are assigned to four columns, and two possible opinions towards GM crops, either in favour or against, that are assigned to two rows. This forms a 4×2 contingency table. The null hypothesis in this case simply states that a person's opinion concerning the use of GM crops is not associated with their type of employment.

In a contingency table test, unlike a goodness-of-fit test, we do not have a theoretical frequency that is expected in advance. Instead, the expected frequencies are determined from the observed sample data based on the null hypothesis that the variables are totally independent of each other and, therefore, that the ratio of frequencies are constant across the categories. In practice, the way the expected values are generated for each category is to determine the ratio between the row totals and then to partition each column total in turn according to this same constant ratio. The simplest way to achieve this is by use of the formula:

$$E_{ij} = \frac{\left(R_i \times C_j \right)}{N}$$

where: E_{ij} = expected frequency for the category in the ith row and jth column of the contingency table;
R_i = total for row i;
C_j = total for column j;
N = total no. of observations.

In Example 14.5, out of a total of 500 persons questioned there are a total of 253 in favour of GM crops while the total number of persons who are scientists is 58. Therefore, if opinion on GM crops is independent of the type of employment, it would be expected that the number of those in favour of GM crops who are also employed as scientists should be equal to: $(253 \times 58)/500 = 29.35$. A similar calculation is performed for each category.

Once calculated, the expected frequency of each variable can be shown in the contingency table and a chi-squared analysis then performed where the value for the chi-squared statistic is given by:

$$\chi^2_{observed} = \Sigma \frac{(O-E)^2}{E}$$

To determine the degrees of freedom for a contingency table chi-squared test, we now have to consider how many constants from the data were used to generate each expected value. To generate each expected value we need to know the total sample size N, the total of all but one of the rows and the total of all but one of the columns (the totals of the last row and last column are then known by default). The degrees of freedom, as in a goodness-of-fit test, are then the number of categories (k) less the number of data constants employed. Fortunately this boils down to a conveniently simple formula that produces the same solution:

$$DF = (r-1) \times (c-1)$$

where: r = no. of row categories;
c = no. of column categories.

14.4.2 A special case: the 2 × 2 contingency table

The contingency table containing just two rows and columns, i.e. a 2 × 2 table, while being the smallest possible table that can occur, is commonly encountered. It can be shown that for a 2 × 2 table an alternative formula for calculation of χ^2 is available that avoids the necessity of calculating expected frequencies separately:

$$\chi^2_{observed} = \frac{N \times (O_{1,1}.O_{2,2} - O_{1,2}.O_{2,1})^2}{(C_1.C_2.R_1.R_2)}$$

where: O_{ij} = the observed frequency value in column i, row j;
C_i = total of column i;
R_j = total of row j;
N = total number of observations.

Finally, the observed chi-squared value is compared with the appropriate critical value obtained from the χ^2 table (see Appendix: Table A14) at the chosen level of probability. If $\chi^2_{observed}$ is equal to or greater than $\chi^2_{critical}$, then the null hypothesis can be rejected and a significant association can be concluded to exist between the categories at the level of probability at which the test was conducted.

A fully calculated example of a chi-squared test of association is shown in Example 14.5.

Example 14.5. A chi-squared test of association (contingency table test).

The association between opinion of a sample of people ($n = 500$) on the use of genetically modified (GM) crops and their type of employment.

Opinion regarding use of GM crops	Type of employment				
	Scientist/ technical	Office worker	Manual worker	Unemployed	Total
In favour	39	72	105	37	253
(Expected)	(29.35)	(71.85)	(115.87)	(35.93)	
Against	19	70	124	34	247
(Expected)	(28.65)	(70.15)	(113.13)	(35.07)	
Total	58	142	229	71	500

Null hypothesis: the frequency of opinion regarding use of GM crops is not associated with type of employment.

Calculation of expected frequencies of each category:

Category: *in favour × scientist*   Category: *in favour × office worker*

$$E = \frac{(R \times C)}{N} = \frac{(253 \times 58)}{500} = 29.35 \quad ; \qquad E = \frac{(R \times C)}{N} = \frac{(253 \times 142)}{500} = 71.85$$

Category: *against × scientist*

$$E = \frac{(R \times C)}{N} = \frac{(247 \times 58)}{500} = 28.65 \quad ; \qquad \text{etc.}$$

(All the expected frequencies are shown in parentheses in the contingency table.)

Calculation of chi-squared:

$$\chi^2_{observed} = \Sigma \frac{(O - E)^2}{E} = \frac{(39 - 29.35)^2}{29.35} + \frac{(72 - 71.85)^2}{71.85} + \ldots + \frac{(34 - 35.07)^2}{35.07} = 8.56$$

Degrees of freedom:

DF $\quad = \quad (r - 1)(c - 1) \quad = \quad (2 - 1)(4 - 1) \quad = \quad 3$

Comparison with theoretical chi-squared:

The theoretical value for χ^2 read from tables at $P = 0.05$ (DF = 3) = 7.81
$\chi^2_{observed} >$ critical χ^2; therefore H_0 is rejected.

continued

Example 14.5. Continued.

Conclusion: the frequency of opinion held by a sample of people on the use of genetically modified crops is associated with their type of employment ($P < 0.05$). In other words, the likelihood of a person being 'for' or 'against' the use of GM crops depends on their type of employment.

14.4.3 Subdivision of a chi-squared contingency table test

In section 14.3.2, it was seen how a Goodness-of-fit test could be progressed by subdividing the analysis in order to establish the specific location of any discovered discrepancies between the observed and expected frequencies. A similar type of procedure can also be performed for a contingency table analysis. For example, the analysis of the contingency table in Example 14.5 revealed that the distribution of opinion on use of GM crops was associated with a person's employment. The question arises whether this is the case for all types of employment inspected, or if it is the case for only one of the types of employment. By inspection of the data in Example 14.5, the largest proportional discrepancy between observed and expected frequencies is for those persons employed as scientists. Therefore a subdivision of the analysis is performed by removing the data for the scientist employment category and performing the chi-squared test on the remaining three employment categories. When this is undertaken it is found that there is no association between opinions held on GM crops and the remaining types of employment; therefore it is concluded that it is just those persons who work as scientists whose opinion on the use of GM crops is significantly related to their type of employment. The full procedure is illustrated in Example 14.6.

Example 14.6. Subdivision of a chi-squared test of association. The association between opinion of a sample of persons on the use of GM crops and their type of employment.

Subdivision of chi-squared undertaken by removal of the scientist category from the data analysis shown in Example 14.5.

Opinion regarding use of GM crops	Type of employment				Total
	(Scientist/ technical)*	Office worker	Manual worker	Un-employed	
In favour	(39)	72	105	37	214
(Expected)		(68.75)	(110.87)	(34.38)	
Against	(19)	70	124	34	228
(Expected)		(73.25)	(118.13)	(36.62)	
Total		142	229	71	442

*Data for the scientist category removed from the chi-squared analysis.

continued

Chapter 14

Example 14.6. Continued.

Null hypothesis: the frequency of opinion regarding use of GM crops is not associated with whether a person is employed as an office worker or as a manual worker or is unemployed.

Calculation of expected frequencies:

Category: *in favour × office worker*   Category: *in favour × manual worker*

$$E = \frac{(R \times C)}{n} = \frac{(214 \times 142)}{442} = 68.75 \quad ; \qquad E = \frac{(R \times C)}{n} = \frac{(214 \times 229)}{442} = 110.87$$

Category: *against × office worker*

$$E = \frac{(R \times C)}{n} = \frac{(228 \times 142)}{442} = 73.25 \quad ; \qquad \text{etc.}$$

(All the expected frequencies are shown in parentheses in the contingency table.)

Calculation of chi-squared:

$$\chi^2_{observed} = \Sigma \frac{(O-E)^2}{E} = \frac{(72-68.75)^2}{68.75} + \frac{(105-110.87)^2}{110.87} + \dots + \frac{(34-36.62)^2}{36.62} = 1.29$$

Degrees of freedom:

DF   = $(r-1)(c-1)$ = $(2-1)(3-1)$ =   2

Comparison with critical chi-squared:

The critical value for χ^2 read from tables at $P = 0.05$ (DF = 2) = 5.99.
$\chi^2_{observed}$ < critical χ^2; therefore H_0 is accepted.

Conclusion: the frequency of opinion on the use of GM crops is not associated with whether persons are employed as office workers or as manual workers or are unemployed. The association of opinion on GM crop use with types of employment is due to the frequency distribution of scientists alone. It appears from this survey that the probability of being in favour of GM crops is greater if the person works as a scientist.

14.5   Chi-squared Test for Heterogeneity

Data are often collected for the same experiment at either spatially separated locations or at different times, or possibly both. For example, the effect of a molluscicide on the death rate of slugs may have been recorded in four separate experiments using the same protocol but conducted at different times. The investigator will therefore wish to know whether the data can be aggregated for the purposes of analysis, thereby increasing the number of replicates, or whether the data for each experiment have to be analysed separately. If the data concerned are frequency data, a chi-squared test can be performed to determine whether different sets of data can be considered to be derived from the same population, i.e. they are **homogeneous**, and may therefore be combined, or whether there are real differences between the samples, i.e. they are **heterogeneous**, and must be analysed separately.

A chi-squared test works in this case because the summation or subtraction of two χ^2 values yields a further valid χ^2 value. First, the χ^2 value and degrees of freedom (DF) for each sample are determined separately. A total χ^2 and total DF value are then obtained by summing the individual sample χ^2 and DF values. The data from the different samples are then pooled and the overall χ^2 and DF values for the pooled data determined. The heterogeneity chi-squared test statistic is finally obtained by subtracting the pooled data χ^2 from the total χ^2. Similarly the DF is obtained by subtracting the pooled DF from the total DF. The heterogeneity χ^2 is tested for significance by comparing with the critical χ^2 value read from the χ^2 probability table at the calculated DF and appropriate level of probability (normally $\alpha = 0.05$). The null hypothesis states that there is no difference between the samples. Acceptance of the null hypothesis leads to the conclusion that the samples are homogeneous and may be pooled for subsequent analysis, while rejection of the null hypothesis implies that the samples are heterogeneous and must be treated separately.

A fully calculated contingency table heterogeneity test is shown in Example 14.7.

Example 14.7. A chi-squared contingency table heterogeneity test.

The effect of a molluscicide on the survival of treated slugs in four separate experiments.

The number of slugs still alive after a precise period of time following treatment with the same molluscicide dose was recorded in four separate experiments. A chi-squared test is performed to inspect whether the four sets of data are heterogeneous and can be combined for further analysis. The data are arranged in a set of contingency tables and the expected frequencies and chi-squared values determined.

Experiment 1

Molluscicide treatment	No. of slugs dead/alive		
	Dead	Alive	Total
Treated	9	15	24
Untreated	15	10	25
Total	24	25	49

$\chi^2 = 2.481$ (DF = 1)

continued

Example 14.7. Continued.

Experiment 2

Molluscicide treatment	No. of slugs dead/alive		Total
	Dead	Alive	
Treated	13	12	25
Un-treated	18	7	25
Total	31	19	50

$\chi^2 = 2.122$ (DF = 1)

Experiment 3

Molluscicide treatment	No. of slugs dead/alive		Total
	Dead	Alive	
Treated	12	13	25
Untreated	17	8	25
Total	29	21	50

$\chi^2 = 2.052$ (DF = 1)

Experiment 4

Molluscicide treatment	No. of slugs dead/alive		Total
	Dead	Alive	
Treated	10	14	24
Untreated	16	9	25
Total	26	23	49

$\chi^2 = 2.452$ (DF = 1)

Null hypothesis: the data from the four samples are homogeneous.

Sum of chi-squares for all four experiments = 2.481 + 2.122 + 2.052 + 2.452
= 9.107 (DF = 4)

Table of pooled data:

Molluscicide treatment	No. of slugs dead/alive		Total
	Dead	Alive	
Treated	44	54	98
Untreated	66	34	100
Total	110	88	198

Pooled data $\chi^2 = 8.926$ (DF = 1)

continued

Example 14.7. Continued.

Heterogeneity chi-squared test statistic:

The heterogeneity chi-squared is equal to the sum of the sample chi-squared values minus the pooled chi-squared:

$$\chi^2_{observed} = \quad 9.107 - 8.926 \quad = \quad 0.181$$

The degrees of freedom are equal to the sum of the sample DF values – the pooled DF:

$$DF = \quad 4 - 1 \quad = \quad 3$$

Comparison with critical chi-squared value:

The critical value for χ^2 read from tables (at $\alpha = 0.05$ and DF = 3) = 7.815

$\chi^2_{observed} <$ critical χ^2; therefore H_0 is accepted.

Conclusion: the four data samples are homogeneous and therefore may be pooled for further analysis.

14.6 Limitations and Corrections in the Use of Chi-squared Tests

As with all statistical tests, there is a range of assumptions that limits the use of significance tests based on the chi-squared statistic.

14.6.1 The use of chi-squared tests when data are in the form of proportions or percentages: a common error

Determination of chi-squared based on data that have been converted to proportions or percentages does not take into account the sample size and is, therefore, invalid. Consider, for example, two cases where two categories have observed and expected frequencies of 4:6 and 400:600 respectively. In each case, were the frequencies to be expressed as percentages or proportions, the ratio of observed to expected would be exactly the same, i.e. 2:3, and each case would yield the same chi-squared value and produce the same probability of significance. This is plainly ludicrous since, by common sense, it is much more likely that an observed frequency of 4 fits an expected frequency of 6 (the difference is only 2) while an observed frequency of 400 is most unlikely to fit an expected frequency of 600 (the difference is 200). Therefore *chi-squared analyses must only be performed on the original count data and not on data that have been converted to proportions or percentages.*

14.6.2 Use of chi-squared tests with small sample sizes

In theory, chi-squared tests become increasingly unreliable as the sample size decreases. In practice, this is not a problem unless the expected values become particularly small. Once expected values start to approach 1, the mathematics of the test make it almost inevitable that Type I statistical errors will occur. Although a rule of thumb, it is generally recommended that *chi-squared testing is not performed if any of the expected frequencies are less than 1 or no more than one-fifth of expected frequencies are less than 5*.

If a particular data set does not conform to these guidelines, the problem may be addressed by combining categories that have small expected values, although this will of course limit the usefulness of any conclusions finally reached. It is much better to ensure that the problem does not arise in the first place by deploying sufficiently large samples.

(It may be noticed that in Example 14.4, where a goodness-of-fit test is used to inspect whether a data set adheres to a Poisson data distribution, these guidelines have apparently been broken in that a large number of the frequencies predicted by the Poisson distribution equation are <5 and a few are actually <1. This is one of those awkward occasions in statistics when there is no real satisfactory solution. It would be possible to combine categories but this would result in a loss of resolution in the data distribution, while increasing the sample size sufficiently to provide expected values of >5 would be totally impractical. The Kolmogorov–Smirnov (K-S) test is an alternative goodness-of-fit test that might be employed under these circumstances (see section 14.7) but there is not universal agreement upon its validity. It is suggested therefore that the chi-squared goodness-of-fit test is employed as described but where the calculated value of χ^2 is close to the critical value used for rejecting the null hypothesis the researcher should then be guarded in drawing definitive conclusions.)

14.6.3 Correction for continuity when degrees of freedom and/or expected values are low

The calculated chi-squared values obtained from nominal data in a goodness-of-fit test belong to a discontinuous data distribution. However, the chi-squared frequency distribution used to produce the critical chi-squared values is a continuous distribution (as explained in section 14.2). This mismatch leads to a potential error in estimating the probability values and increases the chances of making a Type I statistical significance error. Fortunately, this error is very small in all cases except where expected values or the degrees of freedom are low. To overcome this in goodness-of-fit tests, where DF = 1 and/or where expected values in a cell of the table are <5, it is generally recommended to employ a corrective procedure called **Yates's correction**. This simply involves reducing the deviation of each observed frequency from the expected frequency by a value of 0.5 before squaring. Thus:

$$\chi^2 = \Sigma \frac{(O - E - 0.5)^2}{E}$$

In Example 14.1 the application of Yates's correction leads to a reduction of the observed value χ^2 from 2.61 to 1.71. This does not alter the conclusion in this particular example, but there will be cases where application of Yates's correction will reduce the calculated χ^2 value sufficiently for the null hypothesis to be accepted when it would otherwise have been rejected.

The use of Yates's correction is another of those areas in data analysis where there remains some considerable disagreement among statisticians. It may be pointed out, for example, that in a 2 × 2 contingency table where $(O_{1,1} \times O_{2,2}) - (O_{1,2} \times O_{2,1})$ is equal to or less than $n/2$ then application of Yates's correction would actually increase rather than decrease $\chi^2_{observed}$ and thereby increase the potential error. Although some complex procedures have been devised to address this problem, the issue has not been satisfactorily resolved. Consequently, it is probably best to avoid routine application of any correction to a 2 × 2 contingency table and simply accept the increased possibility of a Type I statistical significance error occurring in this particular situation.

14.7 Kolmogorov–Smirnov Goodness-of-fit Test

The Kolmogorov–Smirnov (K-S) goodness-of-fit test is a non-parametric test that, under certain circumstances, may be preferred to a chi-squared goodness-of-fit test. The K-S test is used to compare the observed cumulative frequency of a single variable with an expected or theoretical cumulative frequency and where the categories can be ordered according to the magnitude of a particular criterion. For example, the number of egg-laying hens in a sample group which attain their maximum productivity may be compared when housed in different pens maintained under increasing diurnal light level.

To perform a K-S goodness-of-fit test, the observed categories are first ordered according to magnitude and the cumulative frequency determined. The expected cumulative frequencies are found under the null hypothesis that there is no difference in frequency between categories. The largest absolute difference (d_{max}) between the observed and expected cumulative frequencies for each category then provides the test statistic, which may be denoted $d_{max(observed)}$. The value of $d_{max(observed)}$ is subsequently compared with the appropriate critical value obtained from the probability distribution of d_{max} (see Appendix Table A15); where $d_{max(observed)} \geq d_{max(critical)}$ the null hypothesis may be rejected at the level of probability at which the distribution was consulted. It should be noted that the required value of $d_{max(critical)}$ is read from the probability distribution table at the number of categories k and the total number of observations, N, and it is necessary that N is an exact multiple of k. The example referred to above concerning egg production in chickens housed in pens with differing light levels is used to illustrate the K-S test in Example 14.8.

Example 14.8. A Kolmogorov–Smirnov (K-S) goodness-of-fit test.

The effect of luminosity on egg production in hens.

A sample of 50 egg-laying hens (N = 50) was randomly housed in a series of pens that were maintained under differing luminosity during equal photoperiods. For each pen the number of the hens that attained their maximum productivity (maximum number of eggs laid per week) were recorded. A Kolmogorov–Smirnov (K-S) goodness-of-fit test was performed to examine whether distribution of maximum productivity among the hens was associated with the different pens.

continued

Example 14.6. Continued.

	Pen 1 (5 lux)	Pen 2 (20 lux)	Pen 3 (50 lux)	Pen 4 (100 lux)	Pen 5 (200 lux)	N
	Number of hens attaining maximum productivity in different pens (pens ordered in increasing luminosity)					
Observed frequency	4	6	15	12	13	50
Observed cumulative frequency	4	10	25	37	50	
Expected frequency	10	10	10	10	10	
Expected cumulative frequency	10	20	30	40	50	
d	6	10	5	3	0	

Null hypothesis: there is no difference in the frequency of chickens attaining maximum productivity among the five pens with differing luminosity.

$d_{max(observed)}$ = the maximum value of d = 10

$d_{max(P = 0.05, \, k = 5, \, N = 50)}$ = 9

$d_{max(observed)} > d_{max(P = 0.05, \, k = 5, \, N = 50)}$; therefore H_0 is rejected.

Conclusion: there is a significant difference in the frequency of hens attaining maximum productivity between pens with differing luminosity ($P < 0.05$).

The K-S goodness-of-fit test is more powerful than the equivalent chi-squared test, i.e. it is able to identify smaller differences in observed frequency as significant, particularly where both N and the expected cumulative frequency values are relatively small. In Example 14.8 the K-S goodness-of-fit test identified a significant difference among categories; however, a chi-squared test performed on the same data produces a non-significant result ($\chi^2_{observed}$ = 9.00: $P > 0.05$).

The K-S goodness-of-fit test can also be employed to examine frequencies falling within class intervals for a continuous variable, for example, the number of earthworms present at different depths in the soil where the null hypothesis is that earthworms are uniformly distributed from the soil surface to a particular maximum depth. The test for continuous data will not be described here but see, for example, Sokal and Rohlf (1994) and Zar (2009) for further description. A particularly common and traditional use of the K-S test is to examine whether the observed values for a measured variable adhere to a particular theoretical frequency distribution, in particular, the normal distribution. The test operates by comparing the cumulative frequency of the data within a series of class intervals, with the expected cumulative frequency based on the standard proportions of the normal distribution, and a P value is determined based on the largest discrepancy. This test is often offered by computer software packages for testing normality of the data prior to performance of a parametric statistical analysis. There are, however, some serious theoretical criticisms to the use of this test to inspect for normality, and its employment for this purpose is probably best avoided.

15 Performing Statistical Analyses Using Computer Packages and Presenting Results

Compute: reckon (number or amount). Hence computer, calculator, electronic calculating machine.

(Concise Oxford English Dictionary)

- Introduction to the use of computer software for statistical analysis.
- Using the Microsoft Excel®1 spreadsheet software statistical functions.
- Using the Microsoft Excel Analysis Toolpak.
- Using GenStat: an example of a modern-day statistics software package.
- Rules and techniques for presenting the results of statistical analyses of experiments.

15.1 Introduction

The previous chapters of this book have presented some basic explanation of the principles and procedures for a range of the statistical analyses most commonly employed in agricultural and horticultural research. As explained previously, an understanding of the basic principles of statistical analyses is very important in allowing valid interpretation of the outcomes and the recognition and rectification of any errors that occur in the analytical procedures. In the present day it is very rare, however, that the researcher needs to calculate these analyses from first principles without the aid of computer power. A number of basic statistical procedures can be performed rapidly using modern calculators or functions within spreadsheet programmes such as Microsoft Excel® spreadsheet software. To undertake more extensive statistical analyses a very wide range of statistics software packages is available. These packages range from relatively simple-to-use programs but with limited applicability through to very sophisticated programs that enable highly complex experimental designs to be modelled and analysed. Two software packages will be introduced here; the relatively simple Microsoft Excel® spreadsheet software 'add-in' package called Analysis Toolpak and the advanced statistics package 'GenStat' which was originally designed to support agricultural research. It is important that the reader understands that these are only being discussed here as examples of the genre and the reader should not treat the following as either recommendations of the product or as comprehensive working manuals.

The output from statistical packages is, however, rarely in a format suitable for immediate presentation in final reports, research theses and publishable papers;

therefore the final section of this chapter discusses the basic rules and some of the techniques for the formal reporting of the results of statistical analyses.

15.2 Calculation of Basic Statistics using the Microsoft Excel® Spreadsheet Software Functions

The Microsoft Excel® spreadsheet software program includes a large number of in-built functions, including a range of basic statistical functions, that can be applied to data sets held in an Excel spreadsheet to quickly obtain mathematical transformations and numerical analysis. (In describing the use of these functions, it will be assumed that the reader has a working knowledge of Microsoft Excel® spreadsheet software.) These functions can be obtained by clicking on *Insert* in the main spreadsheet toolbar and selecting *Function* from the drop-down menu to obtain the *Insert Function* menu box (see Fig. 15.1). (Alternatively this may be obtained via the 'function wizard', the location of which will depend on the version of Excel® spreadsheet software being used but is generally located by a button on the screen labelled '*fx*'.) The many functions available are grouped into convenient categories, which includes a *Statistical* category. On selecting a category, a menu of functions available is displayed in the *Select a function* box. Each available function has a name presented in upper case characters, which are generally intuitive but in some cases Excel® spreadsheet software does use some non-standard statistical terminology. By scrolling down the menu, the required function can be selected, noting that whenever a function is clicked on with the mouse a brief description of the function is provided beneath the menu. When a function is selected by a double click, or via the *OK* button, a new

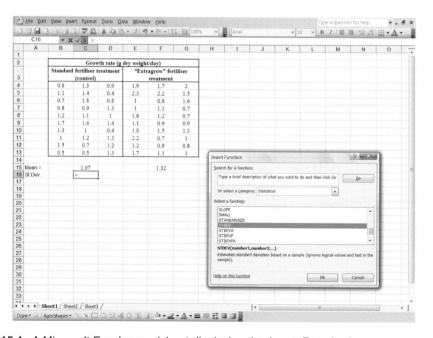

Fig. 15.1. A Microsoft Excel spreadsheet displaying the *Insert Function* box.

box is presented termed the *Function Arguments* box. The spreadsheet location of the data is entered in this box, together with any further criteria required by the specific function selected. On clicking *OK* the numerical solution to the function is transferred to a highlighted cell in the Excel spreadsheet. A typical Excel spreadsheet displaying the *Insert Function* box is illustrated in Fig. 15.1.

Given the large range of functions available, these cannot all be individually described here but once the user has employed one or two functions they do become increasingly intuitive. In Table 15.1 some of the more useful Excel® spreadsheet software mathematical and statistical functions are listed with comments on their use.

Table 15.1. Commonly used Microsoft Excel® spreadsheet software statistical functions.

Function	Description	Further note
ASIN	Returns the arcsin of a numerical value in radians	This is a common transformation in statistics used for % values (see section 4.2.3). Radians can be converted to degrees using the DEGREES function
AVEDEV	Returns the mean of the absolute deviations of a set of data values from their mean: $\Sigma[x-\bar{x}]/n$	
AVERAGE	Returns the arithmetic mean of a data set: $\Sigma x/n$	
BINOMDIST	Returns the binomial distribution probability	
CHITEST	Returns the *P* value for a chi-squared goodness-of-fit test	This does not provide the value of χ^2 itself; the χ^2 value can be obtained via the CHIINV function
DEVSQ	Returns the sum of squares of a set of data: $\Sigma(x - \bar{x})^2$	
FREQUENCY	Returns the number of values that occur within a specified range (class interval)	In the *Function arguments* box the spreadsheet location of the data is entered in the *Data_array* box and the class intervals in the *Bins_array* box
FTEST	Returns the *one-tailed P* value for an *F*-ratio test for comparing the variances of two samples	This test is used to check homogeneity of variances prior to a Student's *t*-test, however, the test should normally be conducted as a two-tailed test (see section 6.4)
LOG10	Returns the base 10 log of a number	This is a common transformation in statistics used for count values (see section 4.2.3)

continued

Table 15.1. Continued.

Function	Description	Further note
MEDIAN	Returns the median of a data set	
NORMSINV	Returns standardized normal deviate (z value) for a given probability	The cumulative probability for a given z value is given by the function NORMSDIST
PEARSON	Returns Pearson's product-moment correlation coefficient for a set of x, y values (r)	The function CORREL performs exactly the same function
POISSON	Returns the Poisson distribution probability	
RAND	Returns a random value ≥ 0 and < 1	The function RANDBETWEEN returns a random value between two specified values
RSQ	Returns the coefficient of determination for a set of x, y values (r^2)	
SLOPE	Returns the slope (b) of the best-fit line through a set of x, y values based on a simple linear regression	The intercept (a) of the regression line with the Y-axis is returned by the function INTERCEPT
STDEV	Returns the standard deviation of a sample (s^2)	
STDEVP	Returns the standard deviation of a population (σ^2)	
SUM	Returns the total of a set of values (Σx)	
SUMSQ	Returns the total of the squares of a set of values (Σx^2)	
TTEST	Returns the P value for a Student's t-test	The *Function arguments* box allows selection of a one- or two-tailed test and also selection of a paired-sample or an independent sample t-test assuming either equal or unequal sample variances
VAR	Returns the variance of a sample (s)	
VARP	Returns the variance of a population (σ)	
ZTEST	Returns the P value for a two-tailed z-test	If the population standard deviation σ^2 is known this can be entered; otherwise σ^2 is estimated by the sample standard deviation s^2

15.3 Calculation of Statistics using the Microsoft Excel® Spreadsheet Software Analysis Toolpak

The Analysis Toolpak is a software package for Microsoft Excel® spreadsheet software that allows data held in an Excel spreadsheet to be easily and rapidly subjected to a range of common statistical analyses. The package produces a wide range of descriptive statistics and calculates a number of standard parametric significant tests including Student's *t*-test, one-way and two-way ANOVA, Pearson's correlation and simple linear regression analysis. It does not, however, handle more advanced factorial designs nor does it perform any non-parametric sample comparison tests. Furthermore, it does use some non-standard statistical terminology that needs to be understood.

The Analysis Toolpak is a so-called 'add-in' package that needs first to be loaded before the Excel software can access it. The procedure for accessing the *Add-ins* menu will depend on the version of Excel® spreadsheet software being used; in general it is obtained via the Excel *Tools* menu. Once the *Add-ins* menu has been located and the *Analysis Toolpak* package selected, the Analysis Toolpak will be loaded and made available by clicking on *Data Analysis* in the main Excel *Tools* menu. (Note that if *Analysis Toolpak* is not listed in the *Add-Ins* menu then the package has not been loaded from the original Excel systems disk. Excel® spreadsheet software will have to be reinstalled from the systems disk in order to obtain it.) The main *Data Analysis* scroll-down menu (illustrated in Fig. 15.2) is displayed which lists the various statistical analyses available.

15.3.1 Using the Excel Analysis Toolpak to calculate descriptive statistics

The Analysis Toolpak allows a range of descriptive statistics to be obtained for multiple samples of data simultaneously, given that each individual data sample is presented in a single column in the Excel spreadsheet.

The *Descriptive Statistics* option is selected in the Analysis Toolpak menu to obtain the *Descriptive Statistics* input box. The spreadsheet cell location of the data is indicated in the *Input Range* box by either typing in the start and end cell locations separated by a colon (e.g. B2:C32) or by highlighting all the cells that contain the data in the spreadsheet using the mouse. If it is required to reserve the first cell in each column for a text label, the small box titled *labels in first row* must be checked; if labels are not included, this box must remain empty. At the bottom of the *Descriptive Statistics* input box the *Summary Statistics* option should be checked and the *confidence limit* and *largest* and *smallest values* selected if required.

In all Analysis Toopak operations, the rather unhelpful default is to produce the statistics output in a fresh spreadsheet. To specify a required location for the display of the output on the existing data spreadsheet, the *Output Range* option must be checked and the cell location entered in the *Output Range* box by either typing in the spreadsheet cell reference or using the mouse to click on an appropriate cell in the spreadsheet. On clicking on *OK*, a descriptive statistics output table is then displayed on the Excel spreadsheet starting at the cell location selected.

An example of the *Descriptive Statistics* input box and output table are shown in Fig. 15.2. (Note that on accessing the *Descriptive Statistics* input box this will obscure the *Data Analysis* menu box.) Note that the Analysis Toolpak package applies no automatic rounding-up procedures and consequently data values may be displayed with an inappropriate large number of significant figures. Rounding up can be achieved through the Excel number format function if required.

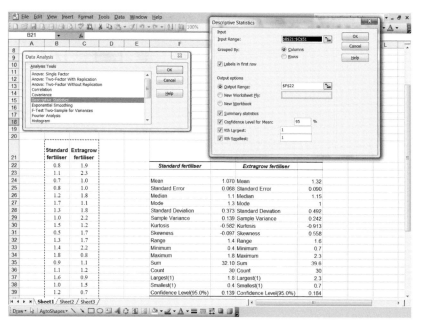

Fig. 15.2. Microsoft Excel® spreadsheet software displaying the Analysis Toolpak menu box, descriptive statistics input box and descriptive statistics output table following appropriate data rounding. The data analysed are from Example 2.3.

15.3.2 Using the Excel Analysis Toolpak to calculate Student's *t*-test

The Analysis Toolpak provides a rapid determination of Student's *t*-test for the analysis of differences either between the means of two independent samples, assuming either equal or unequal variances, or between paired samples. Note that in these cases data for each sample must be entered into the Excel spreadsheet in either a single column or a single row.

In the *Data Analysis* menu the following options are available:

> t-Test: Paired Two Sample for Means
> t-Test: Two-Sample Assuming Equal Variances
> t-Test: Two-Sample Assuming Unequal Variances
> z-Test: Two Sample for Means

Following selection of the required option an appropriate input work box is displayed. (Note that there is also an option for choosing a *z*-test if samples are sufficiently large, i.e. *n* > 30.) The spreadsheet cell locations for the data of the two samples are entered in the *Variable 1 Range* and the *Variable 2 Range* input boxes respectively. If the first cell for each sample contains a text label, this must be indicated by checking the small *labels* box; otherwise ensure that the *labels* box is empty. A value can be entered in the *Hypothesised Mean Difference* box if appropriate. If the null hypothesis of no difference between means is to be tested, the value zero should be entered, or the box can be left empty since a zero value is the default. If, however, it is required to test whether the observed difference between the sample means differs significantly from a given hypothetical value, then this value is entered here. The required critical level of probability for rejection of the null hypothesis is shown in the α input box; the default value is 0.05 but a different value can be entered if required. The required spreadsheet location for the display of the *t*-test output is entered in the *Output Range* box. The program is then run by clicking *OK*, upon which an output table of the results of the analysis, similar to that shown in Fig. 15.3, will be displayed at the selected cell location.

(a)

t-Test:Two-Sample Assuming Equal Variances

	Extragrow fertilizer	Standard fertilizer
Mean	1.320	1.070
Variance	0.242	0.139
Observations	30	30
Pooled Variance	0.191	
Hypothesized mean difference	0.0	
DF	58	
t Stat	2.218	
P(T<=t) one-tail	0.015	
t Critical one-tail	1.672	
P(T<=t) two-tail	0.030	
t Critical two-tail	2.002	

(b)

t-Test: Paired Two Samle for Means

	Resp. rate in Sept.	Resp. rate in Oct.
Mean	11.47	12.35
Variance	0.653	0.396
Observations	10	10
Pooled Variance	−0.084	
Hypothesized mean difference	0	
DF	9	
t Stat	−2.612	
P(T<=t) one-tail	0.014	
t Critical one-tail	1.833	
P(T<=t) two-tail	0.028	
t Critical two-tail	2.262	

Fig. 15.3. The Excel Analysis Toolpak output tables (following appropriate formatting) for Student's *t*-test: (a) to compare independent samples assuming equal variances (data from Example 6.5); (b) to compare non-independent paired samples (data from Example 6.6).

In the Student's *t*-test output table, the term 't Stat' is the value of the Student's *t*-test statistic (or $t_{observed}$ as it is referred to in this book). The terms 't Critical one-tail' and 't Critical two-tail' are the critical values for *t* at the level of probability defined by α for a one-tail and two-tail test respectively, while 'P(T<=t) one-tail' and 'P(T<=t) two-tail' are the precise probabilities of obtaining the observed difference, or one larger, under the one-tailed and two-tailed null hypotheses respectively. For a two-tailed test where the absolute value of *t* is larger than the critical two-tailed *t* value, the null hypothesis can be rejected and a significant difference between samples claimed ($P < 0.05$). When conducting a one-tailed test, the sign of *t* must be compatible with the direction of the null hypothesis for a significant difference to be claimed. Thus, if the null hypothesis is that the first sample mean is not smaller than the second sample mean, then a negative value for *t* would be necessary for the null hypothesis to be rejected.

Note that the Analysis Toolpak also provides an *F* test for checking equality of sample variances prior to undertaking a Student's *t*-test. Excel® spreadsheet software runs this test as a one-tailed test; however, most statisticians agree that this should be a two-tailed test (see Chapter 6, section 6.4). The results of this *F* test are, therefore, not to be relied upon and the test has not been illustrated here.

15.3.3   Using the Excel Analysis Toolpak to calculate analysis of variance tests

The Analysis Toolpak provides a rapid determination of both one-way and two-way ANOVA for the analysis of balanced, fully randomized one- and two-factorial experimental designs. The data to be analysed must first be entered into an Excel spreadsheet in the form of a data table, with replicates for each sample placed in columns. In the case of one-way ANOVA, the top single row can be reserved for sample labels; in the case of two-way ANOVA, both the first column and first row of the table must be reserved for labels. It is important to note that there should be no empty cells within the data table, since these will be read as zero values, and empty cells should not therefore be used to demarcate the data within the spreadsheet.

In the *Data Analysis* menu, the following three ANOVA options are available:

> Anova: Single Factor
> Anova: Two-Factor With Replication
> Anova: Two-Factor Without Replication

Following selection of the required option, an appropriate analysis of variance input box will be displayed. The Excel spreadsheet cell locations holding the data are entered in the *Range Input* box.

For a single-factor ANOVA, it must be indicated, by clicking in the appropriate box, whether the replicate values for samples are entered reading down the columns or across the rows. It must also be indicated, by clicking in the appropriate *labels* box, whether or not the first row or column in the data table holds treatment labels. (It is a good idea to include labels since the labels will then be transferred to the ANOVA output, making this easier to read.)

For a two-factor ANOVA, the Excel® spreadsheet software program assumes labels are present in the first row and first column and if this is not the case an error message will appear and the calculation will cease.

For a two-factor ANOVA with replication, the number of replicates per sample must be indicated in the *rows per sample* box. Note that, unlike in one-way ANOVA, this must be a constant value for all samples and the program cannot handle missing data.

The critical level of probability for rejection of the null hypothesis is entered in the α input box (the default value is 0.05). The cell location on the spreadsheet for display of the ANOVA output is entered in the *Output Range* box and the program then run by clicking *OK*.

An output table of the results of the analysis is displayed in the Excel spreadsheet, starting at the cell location selected. This output includes a table of summary statistics for each treatment sample and the final ANOVA table. The ANOVA table gives the test statistic F for each treatment factor, the treatment interaction where appropriate, the significance P value and the critical value of F at the selected level of probability. Examples of the one-way and two-way ANOVA outputs are shown in Figs 15.4 and 15.5.

In the Excel output ANOVA table, the terms 'between groups' and 'within groups' are used to represent the sources of variation. These terms can, however, be overwritten with more appropriate terms, as in Fig. 15.5.

In the example shown in Fig. 15.5 the *sample* treatment factor is plant density and the *column* treatment factor is irrigation level. The F value for both plant density (*sample*) and irrigation level (*column*) are larger than their respective critical F values; therefore the null hypotheses of no treatment effect may be rejected ($P < 0.01$) for both treatment factors. There is, however, no interaction between treatments ($P = 0.28$).

Anova: Single factor						
SUMMARY						
Groups	Count	Sum	Average	Variance		
25%	3	15.15	5.05	0.4593		
50%	3	17.91	5.97	0.0333		
75%	3	20.73	6.91	0.5493		
100%	3	11.01	3.67	0.1263		
ANOVA						
Source of Variation	SS	df	MS	F	P-value	F crit
Between Groups	17.1612	3	5.7204	19.58706	0.000482	4.066181
Within Groups	2.3364	8	0.29205			
Total	19.4976	11				

Fig. 15.4. The Excel Analysis Toolpak output tables (following appropriate rounding and formatting) for a one-way ANOVA test applied to the data of Example 8.1. The observed between-group F value is larger than the critical F value; therefore the null hypothesis that there is no treatment effect may be rejected ($P < 0.05$).

Anova: Two Factor With Replication					

Summary		Irrigation %			
Plant Density	0	33	66	100	Total
1					
Count	5	5	5	5	20
Sum	68	91	111	89	359
Average	13.6	18.2	22.2	17.8	17.95
Variance	5.8	6.7	6.7	6.7	15.21
2					
Count	5	5	5	5	20
Sum	75	101	107	99	382
Average	15	20.2	21.4	19.8	19.10
Variance	7	1.7	8.8	9.2	11.88
5					
Count	5	5	5	5	20
Sum	51	63	94	83	291
Average	10.2	12.6	18.8	16.6	14.55
Variance	2.7	4.3	8.7	9.3	17.10
Total					
Count	15	15	15	15	
Sum	194	255	312	271	
Average	12.93	17.00	20.80	18.07	
Variance	8.78	14.71	9.17	9.07	

ANOVA						
Source of Variation	SS	df	MS	F	P-value	F crit
Plant Density	223.9000	2	111.9500	17.31	<0.01	3.19
Irrigation	479.3333	3	159.7778	24.71	<0.01	2.80
Interaction	49.9667	6	8.3278	1.29	0.28	2.29
Residual	310.4000	48	6.4667			
Total	1063.6000	59				

Fig. 15.5. The Excel Analysis Toolpak output tables (following appropriate rounding and formatting) for a two-way ANOVA (with replication) test applied to the data of Example 9.1.

15.3.4 Using the Excel Analysis Toolpak to calculate simple linear regression analysis

The Analysis Toolpak provides a rapid determination of the linear regression coefficient based on the method of least squares and associated statistics, and can provide a plot of the regression line. The data to be analysed must be entered into an Excel spreadsheet in two separate columns. The *Regression Analysis* option is selected in the Analysis Toolpak menu to display the regression analysis input box. The spreadsheet cell locations holding the x values (independent variable) and y values (dependent variable) are entered into the *Input x range* box and *Input y range* boxes respectively. If the first cell of each column holds text labels, then this must be indicated by checking the *labels* box; otherwise this box must remain empty. If it is required to force the regression line through the origin at $x = 0$, $y = 0$, this must be indicated by checking the 'Constant is Zero'

SUMMARY OUTPUT							
Regression statistics							
Multiple R	0.91						
R Square	0.82						
Adjusted R Square	0.80						
Standard error	177.92						
Observations	12						
ANOVA							
Source of Variation	Df	SS	MS	F	Significance F		
Regression	1	1,455,916.371	1,455,916.371	45.99	0.000049		
Residual	10	316,564.629	31,656.463				
Total	11	1,772,481.000					
	Coefficients	Standard Error	t Stat	P-value	Lower 95%	Upper 95%	
Intercept	21,137.11	441.91	47.83	<0.01	20,200	22,100	
Regression	−100.90	14.88	−6.78	<0.01	−134.0	−67.8	

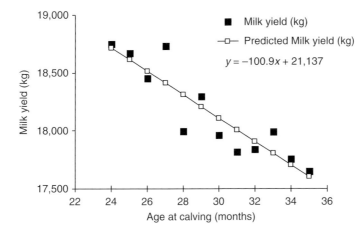

Fig. 15.6. The Excel Analysis Toolpak output tables and scatter plot (following appropriate rounding and formatting) for a linear regression analysis applied to the data of Example 12.1.

box. If it is required to determine the confidence limits of the regression, this must be indicated by checking the appropriate box and entering the confidence level (the default is $P = 95\%$) and, if a scatter plot of the regression line is required, the box labelled *Line Fit Plot* must be checked. A range of other types of plot, including a plot of the residuals, are available and are obtained by checking the appropriate boxes. Finally, the spreadsheet location for the correlation analysis output is entered in the *Output Range* box and the program is then run by clicking *OK*.

An output table of the results of the analysis is produced on the spreadsheet at the cell location specified, as illustrated in Fig. 15.6. The output includes:

1. A table of summary statistics. This gives the two important statistics Pearson's correlation coefficient r (termed *Multiple R*) and the coefficient of determination,

r^2 (termed *R Square*). The term labelled the standard error of the regression, however, is in fact the residual standard deviation (i.e. the equivalent of the $\sqrt{\text{Residual MS}}$). A statistic termed *Adjusted R Square* is also returned; however, this is of little concern here, being of consequence only in cases of multiple regression analysis.

2. An ANOVA table indicating the significance of the regression. The *Significance F* value shown is the probability of obtaining the observed data under the null hypothesis of no linear dependency of y upon x.

3. A table of coefficients: the first column displays the intercept of the regression line, a, and the regression coefficient, b. The table shows Student's t-test statistic and the 95% confidence limits associated with intercept (a) and the regression coefficient (b), as discussed in Chapter 12.

4. A scatter plot showing the values of the dependent variable x predicted by the regression. This graph can be further formatted using the normal Excel® spreadsheet software chart-plotting wizard.

15.4 Calculation of Statistics using Specialized Statistics Software Packages

There are a number of sophisticated statistics packages available today that handle a complete range of statistical analyses, both parametric and non-parametric. In general, there are two major differences between running such packages and running the Microsoft Excel® Analysis Toolpak described above. First, for the majority of statistical applications on most modern statistics software, when entering the data into the spreadsheet the data must be 'column loaded'. That is, all the data must be present in one single column in the spreadsheet, while other columns are used to specify the factor levels to which the data refer. The data are not entered, therefore, in a form that is readily presentable in a formal report. Secondly, the results of an analysis are presented in a separate output window and not within the data spreadsheet itself. However, most of these packages operate well in conjunction with Microsoft Excel® spreadsheet software, allowing data to be freely exchanged between an Excel spreadsheet and the spreadsheet of the statistics package, while the output can subsequently be copied and pasted into Microsoft Word for easy editing and formatting.

It is clearly not possible to describe the use of the full range of statistics software available; however, in the main these all operate in a similar mode and have similar capability. Exactly which package is chosen by individual practitioners will depend on their specific statistics requirements and personal preference. To illustrate the capability of these programs, some detail will now be presented on the use of just one such software package, called 'GenStat', which was originally designed by software engineers at Rothamsted Research, an agricultural research institute in the UK. It is, of course, not possible to present here a comprehensive guide to the use of GenStat; the package itself comes with a very large online manual and there are a number of other published guides to its use, e.g. *GenStat Discovery Edition 3 For Everyday Use* (Buysse *et al.*, 2007), which is freely available on the Internet. In order to demonstrate the mode of data input and the analysis output produced, a range of the worked examples presented in the previous chapters are run here using GenStat (12th edition).

15.4.1 Using GenStat to calculate descriptive statistics

In order to compute a set of descriptive statistics for multiple samples using GenStat, the data sets need to be entered into a GenStat spreadsheet as separate data columns. The formatting can be undertaken in a Microsoft Excel spreadsheet and the spreadsheet then loaded as a file (or copy-pasted) into GenStat, or the data can be directly typed into a GenStat spreadsheet. Since it is generally easier to manipulate and format data within Excel® spreadsheet software, the former option is probably preferable. After opening GenStat and inputting the data, the *Stats* option is selected in the GenStat toolbar to display the statistics drop-down menu. The mouse is pointed at the *Summary Statistics* option to reveal a secondary 'summary statistics' drop-down menu to the side. The GenStat spreadsheet and statistics drop-down selection menus are illustrated in Fig. 15.7.

In order to obtain a standard set of descriptive statistics, the *Summarize Contents of Variates* option is selected by double-clicking and a data input box is then displayed. The spreadsheet column labels that contain the data available for analysis are displayed in the *Available Data* box and by clicking on these the data sets required for analysis are transferred to the *Variables* data box. The statistics required are selected by checking the option boxes below, noting that further statistics are available by clicking the *More statistics* option. Finally, clicking on *Run*

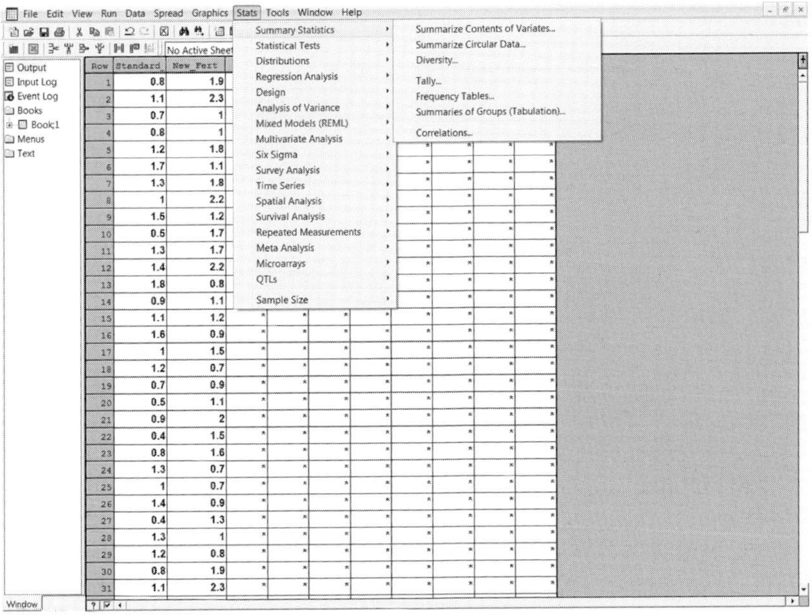

Fig. 15.7. The GenStat spreadsheet and statistics drop-down menus.

results in the calculated statistics being presented in the output window. To display the output window, either *Output* can be selected (by double-clicking) within the windows display tree on the left of the spreadsheet or it can be selected via the *Window* option in the main toolbar. The output can be highlighted and copy-pasted back into a Word document for formatting if required. The GenStat spreadsheet and output window for an analysis of the data of Example 2.3, concerning the growth rate of an untreated and a fertilizer-treated sample of tomatoes, are illustrated in Fig. 15.8.

The reader may have noticed that the values for both the Coefficients of Skewness and Kurtosis are noticeably different when calculated by GenStat and the Excel

(a) GenSat spreadsheet and option boxes

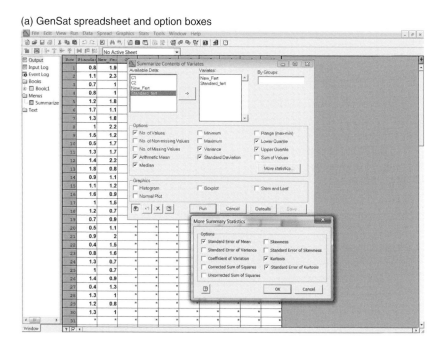

(b) GenStat output

Summary statistics for New_Fert	
Number of values	= 30
Mean	= 1.32
Median	= 1.15
Lower quartile	= 0.9
Upper quartile	= 1.7
Standard deviation	= 0.492
Standard error of mean	= 0.0898
Variance	= 0.242
Kurtosis	= −0.962
Standard error of Kurtosis	= 0.833

Summary statistics for Standard_fert	
Number of values	= 30
Mean	= 1.07
Median	= 1.1
Lower quartile	= 0.8
Upper quartile	= 1.3
Standard deviation	= 0.373
Standard error of mean	= 0.0682
Variance	= 0.139
Kurtosis	= −0.683
Standard error of Kurtosis	= 0.833

Fig. 15.8. (a) GenStat spreadsheet and option boxes for performing summary statistics; (b) copy of summary statistics GenStat output window. Data from Example 2.3.

Analysis Toopak for the same data sets. This is simply because Excel and GenStat use different methods for the determination of these coefficients. Readers will need to consult the respective help menus and manuals for Excel® spreadsheet software and GenStat to ascertain exactly how these statistics have been determined by the software and will then make their own decision on which to employ.

15.4.2   Using GenStat to calculate Student's *t*-test

To perform a Student's *t*-test to compare two samples, the data can be input into a GenStat spreadsheet in two columns exactly as shown previously (Fig. 15.9). Alternatively the data can be entered in one single column and a separate column used to identify which sample each data item belongs to. In the latter case, once the data have been entered into the GenStat spreadsheet, the column holding the sample labels must be identified as a 'factor column' rather than a 'data column'. This is achieved most easily by clicking anywhere in the column and then pressing the right-hand mouse button to obtain an option menu and selecting *Convert to Factor*. To indicate that the column now shows factor labels rather than a variable, an exclamation mark will appear in the column heading.

To run the Student's *t*-test the *Statistical Tests* option is selected from the *Stats* drop-down menu and the *One- and two-sample t-tests* option then selected to display the *T-Tests* input box. Within this box the two variables for comparison by the *t*-test are selected and a choice is made between a one-tailed and a two-tailed test (the latter is the default option). If the data have been entered in a single column with the grouping factors indicated in a separate column, then this must be indicated in the *Data Arrangement* box. An *Options* box provides additional analyses if required, including an *F* test to assess the homogeneity of sample variances and determination of a confidence interval for the difference between the sample means. It also presents a choice for the mode of calculation of variance within the *t*-test, which may be either pooled or determined separately for each sample. (This will depend on whether or not the sample variances are homogeneous, as explained in Chapter 6, section 6.3, 'Analysis of the difference between two small independent samples'.) Finally, clicking on *Run* will result in the outcome of the *t*-test being displayed in the output window. The GenStat spreadsheet and output window for a Student's *t*-test are shown in Fig. 15.9.

Note that in the example shown in Fig. 15.9 a variance comparison *F* test was run which detected that the sample variances were not significantly different ($P = 0.14$). GenStat subsequently determined the Student's *t*-test assuming equal variances. The difference between the means was shown to be significant ($P = 0.030$).

15.4.3   Using GenStat to calculate fully balanced analysis of variance (ANOVA) tests

In order to run ANOVA tests using GenStat, the data must be loaded in single columns with separate factor columns being used to group the data. While the data can be entered directly into a GenStat spreadsheet, it is again recommended that the data are formatted initially in an Excel spreadsheet and then entered into GenStat as a file

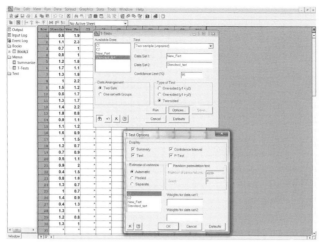

(b) GenSat output window

Two sample t-test

Variates: New_Fert, Standard_fert.

Test for equality of sample variances

Test statistic F = 1.73 on 29 and 29 d.f.
Probability (under null hypothesis of equal variances) = 0.14

Summary

Sample	Size	Mean	Variance	Standard deviation	Standard error of mean
New_Fert	30	1.320	0.2417	0.4916	0.08975
Standard_fert	30	1.070	0.1394	0.3734	0.06817

Difference of means: 0.250
Standard error of difference: 0.113

95% confidence interval for difference in means: (0.02440, 0.4756)

Test of null hypothesis that mean of New_Fert is equal to mean of Standard_fert

Test statistic t = 2.22 on 58 d.f.
Probability = 0.030

Fig. 15.9. (a) GenStat spreadsheet and menu boxes for performing Student's *t*-test for independent samples; (b) copy of *t*-test GenStat output window. Data are from Example 6.5.

(or by copy-pasting if preferred). It must then be ensured that all columns that hold factor grouping labels rather than data are converted to factor columns, this being most readily achieved by using the right-hand mouse button to obtain an option menu and selecting *Convert to Factor*.

To perform a simple fully randomized one-way ANOVA upon the data originally illustrated in Example 8.1, the GenStat spreadsheet would be formatted as shown in Fig. 15.10(a). Note that the first column holds the grouping factor (i.e. level of irrigation), which is indicated by the red exclamation mark in the column label; the second column holds the data (i.e. stem girth). The *One- and Two-way ANOVA* data input box is then obtained via the *Analysis of Variance* option in the *Stats* menu and the *One-way ANOVA* option selected. The column label containing the data is then entered in the *Y-variate box* and the column label containing the factor groupings entered in the *Treatments* box. If required, an *Options* box can be selected which then provides additional statistics in addition to the one-way ANOVA, such as the least significant difference (LSD) and a range of other multiple comparison tests. The

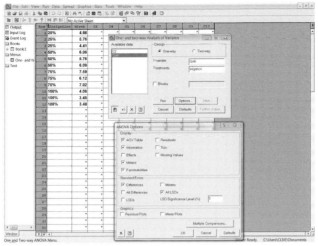

(a) GenSat spreadsheet and option boxes

(b) GenSat output window

Analysis of variance

Variate: Girth

Source of variation	d.f.	s.s.	m.s.	v.r.	F pr.
Irrigation	3	17.1612	5.7204	19.59	<.001
Residual	8	2.3364	0.2920		
Total	11	19.4976			

Information summary

All terms orthogonal, none aliased

Tables of means

Variate: Girth

Grand mean 5.40

Irrigation	100%	25%	50%	75%
	3.67	5.05	5.97	6.91

Standard errors of differences of means

Table	Irrigation
rep.	3
d.f.	8
s.e.d.	0.441

Least significant differences of means (5% level)

Table	Irrigation
rep.	3
d.f.	8
l.s.d.	1.0181

Fig. 15.10. (a) GenStat spreadsheet and menu boxes for performing one-way ANOVA for independent samples; (b) copy of the one-way ANOVA GenStat output window. Data from Example 8.1.

analysis is undertaken by clicking on *Run* and the results of the analysis are then displayed in the GenStat Output Window as illustrated in Fig. 15.10(b).

The spreadsheet format required for performing two-way ANOVA follows the same principles but becomes more complicated by the fact that a further factor column is required to group the data according to the second factor. For example, the experiment to examine the effect of irrigation level on tree girth illustrated above may have been performed at two different plant densities, 'low' and 'high'. In this case the data would be formatted in a GenStat spreadsheet as shown in Fig. 15.11.

To perform two-way ANOVA in GenStat, the same procedure is followed to obtain the ANOVA data input box as detailed for one-way ANOVA above. *Two-way ANOVA* is selected in the design option box, the column label containing the data is entered in the *Y-variate* box and the two treatment factor labels entered in the *Treatment 1* and *Treatment 2* boxes respectively. By clicking on *Options*, additional

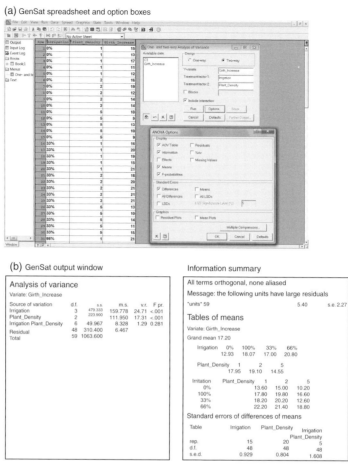

(a) GenSat spreadsheet and option boxes

(b) GenSat output window

Analysis of variance

Variate: Girth_Increase

Source of variation	d.f.	s.s.	m.s.	v.r.	F pr.
Irrigation	3	479.333	159.778	24.71	<.001
Plant_Density	2	223.900	111.950	17.31	<.001
Irrigation Plant_Density	6	49.967	8.328	1.29	0.281
Residual	48	310.400	6.467		
Total	59	1063.600			

Information summary

All terms orthogonal, none aliased

Message: the following units have large residuals

units 59   5.40   s.e. 2.27

Tables of means

Variate: Girth_Increase

Grand mean 17.20

Irrigation	0%	100%	33%	66%
	12.93	18.07	17.00	20.80

Plant_Density	1	2	5
	17.95	19.10	14.55

Irritation	Plant_Density	1	2	5
0%		13.60	15.00	10.20
100%		17.80	19.80	16.60
33%		18.20	20.20	12.60
66%		22.20	21.40	18.80

Standard errors of differences of means

Table	Irrigation	Plant_Density	Irrigation Plant_Density
rep.	15	20	5
d.f.	48	48	48
s.e.d.	0.929	0.804	1.608

Fig. 15.11. (a) GenStat spreadsheet and menu boxes for performing two-way ANOVA for independent samples; (b) copy of ANOVA GenStat output window. Data from Example 9.1.

statistics, including LSD values, may be obtained. Finally, by clicking on *Run*, the result of the analysis is obtained, which is displayed in the GenStat Output Window.

The GenStat output for a two-way ANOVA is shown in Figure 15.11(b) below. The *F* probabilities indicate that there is a significant irrigation effect ($P = 0.036$) but no significant effect of plant density or interaction. Note that GenStat also provides a warning message where particular data items may not adhere to the assumptions required of the analysis; in this case, one particular value appears to show a large residual, i.e. a major departure from the treatment mean.

One-way or two-way ANOVA can be performed on a randomized complete block (RCB) designed experiment quite simply using GenStat. In this case the block groupings are indicated in a further factor column in the spreadsheet, and in the ANOVA data input box the column label containing the block groupings is entered into the *Blocks* box. The ANOVA for an RCB design, in which the effect of the blocks is separated out from the treatment effects, is then computed.

15.4.4 Using GenStat to calculate analysis of variance (ANOVA) tests for more complex factorial designs

In GenStat ANOVA can be determined in a rather more sophisticated manner which employs a general linear model (GLM) approach, which then allows more complex designs to be handled and will provide an analysis of unbalanced factorized experiments where blocks or individual data items are missing (see Chapter 8, section 8.5). Illustrations of the analyses of some of the more standard types of design will be provided here; however, detailed explanation on how to achieve the more complex analyses is beyond the scope of this short introduction to GenStat and the reader is referred to published GenStat manuals for further guidance to these procedures (e.g. McConway *et al.*, 1999).

After selecting *Stats* in the GenStat toolbar and *Analysis of Variance* from the drop-down menu, there are some alternatives to simply choosing *One- and Two-way ANOVA*. One of these alternatives is to select *General*, which produces a new *Analysis of Variance* input box. This box allows the selection of different factorial experimental designs and specification of the treatment and block structures. This enables, for example, the analysis of variance of three-way factorial designed experiments, randomized complete block (RCB) designs, split-plot and Latin square designs and allows the interactions required for inspection to be specified.

Analysis of a two-factor randomized complete block (RCB) design (with missing data)

The data are column loaded into the GenStat spreadsheet as before and treatment factor and block groupings indicated in adjacent factor columns. In cases where there are missing data values, the appropriate cells in the column holding the data are simply left empty; GenStat will then automatically enter an asterisk into these cells to indicate missing values. To conduct an ANOVA for a two-factor RCB, the option *General* is selected from the *Analysis of Variance* menu to display the analysis of variance input box. From here a number of types of design are available, but to input a design model that allows for missing data the default *General Analysis of Variance* is selected. The column label holding the data is indicated in the *Y-variate* box in the normal way. A model of the treatment structure is then entered into the *Treatment Structure* box. This simply involves inputting the column labels holding the factor groupings to be analysed separated by an operator symbol that describes how the factors should be treated within the ANOVA. Thus, if two factors A and B were to be analysed for their main effects without determining the interaction they would be shown as *Factor A + Factor B*; if the interaction was to be determined this would become *Factor A * Factor B*. (The function of other operators available can be found in the GenStat help menu.) Similarly, the column label holding the block groupings are input into the *Block Structure* box. The analysis of variance input box also allows for the selection of contrasted factors and covariates; these, however, involve an advanced level of analysis that has not been covered in this book (if no such factors have been identified, these selection boxes are simply left empty). A further option box allows further multiple comparison analyses to be performed and also, very importantly, allows the existence of missing data to be indicated by simply clicking in the *Missing Values* box. On clicking *Run*, the program will apply a generalized linear model to predict the missing data values and conduct a two-way ANOVA.

The GenStat spreadsheet, data input boxes and output window for this type of analysis are illustrated in Fig. 15.12. The data used for this example are taken from Example 10.3 but two data items have been removed to represent missing data; these are indicated by the asterisk signs in rows 6 and 12 of the spreadsheet. The output shows the estimates for these missing values while the degrees of freedom in the ANOVA table are appropriately reduced. The ANOVA indicates a significant effect of both treatment factors, i.e. fertilizer ($P = 0.003$) and variety ($P < 0.001$), on yield and a significant interaction of these two factors ($P = 0.031$).

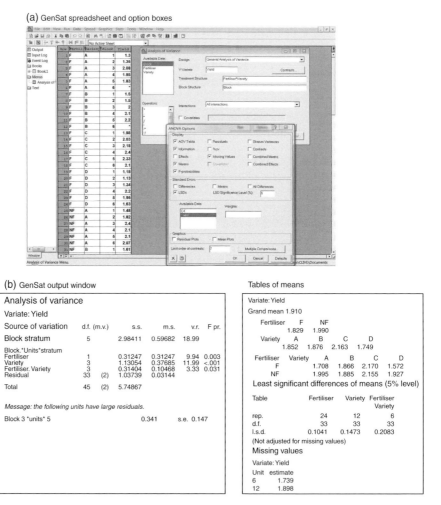

Fig. 15.12. (a) GenStat spreadsheet and menu boxes for performing a two-factor randomized complete block ANOVA with missing data items; (b) copy of ANOVA GenStat output window. Data are from Example 10.3 modified by removing two data items.

Analysis of a Latin square design

Analysing fully balanced Latin square designs with GenStat is relatively straightforward. The option *General* is selected from the *Analysis of Variance* menu to display the analysis of variance input box and *Latin Square* then selected from the design option box. The labels of the columns holding the data, the main treatment factor, the identified row factor and the identified column factor are entered in the *Y-variate* box, the *Treatment* box, the *Rows* box and the *Columns* box, respectively. Other output options, including multiple comparison tests, are obtained via the *Options* box. The GenStat spreadsheet, data input boxes and output window for this type of analysis are illustrated in Fig. 15.13, using the data from Example 10.4. This example concerned the effect of diet supplements on the increase in body weight of lambs. The lambs came from five different breeding lines and were housed in five different pens and to account for these external influences they are treated as row and column factors in a Latin square design. Since we are not really interested in whether the breeding lines and pens produce a significant effect, the ANOVA table produced by GenStat does not report a P value for these, although this can be easily determined by the user if required; however, the effect of the diets is shown to be highly significant ($P < 0.001$).

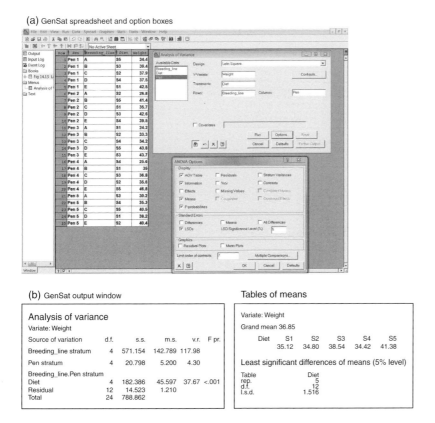

Fig. 15.13. (a) GenStat spreadsheet and menu boxes for performing Latin square ANOVA for independent samples; (b) copy of the ANOVA GenStat output window. Data from Example 10.4.

Analysis of a split-plot design

The *Split-Plot Design* option is selected from within the analysis of variance option and the columns holding the data, blocks, main plots (i.e. whole plots) and subplots are subsequently indicated in the appropriate input boxes. In addition, a model describing the experimental design needs to be given in the *Treatment Structure* input box. In the example shown in Fig. 15.14, a day-length extension treatment and variety are the main plot and subplot treatment factors within a blocked split-plot design and the interaction between these treatments is of interest. The model is therefore entered as:

$$\text{Day extension} + \text{Variety} + (\text{Day extension.Variety})$$

This facilitates computation of the significance of both the factors and their interaction independently. (Note that in GenStat use of an asterisk provides a short cut to writing a model containing treatment interaction in full so that the model above

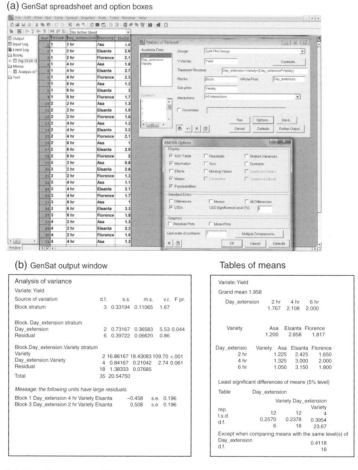

(a) GenSat spreadsheet and option boxes

(b) GenSat output window

Analysis of variance

Variate: Yield

Source of variation	d.f.	s.s.	m.s.	v.r.	F pr.
Block stratum	3	0.33194	0.11065	1.67	
Block.Day_extension stratum					
Day_extension	2	0.73167	0.36583	5.53	0.044
Residual	6	0.39722	0.06620	0.86	
Block.Day_extension.Variety stratum					
Variety	2	16.86167	18.43083	109.70	<.001
Day_extension.Variety	4	0.84167	0.21042	2.74	0.061
Residual	18	1.38333	0.07685		
Total	35	20.54750			

Message: the following units have large residuals.

Block 1 Day_extension 4 hr Variety Elsanta	−0.458	s.e. 0.196
Block 3 Day_extension 2 hr Variety Elsanta	0.508	s.e. 0.196

Tables of means

Variate: Yield

Grand mean 1.958

Day_extension	2 hr	4 hr	6 hr
	1.767	2.108	2.000

Variety	Asa	Elsanta	Florence
	1.200	2.858	1.817

Day_extensio	Variety	Asa	Elsanta	Florence
2 hr		1.225	2.425	1.650
4 hr		1.325	3.000	2.000
6 hr		1.050	3.150	1.800

Least significant differences of means (5% level)

Table	Day_extension		
			Variety Day_extension
			Variety
rep.	12	12	4
l.s.d.	0.2570	0.2378	0.3954
d.f.	6	18	23.67

Except when comparing means with the same level(s) of

Day_extension		0.4118
d.f.		18

Fig. 15.14. (a) GenStat spreadsheet and menu boxes for performing split-plot ANOVA for independent samples; (b) copy of ANOVA GenStat output window. Data from Example 10.5.

could alternatively be entered as Day extension*Variety.) The interactions to be displayed are selected in the *Interactions* box; however, it is usually simplest to select the default of all interactions. Note that GenStat does not display the significance of the block effect since it is assumed this is of no interest; however, this can be readily determined using the appropriate block and residual DF if required.

Analysis of a nested design

The simplest way to analyse a nested design in ANOVA is to select the *General Analysis of Variance* option and to indicate the nested structure within the *Treatment Structure* input box. In the analysis shown in Fig. 15.15, the water content of leaves

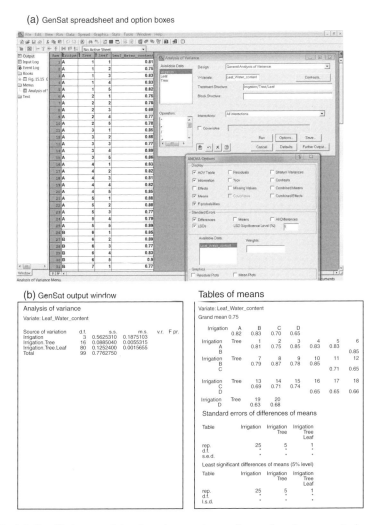

Fig. 15.15. (a) GenStat spreadsheet and menu boxes for performing a nested (hierarchical) ANOVA with one main factor; (b) copy of ANOVA GenStat output window. Data from Example 10.6.

is determined for samples of leaves from different trees nested within different irrigation treatments. The model is thus entered as: Irrigation/Tree/Leaf.

In the output window the mean squares are displayed in an ANOVA table but, since there is not a separated residual MS value, the variance ratios in this example are left undetermined by GenStat. The user will need to complete the calculation depending on the most appropriate way to determine the variance ratios, which in turn will depend on whether the categories of the nested factor(s) are fixed or random (see section 9.6, 'Analysis of nested designs', for discussion of this issue). It is more common for the categories of the nested treatment factors to be random, in which case the *F*-ratios are determined by comparing the MS for each factor with that of the factor immediately below it in the design hierarchy. Therefore, in the analysis of the irrigation experiment illustrated in Fig. 15.15, where both the leaves within trees and the trees within the irrigation plots are selected randomly, the *F* values would be obtained by:

$$F \text{ main factor} = \frac{\text{MS irrigation}}{\text{MS irrigaton.trees}} \text{ and}$$

$$F \text{ nested factor} = \frac{\text{MS irrigation.trees}}{\text{MS irrigaton.trees.leaves}}$$

Where, however, the nested factors are fixed, then both the main treatment and all nested treatment mean squares are compared with the treatment MS lying at the bottom of the design hierarchy.

15.4.5 Using GenStat to perform linear regression analysis

In order to fit a simple linear regression analysis between two variables, data are simply entered into a GenStat spreadsheet in two data columns; again, this may be undertaken by importing an Excel® spreadsheet software file or typing values directly into the GenStat spreadsheet. The *Stats* option is selected in the GenStat toolbar and *Regression Analysis* selected from the subsequent drop-down menu. A menu for a large range of regression procedures appears; for simple linear regression click on *Linear Models* to obtain the linear regression data input box. In the linear regression box select simple linear regression and enter the column heading for the dependent variable (plotted on the *y*-axis) in the *Response Variate* box and the column heading for the independent variable (plotted on the *x*-axis) in the *Explanatory Variable* box. Click on options to select a range of further statistics if required, including a scatter plot of the data, and click *Run* to perform the analysis.

If a plot of the data has been requested, this will appear initially in the GenStat graphics window. In addition to plotting the fitted regression line, the program also plots by default the upper and lower 95% regression confidence limits. The graph can be edited within GenStat and saved for subsequent importing into word processed documents. The numerical outcome of the regression analysis is displayed in the GenStat output window in the normal way. The GenStat output for a simple linear regression analysis performed on the data presented in Example 12.1 is shown in Fig. 15.16.

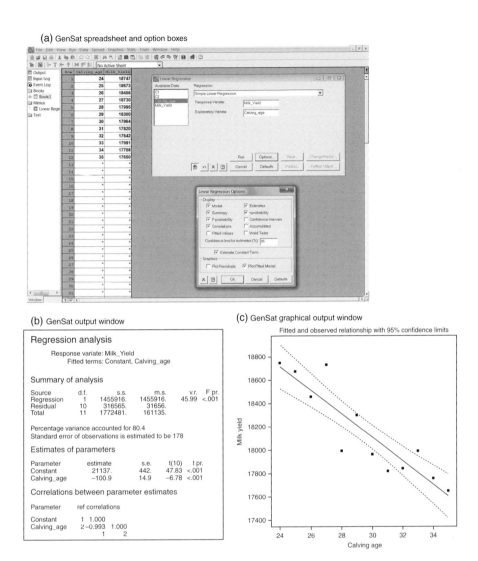

Fig. 15.16. (a) GenStat spreadsheet and menu boxes for performing a simple linear regression analysis; (b) copy of linear regression GenStat output window; (c) copy of linear regression GenStat graphical output window. Analysis was performed on data presented in Example 12.1. Note that in the GenStat output window the term 'percentage variance accounted for' is the adjusted r^2 value.

GenStat is, of course, capable of performing more complex regression analyses, including multiple linear regression and polynomial regression analyses of the types described in Chapter 13 (see sections 13.2 and 13.3.2). Examples of the GenStat spreadsheets, data input and output windows for these analyses are shown in Figs 15.17 and 15.18. Note in Fig. 15.17 that GenStat output is again helpful in indicating potential problems in the analysis by displaying a warning message. In this case, one of the data values in the multiple regression is shown to have a high leverage and would therefore be expected to have an unduly large influence on the regression coefficient.

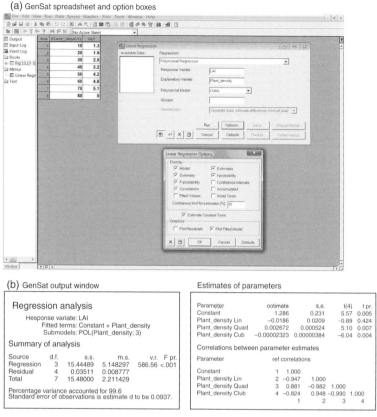

(a) GenSat spreadsheet and option boxes

(b) GenSat output window

Regression analysis

Response variate: LAI
Fitted terms: Constant + Plant_density
Submodels: POL(Plant_density; 3)

Summary of analysis

Source	d.f.	s.s.	m.s.	v.r.	F pr.
Regression	3	15.44489	5.148297	586.56	<.001
Residual	4	0.03511	0.008777		
Total	7	15.48000	2.211429		

Percentage variance accounted for 99.6
Standard error of observations is estimate d to be 0.0937.

Estimates of parameters

Parameter	ootimate	s.e.	t(4)	t pr.
Constant	1.286	0.231	5.57	0.005
Plant_density Lin	−0.0186	0.0209	−0.89	0.424
Plant_density Quad	0.002672	0.000524	5.10	0.007
Plant_density Cub	−0.00002323	0.00000384	−6.04	0.004

Correlations between parameter estimates

Parameter	ref	correlations			
Constant	1	1.000			
Plant_density Lin	2	−0.947	1.000		
Plant_density Quad	3	0.881	−0.982	1.000	
Plant_density Club	4	−0.824	0.948	−0.990	1.000
		1	2	3	4

Fig. 15.17. (a) GenStat spreadsheet and menu boxes for performing a multiple linear regression analysis; (b) copy of multiple linear regression GenStat output window. Analysis was performed on data presented in Example 13.1.

15.5 Reporting the Outcomes of the Statistical Analysis of Experimental Data

15.5.1 The ground rules

It is an obvious statement, but worth reiteration none the less, that no matter how good the research work is, unless it is reported clearly and accurately the research results will be neither acknowledged nor accepted by the research community and the research will neither be published by reputable research journals nor facilitate a successful outcome to the examination of a research thesis. There are several publications that describe appropriate ways for presenting quantitative data in tabular and graphical mode (see, for example, Cleveland, 1993; Tufte, 2001), while editors will invariably provide detailed guidance to the specific requirements for formal publication in specific journals and books. The requirements for reporting the outcomes of statistical analyses may, however, be less well understood. Therefore, in the final section of this book some guidance and rules are provided for presenting statistical output which, it is hoped, will assist the readers' journey along the road to successful research reporting.

(a) GenSat spreadsheet and option boxes

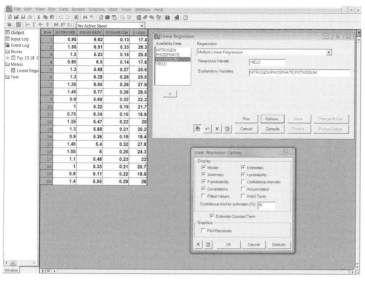

(b) GenSat output window

Regression analysis

Response variate: YIELD
Fitted terms: Constant, NITROGEN, PHOSPHATE, POTASIUM

Summary of analysis

Source	d.f.	s.s.	m.s.	v.r.	F pr.
Regression	3	250.03	83.344	27.53	<.001
Residual	16	48.43	3.027		
Total	19	298.47	15.709		

Percentage variance accounted for 80.7
Standard error of observations is estimated to be 1.74.

Message: the following units have high leverage.

Unit	Response	Leverage
3	20.80	0.43

Estimates of parameters

Parameter	estimate	s.e.	t(16)	t pr.
Constant	4.14	4.77	0.87	0.398
NITROGEN	4.44	2.75	1.61	0.126
PHOSPHATE	0.502	0.723	0.69	0.498
POTASSIUM	47.0	11.2	4.21	<.001

Correlations between parameter estimates

Parameter	ref	correlations			
Constant	1	1.000			
NITROGEN	2	−0.312	1.000		
PHOSPHATE	3	−0.910	0.078	1.000	
POTASSIUM	4	0.048	−0.789	−0.065	1.000
		1	2	3	4

(c) GenSat graphical output window

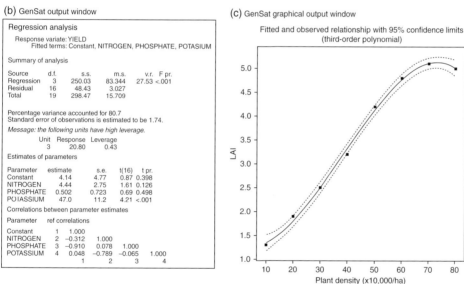

Fitted and observed relationship with 95% confidence limits (third-order polynomial)

Fig. 15.18. (a) GenStat spreadsheet and menu boxes for performing a polynomial (third-order) linear regression analysis; (b) copy of polynomial regression GenStat output window; (c) copy of polynomial regression GenStat graphical output window. Analysis was performed on data presented in Example 13.4.

To begin with, it will be as well to recall the primary purposes of applying statistical data analyses to research data. These were discussed in Chapter 1 and may be summarized as:

1. summarizing large data sets to make then digestible;
2. providing a measure of sample reliability;
3. providing an objective approach to drawing conclusions from observed data.

In a sense, the application of statistics ensures that all researchers across all disciplines apply the same level of objectivity and discernment in the evaluation of their research results, thus enabling valid comparisons to be made between the outcomes from different researchers. If this were not the case, then research groups examining the same problem and obtaining the same research results but applying different levels of objective analysis to the process of evaluation may well arrive at contrary conclusions. At the same time, as shown in the illustrative figures in the previous section, modern statistical software packages produce a wealth of statistical output, a large proportion of which it is not normally essential to report. Deciding, however, on which statistical information to include and which to exclude can be a problematic exercise. In general, there will be a minimum level of statistical information that needs to accompany the presentation of any set of research results that will enable the reader to assess the validity of the conclusions reached and, indeed, allow the reader, should they wish, to make their own judgement on their meaning. Put in a nutshell, what is vital is that the researcher provides evidence that an appropriate analysis has been correctly applied.

15.5.2 Reporting descriptive statistics and sample error

Usually it is sufficient to represent a large set of data by a measure of the central tendency, e.g. the mean, together with a measure of its variability, e.g. the standard deviation. Occasionally, reporting the actual individual values may be important but, more commonly and especially in the case of research journals where space is at a premium, published data tables simply give the treatment means (or, where appropriate, the medians). It is more common to find full data sets presented in dissertations and theses, but even then these will often be confined to an appendix and only the descriptive statistics presented in the results section proper. While this is quite acceptable practice and will invariably make the results more easily read and digested, it does of course make it even more vital that the appropriate statistics are reported.

Where the presented means represent samples rather than whole populations, measures of sample size and reliability become essential. As explained in Chapter 2 (section 2.4), under these circumstances reporting either the standard error (SE) or the confidence limits (CL) as a measure of sample reliability is a minimum requirement. It is, of course, essential to make it clear whether the ± values attached to the means are SEs or CLs, there being absolutely no need to report both. Figures 2.1 and 4.6 illustrated the use of error bars on graphical plots to present these statistics, and a further example of their tabular and graphical presentation is shown in Fig. 15.19.

In cases where a standard parametric statistical analysis such as ANOVA has been applied to a fully randomized experimental design, there is actually no need to show the individual sample standard errors since the analysis has demanded the assumption of homogeneous sample variances. In such cases, display of the residual standard deviation (i.e. $\sqrt{MS_{residual}}$) obtained from the ANOVA will suffice, since this represents the pooled sample standard deviation and as such gives the best estimate of the population standard deviation σ. Alternatively, where the author wishes to direct more attention to the comparison between sample means, the standard error of the difference between means (SED; given by $\sqrt{(2MS_{residual})/n}$) may be plotted as a single error bar on a graph. In both cases, this may then have the virtue of improving the clarity of a complex graph by allowing the removal of the individual sample

Table 1. Mean grain yield (g/m²) and standard errors for four spring-sown wheat cultivars grown in Mn-deficient soil with and without Mn fertilizer application. Mean values not sharing similar letter indices are significantly different ($P < 0.05$) based on a Tukey HSD multiple comparison test following one-way ANOVA ($n = 8$ for all samples).

Cultivar	Mean grain yield (g/m²) ± SE	
	+ Mn fertilizer	– Mn fertilizer
Paragon	237.1 ± 8.1^d	91.1 ± 5.2^a
Maris Butler	217.9 ± 5.6^d	153.3 ± 5.0^{bc}
Sappo	200.2 ± 7.0^{cd}	93.7 ± 3.1^a
Maris Dove	198.3 ± 6.6^{cd}	107.0 ± 2.8^{ab}

Tukey MCD$_{(P = 005)}$ = 50.7

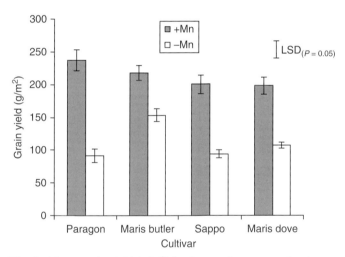

Fig. 1. Mean grain yield (g/m²) for four spring-sown wheat cultivars grown in Mn-deficient soil with and without Mn fertilizer application. Error bars represent ± 95% confidence intervals around sample means; the LSD$_{(P = 0.05)}$ is shown as a single vertical bar ($n = 8$ for all samples).

Fig. 15.19. Examples of tabular and graphical presentation of sample means with accompanying appropriate statistics.

error bars. Where it is desired to make more specific statistical comparisons between the means of equal-sized samples, the least significant difference between means (LSD; given by Student's t × SED) is often presented as illustrated in Fig. 15.19. This practice does, however, rather invite the reader to use the LSD to make all possible paired comparisons between sample means, whereas such comparisons should be confined to only a few preselected treatment means, as explained in section 7.5, 'Least significant difference test'. Alternatively the mininmum critical difference (MCD) arising from other multiple comparison tests, such as the Tukey test, might be displayed, although it will be imperative to make it clear in the figure/table legend exactly which type of MCD value is being presented.

15.5.3 Reporting the outcome of significance tests

While readers, journal editors and thesis examiners will not in general relish having to plough their way through a plethora of reported statistics within a results section in order to reach the major conclusions, they will need to be assured that appropriate and robust data analyses have been applied. The methods section of research reports should, of course, state clearly the statistical methods that have been employed. In the case of straightforward experimental designs, this may consist of no more than a statement of which statistical test was used. In the case of a single-factor fully randomized experiment, for example, it will be necessary to state only that one-way analysis of variance was applied, giving the critical P value used to identify significant main effects and, if appropriate, the method of multiple comparison used to separate significantly different sample means. In the case of more complex factorial designs, such as split-plot and nested designs, it will usually be necessary to describe the statistical model on which the employed ANOVA was based, since it cannot be assumed that the F-ratios are always given by $MS_{treatment}/MS_{residual}$. This is particularly important where a mixed model is involved in which different treatment factors may have been applied at fixed and random levels.

Reporting the results of ANOVA tests

Given that the output from computer analyses invariably provides far more statistical information than it is necessary to present in the results section, we need to decide which statistics to report and the best format for reporting them. In the case of ANOVA tests, the outcomes are conventionally displayed in ANOVA tables that have a highly consistent format and it is a simple matter to copy these straight from the computer statistic output into the results section of a report. While the presentation of the relevant ANOVA table immediately below a graphical display of the treatment means is an established way of presenting results in a thesis, journal editors today invariably resist this because of economy of page space. Full ANOVA tables are usually only presented in science journals in cases where complex non-standard experimental designs have been employed and where it becomes important to show clearly how the ANOVA was conducted. For simple one- and two-way ANOVA, where the F-ratios for each treatment factor and interaction are given by $MS_{treatment}/MS_{residual}$, it is usually sufficient to provide just the F-ratio together with the treatment and residual degrees of freedom and the determined P value. For example the results of the two-way ANOVA applied to the data displayed in Fig. 15.19 may be reported as:

	F-ratio	*P*
Cultivar	$F_{3,32} = 2.098$	0.12
Mn fertilizer	$F_{1,32} = 13.519$	<0.001
Cultivar × Mn fertilizer interaction	$F_{3,32} = 3.383$	0.03

The amount of statistical information given here is reduced to such an extent compared to a full ANOVA table that, if desired, it could be reported quite clearly in the text rather than in a tabular form. For example, the appropriate text might read:

'Differences between cultivars were not significant ($F_{3, 32}$ = 2.098, P = 0.12); however, Mn fertilizer treatment was significant ($F_{1, 32}$ = 13.519, P = <0.001), as was the cultivar × Mn fertilizer interaction effect ($F_{3, 32}$ = 3.383, P = 0.03).'

Note that, since computer programs will generate accurate P values, unless the value is very small indeed, it is far better to report these accurately rather than in the form '$P < 0.05$'. This gives the reader the opportunity of applying their own critical significance level and, if they have a mind to do so, coming to their own opinion on whether to accept or reject the null hypothesis.

Where a multiple comparison procedure has been used to identify significantly different sample means following an ANOVA, a common method for displaying the outcome is to display the means with a label (usually a letter) such that non-significantly different means share the same label. This can be done either within a table, as illustrated in Table 1 of Fig. 15.19, or by labelling columns on a bar chart. It is, of course, essential that the table or figure legend clearly states both the multiple comparison test used and the critical level of probability employed to determine significant differences. Alternatively, a grid showing the differences between all sample means may be formatted and those differences that are identified as being significant highlighted, as illustrated in Example 8.4. This may be a useful approach where there are a large number of sample means to compare and the pattern of significant differences is rather complex. However, where the design is relatively simple, it may suffice simply to cite the actual value of the minimum critical difference derived from the multiple comparison test in the table or figure legend and identify in the text the key significant differences that arise. For example, following on from the cited text shown above, it might be stated that:

The wheat cultivar cv. Maris Butler when grown under Mn-deficient conditions displayed a significantly greater mean grain yield than the modern cultivar cv. Paragon (153.3 ± 5.0 vs. 91.1 ± 5.2; $\text{LSD}_{P = 0.05}$ = 42.5).

Reporting the results of linear regression analyses

Computer output for regression analyses are particularly detailed; see, for example, the output derived for a simple linear regression from the Excel Analysis Toolpak and GenStat packages illustrated in Figs 15.6 and 15.16 respectively. Therefore, when formally presenting the outcome of a regression analysis, we again need to consider what is the minimal information necessary to report in order to demonstrate that an appropriate analysis has been correctly applied. For a simple linear regression, clearly the main information required is the parameters of the fitted linear model (i.e. the estimated slope and intercept) and the significance of the regression. The P value for a regression can be determined either by an F-ratio or a t value, but there is no need to quote both, especially as in the case of a simple linear regression between two variables the F-ratio is equal to t^2. In addition, the coefficient of determination, r^2, will give a measure of the extent to which the variation in the responding variable is explained by the predictor variable. All this information can be conveniently presented within a text box on a graph or within a simple table if desired. Once a regression line has been fitted to data on a scatter plot, it is common practice to display the confidence interval around the regression line. The determination of confidence limits for a linear regression was explained and an example of their presentation shown in Chapter 12 (see section 12.3.5 and Example 12.3). Modern statistics software packages will produce plots showing the confidence limits around a regression line, as illustrated in the GenStat output in Fig. 15.16 and will even produce the confidence limits for a polynomial regression, as shown in Fig. 15.18.

Reporting the results of other types of analyses

A range of other analyses has been described in this book, including chi-squared tests and non-parametric tests for comparing non-normally distributed samples. In presenting the outcomes for all these, the same rules of thumb apply; sufficient evidence needs to be presented to demonstrate that a valid statistical analysis has been applied, to substantiate the conclusions reached and to give the reader opportunity to evaluate the data for themselves. In general, this will involve reporting at a minimum the value of the test statistic, the sample sizes (or degrees of freedom) and the P value. Which other statistics need reporting may well be decided more by the demands of the journal or book editor for whom one is writing or the thesis examiner whom one is trying to satisfy. Where possible, it is always as well to try and determine the level of statistical reporting required by a particular journal, publisher or examiner in advance of preparing a manuscript.

Note

[1]Excel is a trademark of the Microsoft group of companies.

Appendix: Statistical Tables

Table A1.1   Proportions of the normal distribution. Values are the proportions of the standard normal curve that lie beyond a given value of the normal standardized deviate z.

Table A1.2   Probabilities (P) associated with critical values for the standardized deviates (z) of a normal distribution. Use these values to calculate confidence limits for large single samples $(n > 30)$ and to perform the two-sample comparison z-test.

Table A2   Critical two-tailed Student's t values. Use these values to calculate confidence limits for single samples and to perform the two-sample and paired-sample Student's t-tests.

Table A3   Critical two-tailed U values. Use these values to perform the two-sample Mann–Whitney U test.

Table A4   Critical two-tailed S and T values. Use these values to perform the paired two-sample sign test and the paired two-sample Wilcoxon's signed rank test.

Table A5.1   Critical one-tailed F values ($\alpha = 0.05$). Use these values for performing analysis of variance tests at the 5% significance level.

Table A5.2   Critical one-tailed F values ($\alpha = 0.01$). Use these values for performing analysis of variance tests at the 1% significance level.

Table A5.3   Critical two-tailed F values ($\alpha = 0.05$). Use these values for performing the variance ratio F test for testing homogeneity of two sample variances at the 5% significance level prior to performing Student's t-test.

Table A6   Critical values of F_{max}. Use these values for testing the homogeneity of variances prior to performing ANOVA.

Table A7   Critical Tukey q values. Use these values for performing the Tukey HSD and Student–Newman–Keuls (SNK) multiple comparison tests following ANOVA.

Table A8.1   Critical two-tailed q' values for Dunnet's test. Use these values for performing the Dunnet multiple comparison two-tailed test.

Table A8.2   Critical one-tailed q' values for Dunnet's test. Use these values for performing the Dunnet multiple comparison one-tailed test.

Table A9   Critical H values. Use these values for performing the Kruskal–Wallis non-parametric multiple sample test.

Table A10   Critical Friedman χ_r^2 values. Use these values for performing Friedman's non-parametric multiple sample test with randomized blocks.

Table A11   Critical Q values. Use these values for performing a non-parametric multiple comparison test following the Kruskal–Wallis test.

Table A12   Critical values for Pearson's product-moment correlation coefficient, r. Use these values for testing the significance of a correlation determined by Pearson's product-moment analysis.

Table A13   Critical values for Spearman's rank correlation coefficient, r_s. Use these values for testing the significance of a correlation determined by Spearman's rank correlation analysis.

Table A14   Critical values of chi-squared, χ^2. Use these values for performing chi-squared goodness-of-fit tests and contingency table tests.

Table A15   Table of critical values of d_{max}. Use these values for performing the Kolmogorov–Smirnov goodness-of-fit test for category data.

Table A1.1. Proportions of the normal distribution. Values are the proportions of the standard normal curve that lie beyond a given value of the normal standardized deviate z, where $z = \dfrac{x - \mu}{\sigma}$.

z		+0.01	+0.02	+0.03	+0.04	+0.05	+0.06	+0.07	+0.08	+0.09
0.0	0.500	0.496	0.492	0.488	0.484	0.480	0.476	0.472	0.468	0.464
0.1	0.460	0.456	0.452	0.448	0.444	0.440	0.436	0.433	0.429	0.425
0.2	0.421	0.417	0.413	0.409	0.405	0.401	0.397	0.394	0.390	0.386
0.3	0.382	0.378	0.374	0.371	0.367	0.363	0.359	0.356	0.352	0.348
0.4	0.345	0.341	0.337	0.334	0.330	0.326	0.323	0.319	0.316	0.312
0.5	0.309	0.305	0.302	0.298	0.295	0.291	0.288	0.284	0.281	0.278
0.6	0.274	0.271	0.268	0.264	0.261	0.258	0.255	0.251	0.248	0.245
0.7	0.242	0.239	0.236	0.233	0.230	0.227	0.224	0.221	0.218	0.215
0.8	0.212	0.209	0.206	0.203	0.200	0.198	0.195	0.192	0.189	0.187
0.9	0.184	0.181	0.179	0.176	0.174	0.171	0.169	0.166	0.164	0.161
1.0	0.159	0.156	0.154	0.152	0.149	0.147	0.145	0.142	0.140	0.138
1.1	0.136	0.133	0.131	0.129	0.127	0.125	0.123	0.121	0.119	0.117
1.2	0.115	0.113	0.111	0.109	0.107	0.106	0.104	0.102	0.100	0.099
1.3	0.097	0.095	0.093	0.092	0.090	0.089	0.087	0.085	0.084	0.082
1.4	0.081	0.079	0.078	0.076	0.075	0.074	0.072	0.071	0.069	0.068
1.5	0.067	0.066	0.064	0.063	0.062	0.061	0.059	0.058	0.057	0.056
1.6	0.055	0.054	0.053	0.052	0.051	0.049	0.048	0.047	0.046	0.046
1.7	0.045	0.044	0.043	0.042	0.041	0.040	0.039	0.038	0.038	0.037
1.8	0.036	0.035	0.034	0.034	0.033	0.032	0.031	0.031	0.030	0.029
1.9	0.029	0.028	0.027	0.027	0.026	0.026	0.025	0.024	0.024	0.023
2.0	0.023	0.022	0.022	0.021	0.021	0.020	0.020	0.019	0.019	0.018
2.1	0.018	0.017	0.017	0.017	0.016	0.016	0.015	0.015	0.015	0.014
2.2	0.014	0.014	0.013	0.013	0.013	0.012	0.012	0.012	0.011	0.011
2.3	0.011	0.010	0.010	0.010	0.010	0.009	0.009	0.009	0.009	0.008
2.4	0.008	0.008	0.008	0.008	0.007	0.007	0.007	0.007	0.007	0.006
2.5	0.006	0.006	0.006	0.006	0.006	0.005	0.005	0.005	0.005	0.005
2.6	0.005	0.005	0.004	0.004	0.004	0.004	0.004	0.004	0.004	0.004
2.7	0.003	0.003	0.003	0.003	0.003	0.003	0.003	0.003	0.003	0.003
2.8	0.003	0.002	0.002	0.002	0.002	0.002	0.002	0.002	0.002	0.002
2.9	0.002	0.002	0.002	0.002	0.002	0.002	0.002	0.001	0.001	0.001
3.0	0.001	0.001	0.001	0.001	0.001	0.001	0.001	0.001	0.001	0.001

Table A1.2. Probabilities (P) associated with critical values for the standardized deviates (z) of a normal distribution. P is the probability of a randomly selected data item from a normal distributed population occurring either outside or inside the interval defined by the mean $\pm z$ standard deviations from the mean. Use these values to calculate confidence limits for large single samples ($n > 30$) and to perform the two-sample comparison z–test.

P (outside)	0.1	0.05	0.02	0.01	0.001	0.0001	0.00001
P (inside)	0.9	0.95	0.98	0.99	0.999	0.9999	0.99999
z	1.645	1.960	2.326	2.576	3.291	3.891	4.417

Table A2. Critical two-tailed Student's *t* values. Use these values to calculate confidence limits for single samples ($n \leq 30$) and to perform two-sample and paired-sample Student's *t*-tests. (For one-tailed Student's *t*-tests consult the table at twice the critical probability required.)

DF	Probability (α)					
	0.1	0.05	0.02	0.01	0.002	0.001
1	6.314	12.706	31.821	63.656	318.289	636.578
2	2.920	4.303	6.965	9.925	22.328	31.600
3	2.353	3.182	4.541	5.841	10.214	12.924
4	2.132	2.776	3.747	4.604	7.173	8.610
5	2.015	2.571	3.365	4.032	5.894	6.869
6	1.943	2.447	3.143	3.707	5.208	5.959
7	1.895	2.365	2.998	3.499	4.785	5.408
8	1.860	2.306	2.896	3.355	4.501	5.041
9	1.833	2.262	2.821	3.250	4.297	4.781
10	1.812	2.228	2.764	3.169	4.144	4.587
11	1.796	2.201	2.718	3.106	4.025	4.437
12	1.782	2.179	2.681	3.055	3.930	4.318
13	1.771	2.160	2.650	3.012	3.852	4.221
14	1.761	2.145	2.624	2.977	3.787	4.140
15	1.753	2.131	2.602	2.947	3.733	4.073
16	1.746	2.120	2.583	2.921	3.686	4.015
17	1.740	2.110	2.567	2.898	3.646	3.965
18	1.734	2.101	2.552	2.878	3.610	3.922
19	1.729	2.093	2.539	2.861	3.579	3.883
20	1.725	2.086	2.528	2.845	3.552	3.850
21	1.721	2.080	2.518	2.831	3.527	3.819
22	1.717	2.074	2.508	2.819	3.505	3.792
23	1.714	2.069	2.500	2.807	3.485	3.768
24	1.711	2.064	2.492	2.797	3.467	3.745
25	1.708	2.060	2.485	2.787	3.450	3.725
26	1.706	2.056	2.479	2.779	3.435	3.707
27	1.703	2.052	2.473	2.771	3.421	3.689
28	1.701	2.048	2.467	2.763	3.408	3.674
29	1.699	2.045	2.462	2.756	3.396	3.660
30	1.697	2.042	2.457	2.750	3.385	3.646
35	1.690	2.030	2.438	2.724	3.340	3.591
40	1.684	2.021	2.423	2.704	3.307	3.551
45	1.679	2.014	2.412	2.690	3.281	3.520
50	1.676	2.009	2.403	2.678	3.261	3.496
60	1.671	2.000	2.390	2.660	3.232	3.460
70	1.667	1.994	2.381	2.648	3.211	3.435
80	1.664	1.990	2.374	2.639	3.195	3.416
90	1.662	1.987	2.368	2.632	3.183	3.402
100	1.660	1.984	2.364	2.626	3.174	3.390
110	1.659	1.982	2.361	2.621	3.166	3.381
120	1.658	1.980	2.358	2.617	3.160	3.373
∞	1.645	1.960	2.326	2.576	3.090	3.291

Note that the last row of the table (DF = ∞) gives the values for the distribution of the standardized normal deviate *z*.

Table A3. Critical two-tailed U values. Use these values to perform the two-sample Mann–Whitney U test. (For one-tailed Mann–Whitney test, consult table at twice the critical probability required.)

n_b	$n_a = 1$ $\alpha=$ 0.1	0.05	0.02	0.01	$n_a = 3$ $\alpha=$ 0.1	0.05	0.02	0.01	$n_a = 5$ $\alpha=$ 0.1	0.05	0.02	0.01
1												
2												
3					9							
4					12							
5					14	15			21	23	24	25
6					16	17			25	27	28	29
7					19	20	21		29	30	32	34
8					21	22	24		32	34	36	38
9					23	25	26	27	36	38	40	42
10					26	27	29	30	39	42	44	46
11					28	30	32	33	43	46	48	50
12					31	32	34	35	47	49	52	54
13					33	35	37	38	50	53	56	58
14					35	37	40	41	54	57	60	63
15					38	40	42	43	57	61	64	67
16					40	42	45	46	61	65	68	71
17					42	45	47	49	65	68	72	75
18					45	47	50	52	68	72	76	79
19	19				47	50	53	54	72	75	80	83
20	20				49	52	55	57	75	80	84	87
21	21				52	55	58	60	79	83	88	91
22	22				54	57	60	62	82	87	92	96
23	23				56	60	63	65	86	91	96	100
24	24				59	62	66	68	90	95	100	104
25	25				61	65	68	70	93	98	104	108
26	26				63	67	71	73	97	102	108	112
27	27				66	70	74	76	100	106	112	116
28	28				68	72	76	79	104	110	116	120
29	29				70	74	79	81	107	113	120	124
30	30				73	77	81	84	111	117	124	128
31	31				75	79	84	87	115	121	128	133
32	32				77	82	87	89	118	125	132	137
33	33				80	84	89	92	122	128	136	141
34	34				82	87	92	95	125	132	140	145
35	35				84	89	94	97	129	136	144	149
36	36				87	92	97	100	132	140	148	153
37	37				89	94	100	103	136	144	152	157
38	38				91	97	102	105	140	147	156	161
39	39	39			94	99	105	108	143	151	160	165
40	40	40			96	102	107	111	147	155	164	169

$n_a = 2$

n_b	$\alpha = 0.1$	0.05	0.02	0.01
2				
3				
4				
5	10			
6	12			
7	14			
8	15	16		
9	17	18		
10	19	20		
11	21	22		
12	22	23		
13	24	25	26	
14	25	27	28	
15	27	29	30	
16	29	31	32	
17	31	32	34	
18	32	34	36	
19	34	36	37	38
20	36	38	39	40
21	37	39	41	42
22	39	41	43	44
23	41	43	45	46
24	42	45	47	48
25	44	47	49	50
26	46	48	51	52
27	47	50	52	53
28	49	52	54	55
29	51	54	56	57
30	53	55	58	59
31	54	57	60	61
32	56	59	62	63
33	58	61	64	65
34	59	63	65	67
35	61	64	67	69
36	63	66	69	71
37	64	68	71	73
38	66	70	73	75
39	68	71	75	76
40	69	73	77	78

$n_a = 4$

n_b	$\alpha = 0.1$	0.05	0.02	0.01
4	15	16		
5	18	19	20	
6	21	22	23	24
7	24	25	27	28
8	27	28	30	31
9	30	32	33	35
10	33	35	37	38
11	36	38	40	42
12	39	41	43	45
13	42	44	47	49
14	45	47	50	52
15	48	50	53	55
16	50	53	57	59
17	53	57	60	62
18	56	60	63	66
19	59	63	67	69
20	62	66	70	72
21	65	69	73	76
22	68	72	77	79
23	71	75	80	83
24	74	79	83	86
25	77	82	87	90
26	80	85	90	93
27	83	88	93	96
28	86	91	96	100
29	89	94	100	103
30	92	97	103	107
31	95	100	106	110
32	98	104	110	114
33	101	107	113	117
34	104	111	116	120
35	107	113	120	124
36	110	116	123	127
37	113	119	126	131
38	116	122	130	134
39	118	125	133	137
40	121	129	136	141

$n_a = 6$

n_b	$\alpha = 0.1$	0.05	0.02	0.01
6	29	31	33	
7	34	36	38	
8	38	40	42	48
9	42	44	47	53
10	46	49	52	58
11	50	53	57	64
12	55	58	61	69
13	59	62	66	74
14	63	67	71	79
15	67	71	75	85
16	71	75	80	90
17	76	80	84	95
18	80	84	89	100
19	84	89	94	106
20	88	93	98	111
21	92	97	103	116
22	96	102	108	121
23	101	106	112	126
24	105	111	117	132
25	109	115	121	137
26	113	119	126	142
27	117	124	131	147
28	122	128	135	152
29	126	132	140	157
30	130	137	145	163
31	134	141	149	168
32	138	146	154	173
33	142	150	158	178
34	147	154	163	183
35	151	159	168	188
36	155	163	172	194
37	159	167	177	199
38	163	172	182	204
39	167	176	186	209
40	172	181	191	214

continued

Table A3. Continued.

$n_a = 7$

n_b	$\alpha = 0.1$	0.05	0.02	0.01
7	38	41	43	45
8	43	46	49	50
9	48	51	54	56
10	53	56	59	61
11	58	61	65	67
12	63	66	70	72
13	67	71	75	78
14	72	76	81	83
15	77	81	86	89
16	82	86	91	94
17	86	91	96	100
18	91	96	102	105
19	96	101	107	111
20	101	106	112	116
21	106	111	117	122
22	110	116	123	127
23	115	121	128	132
24	120	126	133	138
25	125	131	139	143
26	129	136	144	149
27	134	141	149	154
28	139	146	154	160
29	144	151	160	165
30	149	156	165	170
31	153	161	170	176
32	158	166	175	181
33	163	171	181	187
34	168	176	186	192
35	172	181	191	198
36	177	186	196	203
37	182	191	202	208
38	187	196	207	214
39	191	201	212	219
40	196	206	217	225

$n_a = 9$

n_b	$\alpha = 0.1$	0.05	0.02	0.01
9	60	64	67	70
10	66	70	74	77
11	72	76	81	83
12	78	82	87	90
13	84	89	94	97
14	90	95	100	104
15	96	101	107	111
16	102	107	113	117
17	108	114	120	124
18	114	120	126	131
19	120	126	133	138
20	126	132	140	144
21	132	139	146	151
22	138	145	153	158
23	144	151	159	164
24	150	157	166	171
25	156	163	172	178
26	162	170	179	185
27	168	176	185	191
28	174	182	192	198
29	179	188	198	205
30	185	194	205	212
31	191	201	211	218
32	197	207	218	225
33	203	213	224	232
34	209	219	231	238
35	215	226	237	245
36	221	232	243	252
37	227	238	250	258
38	233	244	257	265
39	239	250	263	272
40	245	257	270	279

$n_a = 11$

n_b	$\alpha = 0.1$	0.05	0.02	0.01
11	87	91	96	109
12	94	99	104	117
13	101	106	112	126
14	108	114	120	135
15	115	121	128	144
16	122	129	135	152
17	130	136	143	161
18	137	143	151	170
19	144	151	159	178
20	151	158	167	187
21	158	166	174	196
22	165	173	182	204
23	172	180	190	213
24	179	188	198	222
25	186	195	205	230
26	194	203	213	239
27	201	210	221	247
28	208	218	229	256
29	215	225	236	265
30	222	232	244	273
31	229	240	252	282
32	236	247	260	290
33	243	255	267	299
34	250	262	275	307
35	257	269	283	316
36	265	277	290	325
37	272	284	298	333
38	279	291	306	342
39	286	299	314	350
40	293	307	321	359

$n_a = 8$

n_b	$\alpha =$ 0.1	0.05	0.02	0.01
8	49	51	55	57
9	54	57	61	63
10	60	63	67	69
11	65	69	73	75
12	70	74	79	81
13	76	80	84	87
14	81	86	90	94
15	87	91	96	100
16	92	97	102	106
17	97	102	108	112
18	103	108	114	118
19	108	114	120	124
20	113	119	126	130
21	119	125	132	136
22	124	131	138	142
23	130	136	144	149
24	135	142	150	155
25	140	147	155	161
26	146	153	161	167
27	151	159	167	173
28	156	164	173	179
29	162	170	179	185
30	167	175	185	191
31	172	181	191	197
32	178	187	197	203
33	183	192	203	209
34	188	198	208	215
35	194	203	214	221
36	199	209	220	228
37	205	215	226	234
38	210	220	232	240
39	215	226	238	246
40	221	231	244	252

$n_a = 10$

n_b	$\alpha =$ 0.1	0.05	0.02	0.01
10	73	77	81	84
11	79	84	88	92
12	86	91	96	99
13	93	97	103	106
14	99	104	110	114
15	106	111	117	121
16	112	118	124	129
17	119	125	132	136
18	125	132	139	143
19	132	138	146	151
20	138	145	153	158
21	145	152	160	166
22	152	159	167	173
23	158	166	175	180
24	165	173	182	188
25	171	179	189	195
26	178	186	196	202
27	184	193	203	210
28	191	200	210	217
29	197	207	217	224
30	204	213	224	232
31	210	220	232	239
32	217	227	239	246
33	223	234	246	254
34	230	241	253	261
35	236	247	260	268
36	243	254	267	276
37	249	261	274	283
38	256	268	281	290
39	262	275	289	298
40	269	281	296	305

$n_a = 12$

n_b	$\alpha =$ 0.1	0.05	0.02	0.01
12	102	107	113	117
13	109	115	121	125
14	117	123	130	134
15	125	131	138	143
16	132	139	146	151
17	140	147	155	160
18	148	155	163	169
19	156	163	172	177
20	163	171	180	186
21	171	179	188	194
22	179	187	197	203
23	186	195	205	212
24	194	203	213	220
25	202	211	222	229
26	209	219	230	238
27	217	227	239	246
28	225	235	247	255
29	232	243	255	263
30	240	251	264	272
31	248	259	272	280
32	256	267	280	289
33	263	275	289	298
34	271	283	297	306
35	279	291	305	315
36	286	299	314	323
37	294	307	322	332
38	302	315	330	340
39	309	323	339	349
40	317	331	347	358

continued

Table A3. Continued.

$n_a = 13$

n_b	$\alpha = 0.1$	0.05	0.02	0.01
13	118	124	130	135
14	126	132	139	144
15	134	141	148	153
16	143	149	157	163
17	151	158	166	172
18	159	167	175	181
19	167	175	184	190
20	176	184	193	200
21	184	193	202	209
22	192	201	211	218
23	201	210	220	227
24	209	218	229	237
25	217	227	238	246
26	225	236	247	255
27	234	244	256	264
28	242	253	265	273
29	250	261	274	283
30	258	270	283	292
31	267	278	292	301
32	275	287	301	310
33	283	296	310	319
34	291	304	319	328
35	299	313	328	338
36	308	321	337	347
37	316	330	346	356
38	324	338	355	365
39	332	347	363	374
40	341	355	372	384

$n_a = 15$

n_b	$\alpha = 0.1$	0.05	0.02	0.01
15	153	161	169	174
16	163	170	179	185
17	172	180	189	195
18	182	190	200	206
19	191	200	210	216
20	200	210	220	227
21	210	219	230	237
22	219	229	240	248
23	229	239	251	258
24	238	249	261	269
25	247	258	271	279
26	257	268	281	290
27	266	278	291	300
28	276	288	301	311
29	285	297	312	321
30	294	307	322	331
31	304	317	332	342
32	313	327	342	352
33	323	336	352	363
34	332	346	362	373
35	341	356	372	383
36	351	366	382	394
37	360	375	393	404
38	369	385	403	415
39	379	395	413	425
40	388	404	423	435

$n_a = 17$

n_b	$\alpha = 0.1$	.05	0.02	0.01
17	193	202	212	219
18	204	213	224	231
19	214	224	235	242
20	225	235	247	254
21	236	246	258	266
22	246	257	269	278
23	257	268	281	289
24	267	279	292	301
25	278	290	303	313
26	288	301	315	324
27	299	312	326	336
28	309	323	337	348
29	320	333	349	359
30	330	344	360	371
31	341	355	371	382
32	351	366	383	394
33	362	377	394	406
34	372	388	405	417
35	383	399	417	429
36	393	410	428	440
37	404	420	439	452
38	414	431	451	464
39	425	442	462	475
40	435	453	473	487

$n_a = 14$

n_b	$\alpha = 0.1$	0.05	0.02	0.01
14	135	141	149	154
15	144	151	159	164
16	153	160	168	174
17	161	169	178	184
18	170	178	187	194
19	179	188	197	203
20	188	197	207	213
21	197	206	216	223
22	206	215	226	233
23	215	224	235	243
24	223	233	245	253
25	232	243	255	263
26	241	252	264	272
27	250	261	274	282
28	259	270	283	292
29	268	279	293	302
30	276	289	302	312
31	285	298	312	321
32	294	307	321	331
33	303	316	331	341
34	312	325	341	351
35	320	334	350	361
36	329	343	360	370
37	338	353	369	380
38	347	362	379	390
39	356	371	388	400
40	364	380	398	410

$n_a = 16$

n_b	$\alpha = 0.1$	0.05	0.02	0.01
16	173	181	190	196
17	183	191	201	207
18	193	202	212	218
19	203	212	222	230
20	213	222	233	241
21	223	233	244	252
22	233	243	255	263
23	243	253	266	274
24	253	264	276	285
25	263	274	287	296
26	273	284	298	307
27	283	295	309	318
28	292	305	319	329
29	302	315	330	340
30	312	326	341	351
31	322	336	352	362
32	332	346	362	373
33	342	357	373	384
34	352	367	384	395
35	362	377	395	406
36	372	388	405	417
37	382	398	416	428
38	392	408	427	439
39	402	418	437	450
40	412	429	448	461

$n_a = 18$

n_b	$\alpha = 0.1$	0.05	0.02	0.01
18	215	225	236	243
19	226	236	248	255
20	237	248	260	268
21	248	259	272	280
22	260	271	284	292
23	271	282	296	305
24	282	294	308	317
25	293	305	320	329
26	304	317	332	341
27	315	328	344	354
28	326	340	355	366
29	337	351	367	378
30	348	363	379	390
31	359	374	391	403
32	370	386	403	415
33	382	397	415	427
34	393	409	427	439
35	404	420	439	451
36	415	432	451	464
37	426	443	463	476
38	437	456	475	488
39	448	466	486	500
40	459	477	498	512

Table A3. Continued.

$n_a = 19$

n_b	$\alpha =$ 0.1	0.05	0.02	0.01
19	238	248	260	268
20	250	261	273	281
21	261	273	286	294
22	273	285	298	307
23	285	297	311	320
24	296	309	323	333
25	308	321	336	346
26	320	333	348	359
27	331	345	361	371
28	343	357	373	384
29	355	369	386	397
30	366	381	398	410
31	378	393	411	423
32	390	405	423	436
33	401	417	436	448
34	413	429	448	461
35	425	441	461	474
36	436	453	473	487
37	448	465	486	500
38	459	477	498	512
39	471	489	511	525
40	482	502	523	538

$n_a = 20$

n_b	$\alpha =$ 0.1	0.05	0.02	0.01
20	262	273	286	295
21	274	286	299	308
22	286	299	313	322
23	299	311	326	335
24	311	324	339	349
25	323	337	352	362
26	335	349	365	376
27	348	362	378	389
28	360	374	391	403
29	372	387	404	416
30	384	400	418	430
31	396	413	431	443
32	409	425	444	456
33	421	438	457	470
34	433	450	470	483
35	445	463	483	497
36	457	475	496	510
37	469	488	509	523
38	482	501	522	537
39		513	535	550
40		526	548	563

Table A4. Critical two-tailed S and T values. Use S values for performing the sign test for paired samples. Use T values for performing Wilcoxon's signed rank test for paired samples.

n	Critical S values Probability (α)				Critical T values Probability (α)			
	0.1	0.05	0.02	0.01	0.1	0.05	0.02	0.01
5	5				15			
6	6	6			19	21		
7	7	7	7		25	26	28	
8	7	8	8	8	31	33	35	36
9	8	9	9	9	37	40	42	44
10	9	9	10	10	45	47	50	52
11	9	10	10	11	53	56	59	61
12	10	10	11	11	61	65	69	71
13	10	11	12	12	70	74	79	82
14	11	12	12	13	80	84	90	93
15	12	12	13	13	90	95	101	105
16	12	13	14	14	101	107	113	117
17	13	13	14	15	112	119	126	130
18	13	14	15	15	124	131	139	144
19	14	15	15	16	137	144	153	158
20	15	15	16	17	150	158	167	173
21	15	16	17	17	164	173	182	189
22	16	17	17	18	178	188	198	205
23	16	17	18	19	193	203	214	222
24	17	18	19	19	209	219	231	239
25	18	18	19	20	225	236	249	257
26	18	19	20	20	241	253	267	276
27	19	20	20	21	259	271	286	295
28	19	20	21	22	276	290	305	315
29	20	21	22	22	295	309	325	335
30	20	21	22	23	314	328	345	356
31	21	22	23	24	333	349	366	378
32	22	23	24	24	353	369	388	400
33	22	23	24	25	374	391	410	423
34	23	24	25	25	395	413	433	447
35	23	24	25	26	417	435	457	471
36	24	25	26	27	439	458	481	495
37	24	25	27	27	462	482	505	521
38	25	26	27	28	485	506	530	547
39	26	27	28	28	509	531	556	573
40	26	27	28	29	534	556	582	600
45	29	30	31	32	664	692	723	744
50	32	33	34	35	809	841	878	902

The null hypothesis is rejected when observed value of S or T is equal to or greater than the critical value. (Note that some published versions of these tables and computer packages require the observed value to be smaller than the given critical value for rejection of H_0.)

Table A5.1. One-tailed critical F values ($\alpha = 0.05$). Use these values for performing analysis of variance tests at the 5% significance level.

V_2 \ V_1	1	2	3	4	5	6	7	8	9	10	12	15	20	25	30	35	40	45	50	60	70	80	90	100	120	∞
1	161.45	199.50	215.71	224.58	230.16	233.99	236.77	238.88	240.54	241.88	243.90	245.95	248.02	249.26	250.10	250.69	251.14	251.49	251.77	252.20	252.50	252.72	252.90	253.04	253.25	254.32
2	18.51	19.00	19.16	19.25	19.30	19.33	19.35	19.37	19.38	19.40	19.41	19.43	19.45	19.46	19.46	19.47	19.47	19.47	19.48	19.48	19.48	19.48	19.48	19.49	19.49	19.50
3	10.13	9.55	9.28	9.12	9.01	8.94	8.89	8.85	8.81	8.79	8.74	8.70	8.66	8.63	8.62	8.60	8.59	8.59	8.58	8.57	8.57	8.56	8.56	8.55	8.55	8.53
4	7.71	6.94	6.59	6.39	6.26	6.16	6.09	6.04	6.00	5.96	5.91	5.86	5.80	5.77	5.75	5.73	5.72	5.71	5.70	5.69	5.68	5.67	5.67	5.66	5.66	5.63
5	6.61	5.79	5.41	5.19	5.05	4.95	4.88	4.82	4.77	4.74	4.68	4.62	4.56	4.52	4.50	4.48	4.46	4.45	4.44	4.43	4.42	4.41	4.41	4.41	4.40	4.37
6	5.99	5.14	4.76	4.53	4.39	4.28	4.21	4.15	4.10	4.06	4.00	3.94	3.87	3.83	3.81	3.79	3.77	3.76	3.75	3.74	3.73	3.72	3.72	3.71	3.70	3.67
7	5.59	4.74	4.35	4.12	3.97	3.87	3.79	3.73	3.68	3.64	3.57	3.51	3.44	3.40	3.38	3.36	3.34	3.33	3.32	3.30	3.29	3.29	3.28	3.27	3.27	3.23
8	5.32	4.46	4.07	3.84	3.69	3.58	3.50	3.44	3.39	3.35	3.28	3.22	3.15	3.11	3.08	3.06	3.04	3.03	3.02	3.01	2.99	2.99	2.98	2.97	2.97	2.93
9	5.12	4.26	3.86	3.63	3.48	3.37	3.29	3.23	3.18	3.14	3.07	3.01	2.94	2.89	2.86	2.84	2.83	2.81	2.80	2.79	2.78	2.77	2.76	2.76	2.75	2.71
10	4.96	4.10	3.71	3.48	3.33	3.22	3.14	3.07	3.02	2.98	2.91	2.85	2.77	2.73	2.70	2.68	2.66	2.65	2.64	2.62	2.61	2.60	2.59	2.59	2.58	2.54
11	4.84	3.98	3.59	3.36	3.20	3.09	3.01	2.95	2.90	2.85	2.79	2.72	2.65	2.60	2.57	2.55	2.53	2.52	2.51	2.49	2.48	2.47	2.46	2.46	2.45	2.40
12	4.75	3.89	3.49	3.26	3.11	3.00	2.91	2.85	2.80	2.75	2.69	2.62	2.54	2.50	2.47	2.44	2.43	2.41	2.40	2.38	2.37	2.36	2.36	2.35	2.34	2.30
13	4.67	3.81	3.41	3.18	3.03	2.92	2.83	2.77	2.71	2.67	2.60	2.53	2.46	2.41	2.38	2.36	2.34	2.33	2.31	2.30	2.28	2.27	2.27	2.26	2.25	2.21
14	4.60	3.74	3.34	3.11	2.96	2.85	2.76	2.70	2.65	2.60	2.53	2.46	2.39	2.34	2.31	2.28	2.27	2.25	2.24	2.22	2.21	2.20	2.19	2.19	2.18	2.13
15	4.54	3.68	3.29	3.06	2.90	2.79	2.71	2.64	2.59	2.54	2.48	2.40	2.33	2.28	2.25	2.22	2.20	2.19	2.18	2.16	2.15	2.14	2.13	2.12	2.11	2.07
16	4.49	3.63	3.24	3.01	2.85	2.74	2.66	2.59	2.54	2.49	2.42	2.35	2.28	2.23	2.19	2.17	2.15	2.14	2.12	2.11	2.09	2.08	2.07	2.07	2.06	2.01
17	4.45	3.59	3.20	2.96	2.81	2.70	2.61	2.55	2.49	2.45	2.38	2.31	2.23	2.18	2.15	2.12	2.10	2.09	2.08	2.06	2.05	2.03	2.03	2.02	2.01	1.96
18	4.41	3.55	3.16	2.93	2.77	2.66	2.58	2.51	2.46	2.41	2.34	2.27	2.19	2.14	2.11	2.08	2.06	2.05	2.04	2.02	2.00	1.99	1.98	1.98	1.97	1.92
19	4.38	3.52	3.13	2.90	2.74	2.63	2.54	2.48	2.42	2.38	2.31	2.23	2.16	2.11	2.07	2.05	2.03	2.01	2.00	1.98	1.97	1.96	1.95	1.94	1.93	1.88
20	4.35	3.49	3.10	2.87	2.71	2.60	2.51	2.45	2.39	2.35	2.28	2.20	2.12	2.07	2.04	2.01	1.99	1.98	1.97	1.95	1.93	1.92	1.91	1.91	1.90	1.84
21	4.32	3.47	3.07	2.84	2.68	2.57	2.49	2.42	2.37	2.32	2.25	2.18	2.10	2.05	2.01	1.98	1.96	1.95	1.94	1.92	1.90	1.89	1.88	1.88	1.87	1.81
22	4.30	3.44	3.05	2.82	2.66	2.55	2.46	2.40	2.34	2.30	2.23	2.15	2.07	2.02	1.98	1.96	1.94	1.92	1.91	1.89	1.88	1.86	1.86	1.85	1.84	1.78
23	4.28	3.42	3.03	2.80	2.64	2.53	2.44	2.37	2.32	2.27	2.20	2.13	2.05	2.00	1.96	1.93	1.91	1.90	1.88	1.86	1.85	1.84	1.83	1.82	1.81	1.76
24	4.26	3.40	3.01	2.78	2.62	2.51	2.42	2.36	2.30	2.25	2.18	2.11	2.03	1.97	1.94	1.91	1.89	1.88	1.86	1.84	1.83	1.82	1.81	1.80	1.79	1.73
25	4.24	3.39	2.99	2.76	2.60	2.49	2.40	2.34	2.28	2.24	2.16	2.09	2.01	1.96	1.92	1.89	1.87	1.86	1.84	1.82	1.81	1.80	1.79	1.78	1.77	1.71
26	4.23	3.37	2.98	2.74	2.59	2.47	2.39	2.32	2.27	2.22	2.15	2.07	1.99	1.94	1.90	1.87	1.85	1.84	1.82	1.80	1.79	1.78	1.77	1.76	1.75	1.69
27	4.21	3.35	2.96	2.73	2.57	2.46	2.37	2.31	2.25	2.20	2.13	2.06	1.97	1.92	1.88	1.86	1.84	1.82	1.81	1.79	1.77	1.76	1.75	1.74	1.73	1.67
28	4.20	3.34	2.95	2.71	2.56	2.45	2.36	2.29	2.24	2.19	2.12	2.04	1.96	1.91	1.87	1.84	1.82	1.80	1.79	1.77	1.75	1.74	1.73	1.73	1.71	1.65
29	4.18	3.33	2.93	2.70	2.55	2.43	2.35	2.28	2.22	2.18	2.10	2.03	1.94	1.89	1.85	1.83	1.81	1.79	1.77	1.75	1.74	1.73	1.72	1.71	1.70	1.64
30	4.17	3.32	2.92	2.69	2.53	2.42	2.33	2.27	2.21	2.16	2.09	2.01	1.93	1.88	1.84	1.81	1.79	1.77	1.76	1.74	1.72	1.71	1.70	1.70	1.68	1.62
40	4.08	3.23	2.84	2.61	2.45	2.34	2.25	2.18	2.12	2.08	2.00	1.92	1.84	1.78	1.74	1.72	1.69	1.67	1.66	1.64	1.62	1.61	1.60	1.59	1.58	1.51
50	4.03	3.18	2.79	2.56	2.40	2.29	2.20	2.13	2.07	2.03	1.95	1.87	1.78	1.73	1.69	1.66	1.63	1.61	1.60	1.58	1.56	1.54	1.53	1.52	1.51	1.44
60	4.00	3.15	2.76	2.53	2.37	2.25	2.17	2.10	2.04	1.99	1.92	1.84	1.75	1.69	1.65	1.62	1.59	1.57	1.56	1.53	1.52	1.50	1.49	1.48	1.47	1.39
80	3.96	3.11	2.72	2.49	2.33	2.21	2.13	2.06	2.00	1.95	1.88	1.79	1.70	1.64	1.60	1.57	1.54	1.52	1.51	1.48	1.46	1.45	1.44	1.43	1.41	1.32
100	3.94	3.09	2.70	2.46	2.31	2.19	2.10	2.03	1.97	1.93	1.85	1.77	1.68	1.62	1.57	1.54	1.52	1.49	1.48	1.45	1.43	1.41	1.40	1.39	1.38	1.28
120	3.92	3.07	2.68	2.45	2.29	2.18	2.09	2.02	1.96	1.91	1.83	1.75	1.66	1.60	1.55	1.52	1.50	1.47	1.46	1.43	1.41	1.39	1.38	1.37	1.35	1.25
∞	3.84	3.00	2.60	2.37	2.21	2.10	2.01	1.94	1.88	1.83	1.75	1.67	1.57	1.51	1.46	1.42	1.39	1.37	1.35	1.32	1.29	1.27	1.26	1.24	1.22	1.00

For ANOVA V_1 and V_2 are the treatment and residual degrees of freedom respectively.

Table A5.2. One-tailed critical F values ($\alpha = 0.01$). Use these values for performing analysis of variance tests at the 1% significance level.

V_1	1	2	3	4	5	6	7	8	9	10	12	15	20	25	30	35	40	45	50	60	70	80	90	100	120	∞
V_2																										
1	4052	4999	5404	5624	5764	5859	5928	5981	6022	6056	6107	6157	6209	6240	6260	6275	6286	6296	6302	6313	6321	6326	6331	6334	6340	6366
2	98.50	99.00	99.16	99.25	99.30	99.33	99.36	99.38	99.39	99.40	99.42	99.43	99.45	99.46	99.47	99.47	99.48	99.48	99.48	99.48	99.48	99.48	99.49	99.49	99.49	99.50
3	34.12	30.82	29.46	28.71	28.24	27.91	27.67	27.49	27.34	27.23	27.05	26.87	26.69	26.58	26.50	26.45	26.41	26.38	26.35	26.32	26.29	26.27	26.25	26.24	26.22	26.13
4	21.20	18.00	16.69	15.98	15.52	15.21	14.98	14.80	14.66	14.55	14.37	14.20	14.02	13.91	13.84	13.79	13.75	13.71	13.69	13.65	13.63	13.61	13.59	13.58	13.56	13.46
5	16.26	13.27	12.06	11.39	10.97	10.67	10.46	10.29	10.16	10.05	9.89	9.72	9.55	9.45	9.38	9.33	9.29	9.26	9.24	9.20	9.18	9.16	9.14	9.13	9.11	9.02
6	13.75	10.92	9.78	9.15	8.75	8.47	8.26	8.10	7.98	7.87	7.72	7.56	7.40	7.30	7.23	7.18	7.14	7.11	7.09	7.06	7.03	7.01	7.00	6.99	6.97	6.88
7	12.25	9.55	8.45	7.85	7.46	7.19	6.99	6.84	6.72	6.62	6.47	6.31	6.16	6.06	5.99	5.94	5.91	5.88	5.86	5.82	5.80	5.78	5.77	5.75	5.74	5.65
8	11.26	8.65	7.59	7.01	6.63	6.37	6.18	6.03	5.91	5.81	5.67	5.52	5.36	5.26	5.20	5.15	5.12	5.09	5.07	5.03	5.01	4.99	4.97	4.96	4.95	4.86
9	10.56	8.02	6.99	6.42	6.06	5.80	5.61	5.47	5.35	5.26	5.11	4.96	4.81	4.71	4.65	4.60	4.57	4.54	4.52	4.48	4.46	4.44	4.43	4.41	4.40	4.31
10	10.04	7.56	6.55	5.99	5.64	5.39	5.20	5.06	4.94	4.85	4.71	4.56	4.41	4.31	4.25	4.20	4.17	4.14	4.12	4.08	4.06	4.04	4.03	4.01	4.00	3.91
11	9.65	7.21	6.22	5.67	5.32	5.07	4.89	4.74	4.63	4.54	4.40	4.25	4.10	4.01	3.94	3.89	3.86	3.83	3.81	3.78	3.75	3.73	3.72	3.71	3.69	3.60
12	9.33	6.93	5.95	5.41	5.06	4.82	4.64	4.50	4.39	4.30	4.16	4.01	3.86	3.76	3.70	3.65	3.62	3.59	3.57	3.54	3.51	3.49	3.48	3.47	3.45	3.36
13	9.07	6.70	5.74	5.21	4.86	4.62	4.44	4.30	4.19	4.10	3.96	3.82	3.66	3.57	3.51	3.46	3.43	3.40	3.38	3.34	3.32	3.30	3.28	3.27	3.25	3.17
14	8.86	6.51	5.56	5.04	4.69	4.46	4.28	4.14	4.03	3.94	3.80	3.66	3.51	3.41	3.35	3.30	3.27	3.24	3.22	3.18	3.16	3.14	3.12	3.11	3.09	3.00
15	8.68	6.36	5.42	4.89	4.56	4.32	4.14	4.00	3.89	3.80	3.67	3.52	3.37	3.28	3.21	3.17	3.13	3.10	3.08	3.05	3.02	3.00	2.99	2.98	2.96	2.87
16	8.53	6.23	5.29	4.77	4.44	4.20	4.03	3.89	3.78	3.69	3.55	3.41	3.26	3.16	3.10	3.05	3.02	2.99	2.97	2.93	2.91	2.89	2.87	2.86	2.84	2.75
17	8.40	6.11	5.19	4.67	4.34	4.10	3.93	3.79	3.68	3.59	3.46	3.31	3.16	3.07	3.00	2.96	2.92	2.89	2.87	2.83	2.81	2.79	2.78	2.76	2.75	2.65
18	8.29	6.01	5.09	4.58	4.25	4.01	3.84	3.71	3.60	3.51	3.37	3.23	3.08	2.98	2.92	2.87	2.84	2.81	2.78	2.75	2.72	2.70	2.69	2.68	2.66	2.57
19	8.18	5.93	5.01	4.50	4.17	3.94	3.77	3.63	3.52	3.43	3.30	3.15	3.00	2.91	2.84	2.80	2.76	2.73	2.71	2.67	2.65	2.63	2.61	2.60	2.58	2.49
20	8.10	5.85	4.94	4.43	4.10	3.87	3.70	3.56	3.46	3.37	3.23	3.09	2.94	2.84	2.78	2.73	2.69	2.67	2.64	2.61	2.58	2.56	2.55	2.54	2.52	2.42
21	8.02	5.78	4.87	4.37	4.04	3.81	3.64	3.51	3.40	3.31	3.17	3.03	2.88	2.79	2.72	2.67	2.64	2.61	2.58	2.55	2.52	2.50	2.49	2.48	2.46	2.36
22	7.95	5.72	4.82	4.31	3.99	3.76	3.59	3.45	3.35	3.26	3.12	2.98	2.83	2.73	2.67	2.62	2.58	2.55	2.53	2.50	2.47	2.45	2.43	2.42	2.40	2.31
23	7.88	5.66	4.76	4.26	3.94	3.71	3.54	3.41	3.30	3.21	3.07	2.93	2.78	2.69	2.62	2.57	2.54	2.51	2.48	2.45	2.42	2.40	2.39	2.37	2.35	2.26
24	7.82	5.61	4.72	4.22	3.90	3.67	3.50	3.36	3.26	3.17	3.03	2.89	2.74	2.64	2.58	2.53	2.49	2.46	2.44	2.40	2.38	2.36	2.34	2.33	2.31	2.21
25	7.77	5.57	4.68	4.18	3.85	3.63	3.46	3.32	3.22	3.13	2.99	2.85	2.70	2.60	2.54	2.49	2.45	2.42	2.40	2.36	2.34	2.32	2.30	2.29	2.27	2.17
26	7.72	5.53	4.64	4.14	3.82	3.59	3.42	3.29	3.18	3.09	2.96	2.81	2.66	2.57	2.50	2.45	2.42	2.39	2.36	2.33	2.30	2.28	2.26	2.25	2.23	2.13
27	7.68	5.49	4.60	4.11	3.78	3.56	3.39	3.26	3.15	3.06	2.93	2.78	2.63	2.54	2.47	2.42	2.38	2.35	2.33	2.29	2.27	2.25	2.23	2.22	2.20	2.10
28	7.64	5.45	4.57	4.07	3.75	3.53	3.36	3.23	3.12	3.03	2.90	2.75	2.60	2.51	2.44	2.39	2.35	2.32	2.30	2.26	2.24	2.22	2.20	2.19	2.17	2.06
29	7.60	5.42	4.54	4.04	3.73	3.50	3.33	3.20	3.09	3.00	2.87	2.73	2.57	2.48	2.41	2.36	2.33	2.30	2.27	2.23	2.21	2.19	2.17	2.16	2.14	2.03
30	7.56	5.39	4.51	4.02	3.70	3.47	3.30	3.17	3.07	2.98	2.84	2.70	2.55	2.45	2.39	2.34	2.30	2.27	2.25	2.21	2.18	2.16	2.14	2.13	2.11	2.01
40	7.31	5.18	4.31	3.83	3.51	3.29	3.12	2.99	2.89	2.80	2.66	2.52	2.37	2.27	2.20	2.15	2.11	2.08	2.06	2.02	1.99	1.97	1.95	1.94	1.92	1.80
50	7.17	5.06	4.20	3.72	3.41	3.19	3.02	2.89	2.78	2.70	2.56	2.42	2.27	2.17	2.10	2.05	2.01	1.97	1.95	1.91	1.88	1.86	1.84	1.82	1.80	1.68
60	7.08	4.98	4.13	3.65	3.34	3.12	2.95	2.82	2.72	2.63	2.50	2.35	2.20	2.10	2.03	1.98	1.94	1.90	1.88	1.84	1.81	1.78	1.76	1.75	1.73	1.60
80	6.96	4.88	4.04	3.56	3.26	3.04	2.87	2.74	2.64	2.55	2.42	2.27	2.12	2.01	1.94	1.89	1.85	1.82	1.79	1.75	1.71	1.69	1.67	1.65	1.63	1.49
100	6.90	4.82	3.98	3.51	3.21	2.99	2.82	2.69	2.59	2.50	2.37	2.22	2.07	1.97	1.89	1.84	1.80	1.76	1.74	1.69	1.66	1.63	1.61	1.60	1.57	1.43
120	6.85	4.79	3.95	3.48	3.17	2.96	2.79	2.66	2.56	2.47	2.34	2.19	2.03	1.93	1.86	1.81	1.76	1.73	1.70	1.66	1.62	1.60	1.58	1.56	1.53	1.38
∞	6.63	4.61	3.78	3.32	3.02	2.80	2.64	2.51	2.41	2.32	2.18	2.04	1.88	1.77	1.70	1.64	1.59	1.55	1.52	1.47	1.43	1.40	1.38	1.36	1.32	1.00

For ANOVA V_1 and V_2 are the treatment and residual degrees of freedom respectively.

Table A5.3. Two-tailed critical *F* values ($\alpha = 0.05$). Use these values for performing the variance ratio *F* test for testing homogeneity of two sample variances at the 5% significance level prior to conducting Student's *t*-test.

V_2 \ V_1	1	2	3	4	5	6	7	8	9	10	12	15	20	25	30	35	40	45	50	60	70	80	90	100	120	∞
1	647.8	799.5	864.2	899.6	921.8	937.1	948.2	956.6	963.3	968.6	976.7	984.4	993.1	998.1	1001	1004	1006	1007	1008	1010	1011	1012	1013	1013	1014	1018
2	38.51	39.00	39.17	39.25	39.30	39.33	39.36	39.37	39.39	39.40	39.41	39.43	39.45	39.46	39.46	39.47	39.47	39.48	39.48	39.48	39.48	39.49	39.49	39.49	39.49	39.50
3	17.44	16.04	15.44	15.10	14.88	14.73	14.62	14.54	14.47	14.42	14.34	14.25	14.17	14.12	14.08	14.06	14.04	14.02	14.01	13.99	13.98	13.97	13.96	13.96	13.95	13.90
4	12.22	10.65	9.98	9.60	9.36	9.20	9.07	8.98	8.90	8.84	8.75	8.66	8.56	8.50	8.46	8.43	8.41	8.39	8.38	8.36	8.35	8.33	8.33	8.32	8.31	8.26
5	10.01	8.43	7.76	7.39	7.15	6.98	6.85	6.76	6.68	6.62	6.52	6.43	6.33	6.27	6.23	6.20	6.18	6.16	6.14	6.12	6.11	6.10	6.09	6.08	6.07	6.02
6	8.81	7.26	6.60	6.23	5.99	5.82	5.70	5.60	5.52	5.46	5.37	5.27	5.17	5.11	5.07	5.04	5.01	4.99	4.98	4.96	4.94	4.93	4.92	4.92	4.90	4.85
7	8.07	6.54	5.89	5.52	5.29	5.12	4.99	4.90	4.82	4.76	4.67	4.57	4.47	4.40	4.36	4.33	4.31	4.29	4.28	4.25	4.24	4.23	4.22	4.21	4.20	4.14
8	7.57	6.06	5.42	5.05	4.82	4.65	4.53	4.43	4.36	4.30	4.20	4.10	4.00	3.94	3.89	3.86	3.84	3.82	3.81	3.78	3.77	3.76	3.75	3.74	3.73	3.67
9	7.21	5.71	5.08	4.72	4.48	4.32	4.20	4.10	4.03	3.96	3.87	3.77	3.67	3.60	3.56	3.53	3.51	3.49	3.47	3.45	3.43	3.42	3.41	3.40	3.39	3.33
10	6.94	5.46	4.83	4.47	4.24	4.07	3.95	3.85	3.78	3.72	3.62	3.52	3.42	3.35	3.31	3.28	3.26	3.24	3.22	3.20	3.18	3.17	3.16	3.15	3.14	3.08
11	6.72	5.26	4.63	4.28	4.04	3.88	3.76	3.66	3.59	3.53	3.43	3.33	3.23	3.16	3.12	3.09	3.06	3.04	3.03	3.00	2.99	2.97	2.96	2.96	2.94	2.88
12	6.55	5.10	4.47	4.12	3.89	3.73	3.61	3.51	3.44	3.37	3.28	3.18	3.07	3.01	2.96	2.93	2.91	2.89	2.87	2.85	2.83	2.82	2.81	2.80	2.79	2.72
13	6.41	4.97	4.35	4.00	3.77	3.60	3.48	3.39	3.31	3.25	3.15	3.05	2.95	2.88	2.84	2.80	2.78	2.76	2.74	2.72	2.70	2.69	2.68	2.67	2.66	2.60
14	6.30	4.86	4.24	3.89	3.66	3.50	3.38	3.29	3.21	3.15	3.05	2.95	2.84	2.78	2.73	2.70	2.67	2.65	2.64	2.61	2.60	2.58	2.57	2.56	2.55	2.49
15	6.20	4.77	4.15	3.80	3.58	3.41	3.29	3.20	3.12	3.06	2.96	2.86	2.76	2.69	2.64	2.61	2.59	2.56	2.55	2.52	2.51	2.49	2.48	2.47	2.46	2.40
16	6.12	4.69	4.08	3.73	3.50	3.34	3.22	3.12	3.05	2.99	2.89	2.79	2.68	2.61	2.57	2.53	2.51	2.49	2.47	2.45	2.43	2.42	2.40	2.40	2.38	2.32
17	6.04	4.62	4.01	3.66	3.44	3.28	3.16	3.06	2.98	2.92	2.82	2.72	2.62	2.55	2.50	2.47	2.44	2.42	2.41	2.38	2.36	2.35	2.34	2.33	2.32	2.25
18	5.98	4.56	3.95	3.61	3.38	3.22	3.10	3.01	2.93	2.87	2.77	2.67	2.56	2.49	2.44	2.41	2.38	2.36	2.35	2.32	2.30	2.29	2.28	2.27	2.26	2.19
19	5.92	4.51	3.90	3.56	3.33	3.17	3.05	2.96	2.88	2.82	2.72	2.62	2.51	2.44	2.39	2.36	2.33	2.31	2.30	2.27	2.25	2.24	2.23	2.22	2.20	2.13
20	5.87	4.46	3.86	3.51	3.29	3.13	3.01	2.91	2.84	2.77	2.68	2.57	2.46	2.40	2.35	2.31	2.29	2.27	2.25	2.22	2.20	2.19	2.18	2.17	2.16	2.09
21	5.83	4.42	3.82	3.48	3.25	3.09	2.97	2.87	2.80	2.73	2.64	2.53	2.42	2.36	2.31	2.27	2.25	2.23	2.21	2.18	2.16	2.15	2.14	2.13	2.11	2.04
22	5.79	4.38	3.78	3.44	3.22	3.05	2.93	2.84	2.76	2.70	2.60	2.50	2.39	2.32	2.27	2.24	2.21	2.19	2.17	2.14	2.13	2.11	2.10	2.09	2.08	2.00
23	5.75	4.35	3.75	3.41	3.18	3.02	2.90	2.81	2.73	2.67	2.57	2.47	2.36	2.29	2.24	2.20	2.18	2.15	2.14	2.11	2.09	2.08	2.07	2.06	2.04	1.97
24	5.72	4.32	3.72	3.38	3.15	2.99	2.87	2.78	2.70	2.64	2.54	2.44	2.33	2.26	2.21	2.17	2.15	2.12	2.11	2.08	2.06	2.05	2.03	2.02	2.01	1.94
25	5.69	4.29	3.69	3.35	3.13	2.97	2.85	2.75	2.68	2.61	2.51	2.41	2.30	2.23	2.18	2.15	2.12	2.10	2.08	2.05	2.03	2.02	2.01	2.00	1.98	1.91
26	5.66	4.27	3.67	3.33	3.10	2.94	2.82	2.73	2.65	2.59	2.49	2.39	2.28	2.21	2.16	2.12	2.09	2.07	2.05	2.03	2.01	1.99	1.98	1.97	1.95	1.88
27	5.63	4.24	3.65	3.31	3.08	2.92	2.80	2.71	2.63	2.57	2.47	2.36	2.25	2.18	2.13	2.10	2.07	2.05	2.03	2.00	1.98	1.97	1.95	1.94	1.93	1.85
28	5.61	4.22	3.63	3.29	3.06	2.90	2.78	2.69	2.61	2.55	2.45	2.34	2.23	2.16	2.11	2.08	2.05	2.03	2.01	1.98	1.96	1.94	1.93	1.92	1.91	1.83
29	5.59	4.20	3.61	3.27	3.04	2.88	2.76	2.67	2.59	2.53	2.43	2.32	2.21	2.14	2.09	2.06	2.03	2.01	1.99	1.96	1.94	1.92	1.91	1.90	1.89	1.81
30	5.57	4.18	3.59	3.25	3.03	2.87	2.75	2.65	2.57	2.51	2.41	2.31	2.20	2.12	2.07	2.04	2.01	1.99	1.97	1.94	1.92	1.90	1.89	1.88	1.87	1.79
40	5.42	4.05	3.46	3.13	2.90	2.74	2.62	2.53	2.45	2.39	2.29	2.18	2.07	1.99	1.94	1.90	1.88	1.85	1.83	1.80	1.78	1.76	1.75	1.74	1.72	1.64
50	5.34	3.97	3.39	3.05	2.83	2.67	2.55	2.46	2.38	2.32	2.22	2.11	1.99	1.92	1.87	1.83	1.80	1.77	1.75	1.72	1.70	1.68	1.67	1.66	1.64	1.55
60	5.29	3.93	3.34	3.01	2.79	2.63	2.51	2.41	2.33	2.27	2.17	2.06	1.94	1.87	1.82	1.78	1.74	1.72	1.70	1.67	1.64	1.63	1.61	1.60	1.58	1.48
80	5.22	3.86	3.28	2.95	2.73	2.57	2.45	2.35	2.28	2.21	2.11	2.00	1.88	1.81	1.75	1.71	1.68	1.65	1.63	1.60	1.57	1.55	1.54	1.53	1.51	1.40
100	5.18	3.83	3.25	2.92	2.70	2.54	2.42	2.32	2.24	2.18	2.08	1.97	1.85	1.77	1.71	1.67	1.64	1.61	1.59	1.56	1.53	1.51	1.50	1.48	1.46	1.35
120	5.15	3.80	3.23	2.89	2.67	2.52	2.39	2.30	2.22	2.16	2.05	1.94	1.82	1.75	1.69	1.65	1.61	1.59	1.56	1.53	1.50	1.48	1.47	1.45	1.43	1.31
∞	5.02	3.69	3.12	2.79	2.57	2.41	2.29	2.19	2.11	2.05	1.94	1.83	1.71	1.63	1.57	1.52	1.48	1.45	1.43	1.39	1.36	1.33	1.31	1.30	1.27	1.00

For comparing variances of two samples, V_1 is the DF of sample with the larger variance, V_2 is the DF of sample with the smaller variance.